Beginning Algebra

PRENTICE-HALL INTERNATIONAL, INC., London
PRENTICE-HALL OF AUSTRALIA, PTY. LTD., Sydney
PRENTICE-HALL OF CANADA, LTD., Toronto
PRENTICE-HALL OF INDIA PRIVATE LTD., New Delhi
PRENTICE-HALL OF JAPAN, INC., Tokyo

Beginning Algebra

JOHN H. MINNICK
RAYMOND C. STRAUSS

De Anza College
Cupertino, California

PRENTICE-HALL, INC.
Englewood Cliffs, New Jersey

© 1968, 1969 by Prentice-Hall, Inc.
Englewood Cliffs, N.J.

All rights reserved. No part of this book may be reproduced in any form or by any means without permission in writing from the publisher.

Current printing (last digit):
10 9 8 7 6 5 4

13-073494-2
Library of Congress Catalog Card Number 70-76880

Printed in the United States of America

Preface

Beginning algebra books are often written more for their instructors than for the student. Such books begin with boring, symbolic chapters containing lists of axioms which students have difficulty grasping or seeing the use of. They emphasize theorems and their proofs, topics which disinterest and confuse many students. Often they postpone introducing the student to many of the concepts and techniques of algebra until much later in the text. Perhaps worst of all, these books present most topics by first stating a general principle, next proving a theorem, and only then giving examples which the student can understand. This is a logically satisfying, mathematically pleasing approach, but not a pedagogically sound one with typical college students enrolled in their first algebra course.

We have attempted to reverse this order by beginning, whenever possible, with examples. (There are more than 800 examples in the text.) From these we draw general conclusions, and only after the concept is intuitively clear do we attempt to verbalize or symbolize the underlying principle. But in doing this we have not abandoned mathematical rigor. All of the usual axioms for real numbers are found in the text. There are theorems, and some are proved. But the emphasis in *Beginning Algebra* is on learning general principles through examples and developing skills and techniques for problem solving.

This has resulted in some unusual developments. Almost all of the fundamental concepts are presented in Chapter 1. It is not expected that the student will fully understand such subtle ideas as what is meant by saying that two algebraic expressions are equivalent, or what the solution set for an open sentence is, or what a function is, after completing this chapter. But by immediately introducing them in the simple context of one variable with small, finite replacement sets, we not only make the student aware of what the course is about, but we also prepare him for the further treatment of these topics in later chapters. By repeated exposure and the subsequent relearning involved, the student will at each stage strengthen his understanding.

We believe that most of the learning takes place when the student solves a problem related to the subject; or rather when he attempts a problem, makes an error, discovers his error, and *corrects it*. For this reason there are over 2000 exercises in the book, and complete answers, including graphs, are given to half of them. We urge the student to attempt as many of these exercises as his time will allow and always to check his answers. The answers to the even-numbered exercises have been withheld, so that these may be used for required homework, if the instructor desires.

A special feature of the book is the use of color as a teaching device. For example, when a number is added to the left and right expressions of an open sentence, this number is printed in color to emphasize what is happening. Similar use of color in multiplication of the left and right expressions in an open sentence or in the numerator and denominator of a fraction helps to make these important techniques clear.

Another feature is the manner in which the subject matter is divided into sections. As an aid in pacing the course, we have included in each section enough material for one meeting of the class. At times this has meant that several different topics are presented together in one section, and at other times a single topic will continue into the following section. We recognize that pacing will vary, depending on the ability of the students, but the suggested pacing (with an occasional day for review) has succeeded with typical students in several junior colleges in California, where a preliminary edition of the text was class tested. For a semester course meeting four hours per week, all 47 sections can be included with ample time for review and testing. For a shorter course certain sections may be omitted without interrupting the continuity or logical development of the sections which are presented. Details of how this might be accomplished are found in the *Instructor's Manual*, which also includes answers to the even-numbered exercises, some sample test questions, and some general comments about using this text.

We are indebted to several reviewers for their constructive comments and suggestions. We also wish to acknowledge and thank Ruth Anne Fish and Stanley Cotter of Foothill College, C. Louise Gillespie of Los Angeles Valley College, and Thomas C. Crooks of Contra Costa College for their help and cooperation in the class testing and evaluation of the preliminary edition of *Beginning Algebra*.

JOHN H. MINNICK
RAYMOND C. STRAUSS

Contents

ONE   Fundamental Concepts   1

     1.1   Algebraic Expressions   1
     1.2   Open Sentences   7
     1.3   The Function Concept   13
     1.4   Set Operations   18
     1.5   Infinite Sets   23
     1.6   Sets of Points   28

TWO   The Real Numbers   37

     2.1   Five Basic Axioms   37
     2.2   Other Basic Axioms   43
     2.3   Integers: Addition and Multiplication   49
     2.4   Integers: Subtraction and Division   55
     2.5   The Rational Numbers   62
     2.6   The Number Line   72

THREE First Degree Sentences   82

    3.1 Polynomial Expressions   82
    3.2 First Degree Equations   89
    3.3 First Degree Equations, Continued   95
    3.4 First Degree Inequalities   100
    3.5 Graphs of Sets   105
    3.6 Applications   110

FOUR Polynomials   120

    4.1 Natural Number Exponents   120
    4.2 Multiplication of Polynomials   125
    4.3 Factoring Polynomials   131
    4.4 Complete Factoring   137
    4.5 Second Degree Equations   142
    4.6 Second Degree Inequalities   149

FIVE Rational Algebraic Expressions   157

    5.1 Products and Quotients   157
    5.2 Rational Expressions: Sums and Differences   165
    5.3 Sums and Differences, Continued   173
    5.4 Equations and Inequalities   181
    5.5 Applications   189
    5.6 Complex Fractions and Division of Polynomials   197
    5.7 Integral Exponents   207

SIX Expressions with Two Variables   215

    6.1 Sets of Ordered Pairs and Their Graphs   215
    6.2 Equivalent Expressions   225
    6.3 Equations Over $R \times R$ 230
    6.4 Intersections of Solution Sets   239
    6.5 Inequalities Over $R \times R$ 247

SEVEN Functions and Relations — 255

- 7.1 Linear Functions — 255
- 7.2 Nonlinear Functions — 264
- 7.3 Special Graphing Techniques — 271
- 7.4 The Inverse of a Relation — 277

EIGHT Expressions with Radicals — 286

- 8.1 Roots and Radicals — 286
- 8.2 Rational Number Exponents — 292
- 8.3 Simplifying Radical Expressions — 298
- 8.4 Equivalent Radical Expressions — 304

NINE Quadratic Equations — 312

- 9.1 Completing the Square — 312
- 9.2 The Quadratic Theorem — 319
- 9.3 The Discriminant — 326

Answers to Odd-Numbered Exercises — 333

Glossary — 368

Index — 374

SYMBOLS

{ }	designates a set
=	is equal to
≠	is not equal to
<	is less than
≤	is less than or equal to
>	is greater than
≥	is greater than or equal to
∪	union
∩	intersection
×	Cartesian product
$\overline{A}$	complement of A
∅	empty set
∈	is a member of
$\langle a, b \rangle$	an open interval
$[a, b]$	a closed interval
$\langle a, b]$, $[a, b \rangle$	a half closed interval
∞	infinity
(a, b)	an ordered pair
:	such that
$\|a\|$	absolute value of a
N	the set of all natural numbers
I	the set of all integers
F	the set of all rational numbers
Q	the set of all quotient numbers
R	the set of all real numbers
S^{-1}	the inverse of S
$f(x)$	the expression which determines the function f
$f(a)$	If a is some specified number from the domain of f, then $f(a)$ is the corresponding number from the range
$\sqrt{a}$	the principal square root of a
$-\sqrt{a}$	the negative square root of a
$\sqrt[n]{a}$	the principal nth root of a

Beginning Algebra

Fundamental Concepts Chapter ONE

1.1 Algebraic Expressions

The subject of arithmetic concerns itself with numbers and operations. Our study of algebra begins with a review of these concepts. While we shall not usually make an issue of the difference between a number and a numeral, it is appropriate to consider briefly this difference.

A **number** is an abstract, intangible, but carefully defined idea. Like honesty and red it is something we talk about, but cannot actually see or handle. A **numeral** is a symbol which we use to represent the idea of number, just as we use symbols, the words "honesty" and "red," to represent these ideas. The number six can be represented by the numeral $4 + 2$ as well as by other numerals, such as $\frac{12}{2}$, $9 - 3$, VI, 6, $2 \cdot 3$, etc. None of these symbols *is* the number six; all of them only *represent* the number six. Indeed, every number can be represented by many different numerals.*

Some numerals include operational symbols. Thus, the representation of six by $9 - 3$ includes numerals which represent the numbers nine and three as well as the operational symbol $-$ indicating subtraction. For many of the numbers we shall consider there is a unique numeral called the **simplest form** of that number. For the natural numbers the simplest

---

* We shall not attempt to define a number.

2 *Fundamental Concepts*

form will be a numeral without operational symbols; 1, 2, 3, 4, and so on. For numbers which must be represented by fractions, their simplest forms will be fractions in lowest terms.

A **numerical expression** will be a numeral which may include both number and operational symbols. We shall frequently wish to simplify numerical expressions to find the simplest form of the number they represent. When doing so, we must pay careful attention, not only to the numbers and operations indicated, but also to the order in which those operations are performed. For example, the numerical expression $4 + 7 \cdot 2$ would represent the number twenty-two if the operation of addition were performed first, but would represent the number eighteen if the multiplication were performed first. In order to avoid such ambiguous situations, we shall adopt a hierarchy or order of operations.

The operation order we shall use is one which is rather generally agreed upon: multiplications and divisions will precede additions and subtractions. Thus, in simplest form, $4 + 7 \cdot 2 = 18$; $24 \div 8 + 4 = 7$, $10 - 3 \cdot 2 = 4$; $15 + \frac{6}{3} = 17$. In order to be able to write numerical expressions such as $4 + 7 \cdot 2$ in which we wish the operation of addition to precede that of multiplication, we introduce symbols of grouping. These will usually be parentheses, ( ), but we shall also use brackets, [ ], or the bar, _____. Hence, in simplest form, $(4 + 7) \cdot 2 = 22$; $24 \div [8 + 4] = 2$; $\frac{15 + 6}{3} = 7$.

While we have placed multiplication and division on a higher order level than addition and subtraction, we shall not establish a hierarchy between those operations on the same level. Thus, when only additions and subtractions appear in an expression, we agree to perform the operations as encountered from left to right, unless grouping symbols direct otherwise. Then $4 + 7 - 2 + 3 = 12$; $5 + 9 - 8 + 1 = 7$; $5 + (6 - 3) + 7 - 4 = 11$; $14 - (5 + 3) + 9 - 2 = 13$. Similarly, let us move from left to right when only multiplications and divisions appear. Then $4 \cdot \frac{7}{2} = 14$; $50 \div (2 \cdot 5) = 5$; $\frac{8}{4} \cdot 5 = 10$; $5 \cdot (\frac{12}{4}) = 15$.

We may now summarize our conventions concerning the order of operations.

> **First** Perform the operations within symbols of grouping in the prescribed order.
> **Second** Multiply and divide from left to right.
> **Third** Add and subtract from left to right.

Some examples will illustrate the conventions.

1.1 *Algebraic Expressions*

Example 1 Find the simplest form of $2(3 + 4) - 5 + 1$.

Solution $2(3 + 4) - 5 + 1 = 2 \cdot 7 - 5 + 1 = 14 - 5 + 1 = 10$

Example 2 Find the simplest form of $[(2 + 3)4 - 1]2 + 3$.

Solution $[(2 + 3)4 - 1]2 + 3 = [5 \cdot 4 - 1]2 + 3$
$= (20 - 1)2 + 3$
$= 19 \cdot 2 + 3 = 38 + 3 = 41$

Example 3 Find the simplest form of $\frac{3}{4} \div [\frac{1}{2} + 3 \cdot \frac{1}{4}] + 1$.

Solution $\frac{3}{4} \div [\frac{1}{2} + 3 \cdot \frac{1}{4}] + 1 = \frac{3}{4} \div [\frac{1}{2} + \frac{3}{4}] + 1$
$= \frac{3}{4} \div [\frac{2}{4} + \frac{3}{4}] + 1$
$= \frac{3}{4} \div \frac{5}{4} + 1 = \frac{3}{4} \cdot \frac{4}{5} + 1$
$= \frac{3}{5} + 1 = \frac{8}{5}$

In a numerical expression there is no difficulty finding the simplest form of the number represented. But consider the expression $2 \cdot x + 3$. Since it is not clear which number the letter x represents, it is also not clear which number the expression $2 \cdot x + 3$ represents. Because we shall frequently encounter situations similar to this in our study of algebra, let us establish a vocabulary and certain conventions to cover such occurrences.

Whenever a letter is used in an expression, let us call that expression an **algebraic expression.** Thus $4 + x - 3$, $12 \cdot x - 19$, and $3 \cdot x + 2 \cdot x$ are all algebraic expressions. Without further information, the number each algebraic expression represents cannot be determined. If in the above algebraic expressions the letter x is replaced by the numeral 3, the simplest forms of the resulting numerical expressions can be easily found.

$4 + x - 3$ becomes $4 + 3 - 3 = 4$
$12 \cdot x - 19$ becomes $12 \cdot 3 - 19 = 17$
$3 \cdot x + 2 \cdot x$ becomes $3 \cdot 3 + 2 \cdot 3 = 15$

If, on the other hand, we replace x by 2, then

$4 + x - 3$ becomes $4 + 2 - 3 = 3$
$12 \cdot x - 19$ becomes $12 \cdot 2 - 19 = 5$
$3 \cdot x + 2 \cdot x$ becomes $3 \cdot 2 + 2 \cdot 2 = 10$

Clearly the number represented by an algebraic expression depends on the replacement chosen for the letter which appears in it.

Both letters and numerals are placeholders—symbols which take the place of numbers. When a placeholder can represent only a single number we shall call it a **constant.** Thus, 4, 19, $\frac{21}{2}$ are constants; x is also a constant if only one number is allowed as a replacement for it. But often any member of a set of numbers is allowed as a replacement for a letter. In that case we shall call the letter a **variable.** We shall call the set of replacements for any letter its replacement set. Thus, a variable is a placeholder whose replacement set contains more than one number; a constant is a placeholder which can represent only one number. For example, if the replacement set for x is $\{2, 3, 4\}$, then in the expression $3 \cdot x - 4$, x is a variable and 3 and 4 are constants. If the replacement set for x is $\{5\}$, then in the expression $2 + 5 \cdot x$, 2, 5, and x are all constants.

The **value** of an expression is the number it represents. We shall usually express the value in simplest form. Determining the simplest form of the value of a numerical expression has already been explained. But to determine the value of an algebraic expression we must first know what replacement to make for all placeholders. Hence the expression $4 \cdot x + 5$ has the value 33, if the replacement for x is 7, but has the value 45 if x is replaced by 10. If we do not know the replacement for x, we cannot determine the value of $4 \cdot x + 5$.

Several conventions will be followed. In an algebraic expression, symbols of multiplication will often be dropped, if one factor is a literal placeholder. We shall write $4x + 5$ instead of $4 \cdot x + 5$. For any algebraic expression the replacement set for the variable will also be called the **domain** of the expression. The set of all numbers that the expression can represent with the possible replacements of the variable—the set of all values of the expression—will be called the **range** of the expression. Hence, if the replacement set for x is $\{2, 3, 4\}$, the domain of $4x + 5$ is also $\{2, 3, 4\}$, and the range of $4x + 5$ is $\{13, 17, 21\}$. Note that both the domain and range of an expression are *sets* of numbers.

Example 4 Find the value of the expression $5 + 2(x + 1) - 3$ when the placeholder is replaced by $\frac{1}{2}$.

Solution $5 + 2(x + 1) - 3$ becomes $5 + 2\left(\frac{1}{2} + 1\right) - 3 = 5 + 2\left(\frac{3}{2}\right) - 3 = 5 + 3 - 3 = 5$. (Note that the value is 5, a number, not a set of numbers.)

We shall often use the symbolic phrase "$x = 7$" to represent "x is replaced by 7" when the context of a problem makes it clear that this is our intention. The following examples will illustrate this notation. (In Section 1.2 we discuss another use of the symbol $=$.)

Example 5 The domain of the expression $\dfrac{2x + 3}{x}$ is $\{2, \frac{1}{3}\}$. Find its range.

Solution If $x = 2$, then

$$\frac{2x + 3}{x} = \frac{2 \cdot 2 + 3}{2} = \frac{4 + 3}{2} = \frac{7}{2}$$

1.1 Algebraic Expressions

If $x = \frac{1}{3}$, then

$$\frac{2x+3}{x} = \frac{2(\frac{1}{3})+3}{\frac{1}{3}} = \frac{\frac{2}{3}+3}{\frac{1}{3}} = \frac{\frac{11}{3}}{\frac{1}{3}} = 11$$

The range is $\{\frac{7}{2}, 11\}$.

Example 6 The range of the expression $2x + 1$ is $\{2\}$. Find its domain.

Solution We want to find a replacement for x which makes the expression $2x + 1$ have the value 2. Since $2(\frac{1}{2}) + 1 = 1 + 1 = 2$, we see that $\frac{1}{2}$ is such a replacement. That is when $x = \frac{1}{2}$, then $2x + 1 = 2(\frac{1}{2}) + 1 = 2$. Hence $\{\frac{1}{2}\}$ is the domain of the expression.

Sometimes we shall want to represent verbal descriptions by algebraic expressions. The skill needed to make such translations is acquired gradually. Some examples will illustrate the technique involved.

Example 7 Write an expression which represents four more than x.

Solution $x + 4$

Example 8 Write an expression which represents three times a number.

Solution $3x$

Example 9 Write an expression which represents the sum of 4 and twice a number which is three less than x.

Solution $4 + 2(x - 3)$

Example 10 It is estimated that a total of 16 pounds of cooking equipment are needed for a four-man group which plans to backpack into the mountains. Each man must carry six pounds of extra clothing for the trip and two pounds of food for each day of the trip. What is the total weight of all equipment and supplies needed for a trip of x days?

Solution $16 + 4(6 + 2x)$

Exercises 1.1

For Exercises 1–24 find the value of the expression (in simplest form).

1. $3 + 4 \cdot 5$
2. $12 - 2(5)$
3. $3(2 + 4) - 5$
4. $8 + 2(5 + 3)$
5. $(6 + 2)7 - 2$
6. $3 \cdot 5 - 3 + 1$
7. $3(2 + 5) - 4 + 3$
8. $(2 + 9)5 - 3 + 1$

6 Fundamental Concepts

9. $3[2 + 5(4 - 1)]$
10. $[8 - (2 + 1)2]4$
11. $[(5 + 1)3 - 1]3 + 2$
12. $[2 + 6(4 - 1)]4 - 2$
13. $21 + 6 \div 3$
14. $18 - 12 \div 2$
15. $\frac{30}{5} + 1$
16. $8 + 12 \div 3 - 1$
17. $(\frac{3}{4})(\frac{2}{9})$
18. $(\frac{5}{4}) \div (\frac{15}{8})$
19. $\frac{2}{3} + \frac{1}{2}$
20. $\frac{1}{4} + \frac{1}{8} \cdot \frac{2}{3}$
21. $(\frac{1}{2} - \frac{1}{4})2 + 1$
22. $\dfrac{1/3}{5/6}$
23. $5 - \frac{2}{5}$
24. $2 + \frac{3}{8}(4 - \frac{1}{2})$

For Exercises 25–44 an expression and its domain are given. Find its range.

25. $2x + 5$, $\{3\}$
26. $5 + 4y$, $\{2\}$
27. $15 - 3z$, $\{4\}$
28. $21 - 5x + 2$, $\{3\}$
29. $x(2 + x)$, $\{3\}$
30. $y + 3(4y - 1)$, $\{2\}$
31. $(x + 2)(x + 3)$, $\{1\}$
32. $(z + 2)z + 1$, $\{4\}$
33. $3x + x(x + 1)$, $\{1, 2, 3\}$
34. $(2y + 1)y + 4$, $\{0, 2, 4\}$
35. $p + 5(2 + 3p)$, $\{0, 1\}$
36. $3(x + 3)(x + 2)$, $\{1, 2, 3\}$
37. $4 + 3y$, $\{\frac{1}{2}, \frac{1}{5}\}$
38. $r + 2(r + 1)$, $\{\frac{1}{2}, \frac{1}{3}\}$
39. $3x + x(5 + x)$, $\{\frac{1}{2}, \frac{1}{4}\}$
40. $x + (x + 3)(2x)$, $\{\frac{1}{3}, \frac{1}{4}\}$
41. $\dfrac{2z + 1}{z + 3}$, $\{\frac{1}{3}\}$
42. $\dfrac{x + 2}{3x + 2}$, $\{\frac{1}{4}\}$
43. $\dfrac{w(w + 1)}{w + 2}$, $\{\frac{1}{2}\}$
44. $\dfrac{5 + y(y + 2)}{3y + 1}$, $\{\frac{1}{2}\}$

For Exercises 45–52 an expression and its range are given. Find its domain.

45. $3x$, $\{12\}$
46. $2y$, $\{6, 10\}$
47. $4x$, $\{2\}$
48. $5z$, $\{10\}$
49. $x + 3$, $\{10\}$
50. $y - 4$, $\{8\}$
51. $r + 3$, $\{5, 6\}$
52. $x - 4$, $\{10, 15\}$

For Exercises 53–64 write an expression in the variable x as described.

53. Add 3 to x and multiply the result by 4.
54. Subtract three times x from 7.
55. Multiply 5 by 3 more than x.

56. Add 6 to the product of 3 and 5 more than x.
57. Multiply x by the sum of 4 and twice x.
58. Add 9 to the product of x and four more than x.
59. Multiply x by 6 less than x.
60. Add twice x to the product of 3 times x and 4 less than x.
61. A TV repairman charges $15 for a house call plus $8 per hour for the time worked. How much is the charge for a house call that lasts x hours?
62. If golf balls sell for $2 a package, how much change is returned from a ten dollar bill when x packages are purchased?
63. To promote Christmas sales a magazine offers one year subscriptions for $7 with each additional gift subscription only $5. How much will x subscriptions cost?
64. Expenses for a ski trip are estimated to be $15 for transportation, $11 per day for room and board, and $7 per day for the ski lift. What is the total expense for a trip of x days?

1.2 Open Sentences

We have agreed to call certain collections of number, operation, and grouping symbols *numerical expressions*. For example, $4 + (7 - \frac{2}{3})9$, $5 \cdot 8 - 3 \cdot 7 + 11$, and $\frac{2 + 3 \cdot 5}{12 - 2}$ are all numerical expressions. Is $15 - 6 \cdot 2 = 7 - (6 - 2)$ a numerical expression? Note that $15 - 6 \cdot 2$ and $7 - (6 - 2)$ are both numerical expressions. But because of the inclusion of the symbol $=$, we shall not call $15 - 6 \cdot 2 = 7 - (6 - 2)$ a numerical expression. Similarly, we shall not call $3 + 7 \neq 12 - 6$ a numerical expression (even though it contains two such expressions), because it contains the symbol $\neq$ (is not equal to).

When we write $2 + 2 = 4$, we mean that $2 + 2$ and 4 are numerals for the same number. But clearly 10 and 6 are not numerals for the same number. Hence, when we write $3 + 7 = 12 - 6$, we are making a **false numerical statement.** On the other hand both $15 - 6 \cdot 2$ and $7 - (6 - 2)$ are numerals which represent the number whose simplest form is 3, and so $15 - 6 \cdot 2 = 7 - (6 - 2)$ is a **true numerical statement.** Another example of a numerical statement is $4 + 7 \cdot 3 < 33 + 0$. This statement includes the symbol $<$ (is less than), and it is a true numerical statement, since it is true that $25 < 33$.

A numerical statement, then, contains two numerical expressions and a connective, such as $=$, $<$, or $>$ (is greater than), to which one of the labels "true" or "false" can be assigned. Deciding whether a numerical statement is true or false usually involves finding the simplest form of each numerical expression and making the indicated comparison.

Example 1 Is the following a true or false numerical statement?
$$\tfrac{4}{6} + 9 > 11 - 4 \cdot 2$$

Solution True, since $\tfrac{29}{3} > 3$.

Example 2 Is the following a true or false numerical statement?
$$5 + 3 \cdot 2 \neq 4 \cdot 3 - 1$$

Solution False, since the statement says that $11 \neq 11$.

For the two reasons stated below, $2 + x = 12 - 3$ is not a numerical statement. First, note that on the left we have an algebraic, rather than a numerical, expression. Second, we cannot say whether this is a true or false assertion since we do not know what number to replace x by, and so do not know the value of the algebraic expression. But suppose we decide to replace x by 7. We then have the numerical statement $2 + 7 = 12 - 3$, which is true. Suppose, on the other hand, we replace x by 10. We now have $2 + 10 = 12 - 3$, which is false. Similarly, replacing x by 41 yields another false numerical statement $2 + 41 = 12 - 3$. Indeed, each replacement for x yields a numerical statement which is either true or false. But we cannot say that $2 + x = 12 - 3$ is true, nor can we say that it is false. We shall call $2 + x = 12 - 3$ an **open sentence.**

Another example of an open sentence is $2x - 3 > x + 7$. Here both the left and right sides of the sentence are algebraic expressions, and the connective is $>$. But again, we cannot decide whether the sentence is true or false unless we know the replacement for x. If x is replaced by 3, the resulting numerical statement is false. If x is replaced by 12, the resulting statement is true. This is typical of open sentences. When a replacement is chosen for the variable, the open sentence becomes a numerical statement, which is either true or false.

Example 3 If the variable y is replaced by 4 in the open sentence $2y - 3 > y + 7$, is the resulting numerical statement true or false?

Solution If $y = 4$, then $2y - 3 = 2 \cdot 4 - 3 = 5$ and $y + 7 = 4 + 7 = 11$. Then $5 > 11$ is a false numerical statement.

Example 4 If the variable t is replaced by $\tfrac{1}{2}$ in the open sentence $12t + 1 = 14/t$, is the resulting numerical statement true or false?

1.2 Open Sentences

Solution If $t = \frac{1}{2}$, the sentence $12t + 1 = 14/t$ becomes $12\left(\frac{1}{2}\right) + 1 = 14/\frac{1}{2}$, or $7 = 28$, a false statement.

Example 5 If the variable z is replaced by 0 in the open sentence $12 - z = z + 12$, is the resulting statement true or false?

Solution If $z = 0$, then $12 - z = z + 12$ becomes $12 - 0 = 0 + 12$, or $12 = 12$, a true numerical statement.

$2x + 1$ is an example of an algebraic expression. Associated with the variable x is a certain replacement set which we also call the domain of the expression. If the replacement set for x in the above expression is $\{1, 2, 6\}$, then the range of the expression is $\{3, 5, 13\}$. Consider the open sentence $2x + 1 = 5$. It is not possible to assign either label true or false to this open sentence. But if we consider $\{1, 2, 6\}$ as the replacement set for the variable x, then for one member of the replacement set, namely 2, the sentence $2x + 1 = 5$ becomes $2 \cdot 2 + 1 = 5$, a true numerical statement. But for the other members of the replacement set, namely 1 and 6, the sentence becomes $2 \cdot 1 + 1 = 5$ and $2 \cdot 6 + 1 = 5$, both false numerical statements. This illustrates an important property of an open sentence. For some replacements of the variable the sentence becomes a true statement, for others a false statement. Let us call the set of all members of the replacement set which result in a true statement the **solution set** of the open sentence. Then if the replacement set is $\{1, 2, 6\}$ for the open sentence $2x + 1 = 5$, we shall say the solution set is $\{2\}$.

Example 6 If the replacement set for t is $\{1, 4, 5, 7\}$, find the solution set for the open sentence $(t - 1)(t + 1) = 48$.

Solution $(t - 1)(t + 1) = 48$

If $t = 1$, $(1 - 1)(1 + 1) = 48$ is false.
If $t = 3$, $(3 - 1)(3 + 1) = 48$ is false.
If $t = 5$, $(5 - 1)(5 + 1) = 48$ is false.
If $t = 7$, $(7 - 1)(7 + 1) = 48$ is true.

The solution set is $\{7\}$.

Example 7 If the replacement set for v is $\{0, 1, 2\}$, find the solution set for $2v + 4 = 2(v + 2)$.

Solution $2v + 4 = 2(v + 2)$

If $v = 0$, $2 \cdot 0 + 4 = 2(0 + 2)$ is true.
If $v = 1$, $2 \cdot 1 + 4 = 2(1 + 2)$ is true.
If $v = 2$, $2 \cdot 2 + 4 = 2(2 + 2)$ is true.

The solution set is the entire replacement set, $\{0, 1, 2\}$.

Example 8 If the replacement set for x is $\{0, 1\}$, find the solution set for $x + 4 < 2x + 1$.

Solution $x + 4 < 2x + 1$

If $x = 0$, $0 + 4 < 2 \cdot 0 + 1$ is false.
If $x = 1$, $1 + 4 < 2 \cdot 1 + 1$ is false.

The solution set must be empty, so we indicate this by writing $\{\ \}$.

In Example 7 we considered the open sentence $2v + 4 = 2(v + 2)$ and the replacement set $\{0, 1, 2\}$. We saw that the solution set was the entire replacement set. That is, for each replacement of v by a member of its replacement set the value of the algebraic expression $2v + 4$ was the same as the value of the algebraic expression $2(v + 2)$. We shall describe this situation by saying that $2v + 4$ and $2(v + 2)$ are **equivalent algebraic expressions over the domain** $\{0, 1, 2\}$. Consider the expressions $x/2 + 1$ and $(x + 2)/2$. Let us use $\{10, 20\}$ as the replacement set for the variable. When x is replaced by 10, the first expression becomes $10/2 + 1 = 6$, and the second becomes $(10 + 2)/2 = 6$. And when x is replaced by 20, the first becomes $20/2 + 1 = 11$ and the second $(20 + 2)/2 = 11$. The value of each expression is the same whenever we replace the variable by the same member of its replacement set. Thus, $x/2 + 1$ and $(x + 2)/2$ are equivalent algebraic expressions over the domain $\{10, 20\}$.

Example 9 Are $5x + 2$ and $\dfrac{6 + x}{x}$ equivalent algebraic expressions over the domain $\{1, 2, 3\}$?

Solution If $x = 1$, then $5x + 2 = 5 \cdot 1 + 2 = 7$, and
$$\frac{6 + x}{x} = \frac{6 + 1}{1} = 7$$

If $x = 2$, then $5x + 2 = 12$, but
$$\frac{6 + x}{x} = \frac{8}{2} = 4$$

Since $12 \neq 4$, we see that the expressions are not equivalent over the domain $\{1, 2, 3\}$. Note that it is not necessary to even consider 3 as a replacement since, regardless of what the values of the expressions would be with this replacement, the expressions cannot be equivalent over a domain which includes 2.

We have seen that it is possible to form algebraic expressions from certain verbal expressions. Some verbal statements lead to open sentences

1.2 Open Sentences

involving one or more algebraic expressions. The following examples will illustrate this.

Example 10 Form an open sentence from the statement, "The number which is one more than x is 7 less than twice x."

Solution One possibility is $x + 1 = 2x - 7$.

Example 11 Form an open sentence from the statement, "If the cost of a package of chewing gum increases 1 cent, the new price will be 120% of the original."

Solution $x + 1 = 1.2x$ $(120\% = 1.2)$

Exercises 1.2

For Exercises 1–16 determine whether a numerical expression or a numerical statement is given. Find the values (in simplest form) of the numerical expressions, and tell whether the statements are true or false.

1. $3 + 7(4 - 2) = (3 + 7)(4 - 2)$
2. $12 + (3 - 1)7 - 3$
3. $4(3 + 6) - 8 > 4 \cdot 3 + 6 - 8$
4. $\dfrac{3(2 + 4) - 5}{3 + 2 \cdot 5} = 1 + 2 \cdot 3 - \dfrac{12}{2}$
5. $21 + [3 + (8 - 6)2 + 1]$
6. $6(3 + 1) = 6 \cdot 3 + 1$
7. $\dfrac{2 + 3 \cdot 4}{(1 + 2)2 + 1} < 3 + 0 \cdot (4 + 9)$
8. $8 + 2(5 - 1) - 3 > [3(7 + 2) - 4] - 3 + 1$
9. $493 \cdot 736 + 56 = 736 \cdot 493 + 56$
10. $873(265 + 973) = (973 + 265)873$
11. $(376 + 834) \cdot 0 < (376 + 843) + 0$
12. $(511)(263)(0)(846) < (177)(936) - 0$
13. $3(\tfrac{1}{2} + 4) \neq (\tfrac{1}{2})(3 + 4)$
14. $\dfrac{5(6 - 3) + 1}{8 + 2 \cdot 2} - \dfrac{3 + \tfrac{8}{4}}{2(0 + 3) + 9}$
15. $\dfrac{\tfrac{2}{3} + 4}{\tfrac{5}{3}} = 2\tfrac{4}{5}$
16. $(2 + \tfrac{1}{3})(\tfrac{1}{2}) + 1 \neq \tfrac{2}{3} + \tfrac{3}{2}$

For Exercises 17–28 determine whether an algebraic expression or an open sentence is given. Find the ranges of the expressions and the solution sets of the open sentences with replacement sets for the variable as indicated.

17. $5x + 2 = x + 10$, $\{0, 1, 2\}$
18. $3(2x + 1) < 4x + 7$, $\{0, 1, 2\}$
19. $3 + 2x \neq 3x + 2$, $\{0, 1, 2\}$
20. $5(x + 2) - 6 > 2(x + 3) + x$, $\{1, 2, 3\}$
21. $x(3 + 2x) + 3(x + 2x)$, $\{0, 2\}$
22. $x + 2[3 + 4(x + 1)] = 5(x + 5) + 1$, $\{2, 3, 4\}$
23. $4(x + 1) - 3 = x + 3(2 + x)$, $\{0, 1, 2\}$
24. $x(2 + 3x) + 2(x + 1) = x[1 + 3(x + 1)] + 2$, $\{0, 2, 4\}$
25. $1 + 4(x + 1) = 2(x + 3)$, $\{\frac{1}{3}, \frac{1}{2}\}$
26. $5 + 2x < 4(1 + x)$, $\{0, \frac{2}{3}, \frac{5}{2}\}$
27. $\dfrac{x(x + 1)}{2 + x} < \dfrac{1 + 2x}{x}$, $\{1, 2, 3\}$
28. $\dfrac{2(3 + x)}{1 + 3x} + \dfrac{x + 3}{2x + 1}$, $\{0, 1, 2\}$

For Exercises 29–36, two expressions and a domain are given. Determine whether or not the expressions are equivalent over the given domain.

29. $2x + 5x$ and $3x + 4x$, $\{1, 2, 3\}$
30. $3(x + 2)$ and $3x + 2$, $\{0, 1, 2\}$
31. $3(4x)$ and $12x$, $\{1, 2, 3, 4\}$
32. $2 + (3x)$ and $3 + (2x)$, $\{1, 2, 3\}$
33. $3x + 2x$ and $5x$, $\{\frac{1}{2}, \frac{2}{3}\}$
34. $4x + 3$ and $7x$, $\{1, \frac{5}{2}\}$
35. $\dfrac{2x + 1}{2}$ and $x + \frac{1}{2}$, $\{\frac{1}{4}, 2\}$
36. $3(x + 2)$ and $2(x + 1) + (x + 4)$, $\{\frac{1}{3}, 2, \frac{7}{2}\}$

For Exercises 37–44 a verbal description of either an algebraic expression or an open sentence is given. Use x as the variable and write that expression or sentence.

37. The sum of 4 and 3 times x is 5 times x.

38. Three times x is 1 more than twice x.
39. The product of one less than twice x and 3 more than x.
40. The product of 2 and 1 more than x is less than the sum of x and 5.
41. Four more than the quotient of an unknown and three less than that unknown.
42. The product of an unknown and 2 more than that unknown is 6 more than that unknown.
43. The number of girls in an algebra class is unknown, but there are 5 more than twice as many boys as girls. Altogether there are 29 students.
44. Mary will not tell her age, but in six years her age will be $\frac{5}{4}$ as much as it is now.

1.3 The Function Concept

We have seen that the value of an algebraic expression depends on the replacement for the variable. If x is replaced by 0, the value of $2x + 1$ is $2 \cdot 0 + 1 = 0 + 1 = 1$. If x is replaced by 1, the value is 3, and if x is replaced by 2, the value is 5. That is, if the domain of the expression $2x + 1$ is $\{0, 1, 2\}$, then its range is the set $\{1, 3, 5\}$.

Consider the same expression $2x + 1$, but now let the domain of the expression be $\{5, 10\}$. When x is replaced by 5, the value of the expression is 11; when x is replaced by 10, the expression has the value 21. Thus, when the domain is $\{5, 10\}$, the range of the expression is $\{11, 21\}$. And if we use $\{3, 4, 5, 6, 7\}$ as the domain for the expression $2x + 1$, the range is the set $\{7, 9, 11, 13, 15\}$. Note that even though we used the same expression, changing the domain resulted in a change in the range.

Consider, again, the set $\{0, 1, 2\}$ as the replacement set for the variable x, but now let us use the algebraic expression $3x/2$. If x is replaced by 0, the value of the expression is $3 \cdot 0/2 = 0/2 = 0$. If we replace x by 1, the value of the expression is $\frac{3}{2}$, and if 2 is the replacement for x, the expression has the value 3. Thus, the expression $3x/2$ has range $\{0, \frac{3}{2}, 3\}$ when its domain is $\{0, 1, 2\}$. Observe that even though we used the same domain, $\{0, 1, 2\}$, the range is different when we change from the expression $2x + 1$ to the expression $3x/2$.

What we have been observing is generally true. The range of an expression will depend on two things: what algebraic expression we consider, and what its domain is. We can picture this symbolically as follows:

Algebraic expression: domain $\rightarrow$ range

Using our examples, we see that this symbolism becomes

$$2x + 1: \quad \{0, 1, 2\} \to \{1, 3, 5\}$$
$$2x + 1: \quad \{5, 10\} \to \{11, 21\}$$
$$2x + 1: \quad \{3, 4, 5, 6, 7\} \to \{7, 9, 11, 13, 15\}$$
$$3x/2: \quad \{0, 1, 2\} \to \{0, \tfrac{3}{2}, 3\}$$

Let us slightly modify our point of view. The algebraic expression $2x + 1$ can be thought of as establishing a **correspondence** between the set $\{0, 1, 2\}$, its domain, and the set $\{1, 3, 5\}$, its range. We can exhibit this correspondence by writing

$$2x + 1: \quad 0 \to 1$$
$$1 \to 3$$
$$2 \to 5$$

Similarly, the algebraic expression $3x/2$ establishes a correspondence between the sets $\{0, 1, 2\}$ and $\{0, \tfrac{3}{2}, 3\}$ which we can represent by

$$3x/2: \quad 0 \to 0$$
$$1 \to \tfrac{3}{2}$$
$$2 \to 3$$

Indeed any algebraic expression establishes a correspondence between the set called its domain and the set called its range.

Example 1 Show the correspondence between the domain and the range of the algebraic expression $\dfrac{3x + 2}{5}$ when the domain is $\{0, 2, 4, 6\}$.

Solution $\dfrac{3x + 2}{5}: \quad 0 \to \tfrac{2}{5}$
$$2 \to \tfrac{8}{5}$$
$$4 \to \tfrac{14}{5}$$
$$6 \to 4$$

The kind of correspondence established by an algebraic expression between its domain and its range is an example of what is called a **function**, and the domain and range of the expression are also called the domain and range of the function. We shall not attempt to define a function precisely until Chapter 7. Then we shall consider functions which arise out of other contexts or which are not related to a single algebraic expression. But for now we shall consider only those functions which are defined by an expression and its domain.

We often use a letter to name a function. The function determined by the expression $2x + 1$ with domain $\{0, 1, 2\}$ can be named f, and we may

indicate this by writing
$$f: 2x + 1, \{0, 1, 2\}$$

We can use g to name the function determined by the expression $2x + 1$ with domain $\{3, 4, 5, 6, 7\}$ and write
$$g: 2x + 1, \{3, 4, 5, 6, 7\}$$
Similarly
$$h: \frac{3x}{2}, \{0, 1, 2\}$$
refers to the function named h determined by the expression $3x/2$ with domain $\{0, 1, 2\}$.

Consider the function $f: 2x + 1, \{0, 1, 2\}$. By $f(1)$ we mean the member of the range of f that corresponds to the member 1 in the domain. Since replacing the variable by 1 gives $2 \cdot 1 + 1 = 3$, we shall write $f(1) = 3$. Similarly, $f(0) = 2 \cdot 0 + 1 = 1$ and $f(2) = 2 \cdot 2 + 1 = 5$. In the same way, if $g: 2x + 1, \{3, 4, 5, 6, 7\}$, then $g(5) = 2 \cdot 5 + 1 = 11$. If $h: 3x/2, \{0, 1, 2\}$, then $h(2) = 3 \cdot 2/2 = 3$.

Example 2 Let $k: \dfrac{2x - 3}{x + 1}, \{6, 7, 8\}$. Find $k(7)$.

Solution $k(7) = \dfrac{2 \cdot 7 - 3}{7 + 1} = \dfrac{11}{8}$

Example 3 Let $p: \dfrac{4 + 3y}{2y + 1}, \{0, 1\}$. Find the domain and range of p.

Solution The domain of p is actually given, $\{0, 1\}$. To find the range we find
$$p(0) = \frac{4 + 3 \cdot 0}{2 \cdot 0 + 1} = \frac{4}{1} = 4$$
and
$$p(1) = \frac{4 + 3 \cdot 1}{2 \cdot 1 + 1} = \frac{7}{3}$$
The range is, therefore, $\{4, \tfrac{7}{3}\}$.

Sometimes we shall be interested in a problem of the type illustrated in the next example.

Example 4 Let $f: 2x + 5, \{0, 1, 2, 3, 4\}$. What member of the domain corresponds to the number 11 in the range of f?

Solution Since $f(3) = 11$, the number of the domain sought is 3.

16   Fundamental Concepts

We shall usually phrase such problems in this way: "Find an eleven of the function f." Then if the instructions are to "Find a thirteen of f," we see that 4 is the required number, since $f(4) = 13$.

Example 5 Let $g: 3x - 15$, $\{5, 10\}$. Find a zero of g.

Solution Since $g(5) = 0$, then 5 is a zero of g.

Example 6 Let $h: \dfrac{2z - 10}{z - 1}$, $\{5, 8, 10\}$. Find a zero of h. Find a 1 of h.

Solution $h(5) = 0$, $h(8) = \frac{6}{7}$, $h(10) = \frac{10}{9}$. A zero of h is 5, but there is no 1 of h.

Consider the open sentence $3x - 15 = 0$, where the replacement set for x is $\{5, 10\}$. The solution set can be seen to be $\{5\}$. If we compare this with Example 5, we see a strong connection between the problem of finding a zero of the function determined by the algebraic expression $3x - 15$ with domain $\{5, 10\}$ and finding the solution set of the open sentence $3x - 15 = 0$ with replacement set $\{5, 10\}$. Similarly, finding an eleven of the function determined by the expression $2x + 5$ with domain $\{0, 1, 2, 3, 4, 5\}$ is equivalent to the problem of finding the solution set of the open sentence $2x + 5 = 11$ with replacement set $\{0, 1, 2, 3, 4, 5\}$.

Now consider $f: 2x + 5$, $\{2, 3, 4\}$. Since $f(2) = 2 \cdot 2 + 5$, $f(3) = 2 \cdot 3 + 5$, and $f(4) = 2 \cdot 4 + 5$, it seems natural that if b is any member of the domain of f, then we would write $f(b) = 2b + 5$, where the variable x in the expression $2x + 5$ has been replaced by b. And if x is any member of the domain of f, we would write $f(x) = 2x + 5$. Then we can completely describe the function f by writing $f(x) = 2x + 5$, $\{2, 3, 4\}$. Note that $f(2)$, $f(3)$ and $f(4)$ refer to the numbers 9, 11, and 13 in the range of the function f. Further note that $f(x)$ is *not* the name of the function, nor a specific number in the range of the function, but it refers to the expression $2x + 5$.

Example 7 $f(y) = 2 + 3y$, $\{0, \frac{1}{3}, 1\}$.

(a) What is the name of this function?
(b) What is its domain?
(c) Find $f(0)$.
(d) What is its range?
(e) What is the 3 of this function?
(f) Find $f(2)$.

Solution

(a) f (b) $\{0, \frac{1}{3}, 1\}$ (c) $f(0) = 2$ (d) $\{2, 3, 5\}$

1.3 The Function Concept

(e) $\frac{1}{3}$ (f) Impossible, since 2 is not in the domain of f

Exercises 1.3

For Exercises 1–8 show the correspondence between the domain and range of the function as in Example 1.

1. $f\colon 3x + 2$, $\{0, 1, 2\}$
2. $g\colon x(x + 1)$, $\{0, 2, 3\}$
3. $h\colon 5x/3$, $\{0, 1, 3\}$
4. $k\colon y + 2(y + 1)$, $\{0, 2, 3\}$
5. $p(x) = \dfrac{3x - 1}{x + 1}$, $\{1, 2, 3\}$
6. $f(y) = \dfrac{4 + 3y}{y}$, $\{1, 3, 5\}$
7. $g(z) = z/3 + z/2$, $\{0, 2, 6\}$
8. $h(x) = 2 + 3/x - x$, $\{1, 2, 3\}$

For Exercises 9–16 find the indicated number in the range of the function.

9. $g\colon 2(y + 1) - 3$, $\{2, 3, 4\}$; $g(4)$
10. $f(x) = \dfrac{x + 2}{2x + 3}$, $\{0, 1, 2\}$; $f(0)$
11. $g(z) = (z + 2)(z + 1)$, $\{0, 3, 7\}$; $g(3)$
12. $h\colon x + 2(x + 5)$, $\{0, 1\}$; $h(0)$
13. $k(x) = (x + 2)x + 3$, $\{0, 2, 5\}$; $k(2)$
14. $s\colon y + 2(y + 1)$, $\{\frac{1}{2}, 3\}$; $s(\frac{1}{2})$
15. $p(z) = 3z - 2(z + 1)$, $\{3, \frac{7}{2}\}$; $p(\frac{7}{2})$
16. $f(y) = \dfrac{y + 2}{2y + 1}$, $\{1, \frac{3}{2}, 2\}$; $f(\frac{3}{2})$

For Exercises 17–24 find the number in the domain which corresponds to the given number in the range, if it exists, as in Examples 5 and 6.

17. $g(x) = 3(x + 2) - 5$, $\{1, 2, 3\}$; ten
18. $f(y) = 2 + 3y(y + 1)$, $\{0, 1, 2\}$; eight
19. $h(x) = \dfrac{x + 1}{x}$, $\{1, 2, 3\}$; three

18 Fundamental Concepts

20. $k(z) = z + z(z + 1)$, $\{1, 2, 3\}$; eight
21. $f(x) = 3(x + 2) - 12$, $\{2, 3, 4\}$; zero
22. $g(y) = 14 - 2(1 + 3y)$, $\{0, 1, 2\}$; zero
23. $k(x) = (8 - x)x + 3$, $\{0, 8\}$; zero
24. $f(z) = (3 - 2z)(z + 1)$, $\{0, \frac{1}{2}, \frac{3}{2}\}$; zero

For Exercises 25–40 consider the three functions defined as follows and determine the indicated number or set of numbers.

$$f(x) = 2 + 3x, \qquad \{0, 1, 2, 3\}$$
$$g(x) = x + 2(x + 1), \quad \{1, 2, 5\}$$
$$h(x) = x(2 - x), \qquad \{0, 1, 2\}$$

25. $f(3)$
26. $g(2)$
27. $h(1)$
28. $g(1)$
29. $f(0)$
30. Domain of f
31. Range of g
32. Range of h
33. The eight of g
34. The zeros of h
35. The fourteen of f
36. The one of h
37. $7 + f(1)$
38. $3 \cdot f(1)$
39. $g(5) - 10$
40. $f(2) + g(1)$

1.4 Set Operations

We have defined replacement set and solution set and have even represented sets symbolically, but we have not discussed the concept of "set" formally. Let us do so now.

The idea of **set** is too fundamental to attempt to define. Let us merely say that any collection of objects shall be called a set. We have been calling those things which belong to a set the **members** of the set. We may also use the word **element** as a synonym for member. We shall continue to use braces, { }, to set off the members of a set and commas to separate one member from another. We shall use the symbol ∈ to stand for the phrase "is a member of." For example, if we consider the set $\{1, 3, 5, 7, 9\}$, we observe that 1 is a member of this set, 57 is not one of its members, the set contains five elements, and $7 \in \{1, 3, 5, 7, 9\}$.

Often we shall use a single (upper-case) letter to represent a set. If we write $A = \{2, 4, 6, 8\}$, then we mean that A is the set whose members are 2, 4, 6, and 8. When two sets have exactly the same members, we shall call them **equal** and represent this by means of the symbol =. Then we write $\{1, 2, 3, 4, 5\} = \{1, 5, 3, 2, 4\}$ even though the members are not listed in the same order. The sets are equal because they have exactly the same members. Also $\{1, 2, 3, 4, 5\} = \{1, 1, 2, 3, 3, 3, 4, 5, 5, 5\}$ even though the set on the right contains several (unnecessary) repetitions of some of its members. The sets are equal because each contains the numbers 1, 2, 3, 4 and 5 and nothing else.

We have already encountered a set with no members, which we represented by the symbol $\{\ \}$. We shall call this the **empty set** and shall often represent it by the symbol $\varnothing$. Note that $\{0\} \neq \varnothing$, for the sets do not have the same members. The set on the left has a single member, while the set on the right has no members. Sometimes we shall encounter sets which have so many members that it would be quite frustrating to have to list all of them. Consider the following set.

$$\{1, 3, 5, 7, 9, 11, 13, 15, 17, 19, 21, 23, 25, 27, 29, 31, 33, 35, 37, 39\}$$

Fortunately there are several ways to avoid having to write all the members of this set. We might use a verbal description of the set members: "The set of all odd natural numbers less than 40." Another way to describe this set is to use a partial listing of its members: $\{1, 3, 5, 7, \ldots, 39\}$. In using this incomplete listing method we merely omit from our list of members those whose identity can *clearly* be inferred and replace them with three dots. In fact the only time we can use a partial listing is when there is no doubt of the set's membership, and this can occur only when the pattern of membership is clear from a few example members. Thus $\{10, 20, 30, \ldots, 1000\}$ is obviously the set of the first one hundred multiples of 10, but it is not at all clear what set is meant by $\{1, 50, 23, \ldots, 57\}$. Then $\{1, 2, 3, \ldots, 10\} = \{1, 2, 3, 4, 5, 6, 7, 8, 9, 10\}$, but $\{1, 2, 3, \ldots, 10\} \neq \{1, 2, 3, 10\}$.

Let $A = \{1, 2, 3\}$ and $B = \{1, 2, 3, 4, 5\}$. Note that every member of set A is also a member of set B. We shall describe this relationship between sets A and B by saying that A is a **subset** of B. To illustrate further, if $C = \{4, 5, 6\}$ and $D = \{1, 2, 3, \ldots, 20\}$, we see that every member of C is also a member of D, and we say that C is a subset of D. On the other hand, since $E = \{1, 2, 3, 4\}$ contains one member, 4, which is not a member of $A = \{1, 2, 3\}$, we cannot say that E is a subset of A. For P to be a subset of Q two conditions must be met: P must be, first of all, a set, and, at the same time, each member of P must also be a member of Q.

Example 1 Let $A = \{1, 2, 3, 4, 5\}$. Decide whether each of the following is or is not a subset of A.

(a) $\{1, 2, 3\}$ (b) $\{4, 5, 6\}$ (c) $\{1, 2, 3, 4, 5\}$

(d) 1   (e) $\{1, 2, 3, \ldots, 20\}$

Solution

(a) Is (b) Is not (6 is not a member of A)
(c) Is (d) Is not (1 is not a set)
(e) Is not

Let $R = \{1, 2, 3\}$ and $S = \{1, 3, 5, 7\}$. We can form a new set, $\{1, 2, 3, 5, 7\}$, which contains each member of R as well as each member of S, but no other members. We shall call this set the **union** of sets R and S. Similarly, if $P = \{2, 4, 6\}$ and $Q = \{4, 5, 6, 7\}$, then the union of P and Q is the set $\{2, 4, 5, 6, 7\}$. It will be helpful to introduce a symbol to represent the union of two sets. The symbol we shall use is $\cup$. Then, if $M = \{1, 5, 10\}$ and $N = \{5, 10, 15\}$, we shall represent the union of M and N by $M \cup N = \{1, 5, 10, 15\}$.

Example 2 Find:
(a) $\{1, 2, 3\} \cup \{4, 5, 6\}$ (b) $\{4, 5, 10\} \cup \{2, 4, 6\}$
(c) $\{5, 7, 9, 11\} \cup \{7, 9\}$

Solution

(a) $\{1, 2, 3, 4, 5, 6\}$ (b) $\{2, 4, 5, 6, 10\}$ (c) $\{5, 7, 9, 11\}$

Let $R = \{1, 2, 3\}$ and $T = \{1, 3, 5, 7\}$. This time let us form a new set, $\{1, 3\}$, which contains all those members found in set R, and, at the same time, in set T, but no other members. We shall call this set the **intersection** of sets R and T. If $P = \{2, 4, 6\}$ and $Q = \{3, 4, 5\}$, then the intersection of P and Q is the set $\{4\}$. Again, a symbol, $\cap$, will be used to represent the intersection of two sets. Then if $M = \{1, 2, 3, 4\}$ and $N = \{2, 4, 6, 8\}$, we shall represent the intersection of M and N by $M \cap N = \{2, 4\}$.

Example 3 Find:
(a) $\{10, 20, 30, 40\} \cap \{5, 10, 15\}$ (b) $\{1, 2, 3, 4, 5\} \cap \{1, 5\}$
(c) $\{1, 2, 3, \ldots, 20\} \cap \{10, 20, 30\}$ (d) $\{2, 4, 6, 8\} \cap \{1, 3, 5, 7\}$

Solution (a) $\{10\}$ (b) $\{1, 5\}$ (c) $\{10, 20\}$ (d) $\varnothing$

In some discussions we may wish to designate some set as a **universe** and consider subsets of this universe. For example, we might call set $U = \{1, 2, 3, 4, 5\}$ our universe and consider one of its subsets $A = \{2, 3\}$.

1.4 Set Operations

We could then define the **complement** of set A to be the set $\{1, 4, 5\}$. Similarly, if $B = \{1, 3, 5\}$, the complement of B is $\{2, 4\}$. Note that the complement of A contains those members found in U which are not found in A, and the complement of B contains members of U which are not found in B. We shall use the symbol $\overline{C}$ to represent the complement of C. Hence, if $C = \{1, 3\}$, then $\overline{C} = \{2, 4, 5\}$. One word of caution. Unless we know what universe a given set is a subset of, we cannot determine its complement. That is, the question "What is the complement of $D = \{10, 20, 30\}$?" cannot be answered unless we know what universe D is a subset of.

Example 4 Let $U = \{0, 5, 10, 15, 20\}$ be the universe.
  (a) If $A = \{5, 10\}$, find $\overline{A}$ (b) If $B = \{0, 10, 15, 20\}$, find $\overline{B}$
  (c) What set is $\overline{U}$?

  Solution (a) $\{0, 15, 20\}$ (b) $\{5\}$ (c) $\varnothing$

In Example 4(c) we observe that $\overline{U} = \varnothing$ since the set containing all members of $\{0, 5, 10, 15, 20\}$ not in U must be empty. In Example 3(d) the answer was also the empty set. These examples illustrate one reason for even considering a set with no members—it allows us to answer some questions (with a set) where we might otherwise have no available answer. Thus, any two sets have some set as their intersection and any subset of a universe has a complement. Of course, the empty set also creates some interesting results. If $A = \{0, 1, 2, 3\}$, we must conclude that $\varnothing$ is a subset of A, for it is true that $\varnothing$ is a set, and it is true that every member of $\varnothing$ is also a member of A. (For which member of $\varnothing$ is *not* a member of A?) Indeed, by this reasoning we are forced to conclude that the empty set is a subset of *every* set!

Example 5 Let $V = \{3, 6, 9, 12, 15\}$ be the universe and let $A = \{3, 6\}$ and $B = \{3, 12, 15\}$. Find each of the following:
  (a) $A \cup B$ (b) $A \cap B$ (c) $\overline{A} \cup \overline{B}$
  (d) $\overline{A \cap B}$ (e) $\overline{\varnothing}$

  Solution

  (a) $\{3, 6, 12, 15\}$ (b) $\{3\}$
  (c) $\overline{A} \cup \overline{B} = \{9, 12, 15\} \cup \{6, 9\} = \{6, 9, 12, 15\}$
  (d) $\overline{A \cap B} = \overline{\{3\}} = \{6, 9, 12, 15\}$
  (e) $\overline{\varnothing} = \overline{\{\ \}} = \{3, 6, 9, 12, 15\} = V$

Exercises 1.4

For Exercises 1–24 answer true or false.

1. $\{1, 2, 3\} = \{3, 2, 1\}$
2. $\{2, 3, 4\}$ is a subset of $\{3, 4\}$
3. $4 \in \{2, 3, 4\}$
4. 12 is a member of $\{2, 4, 6, \ldots, 20\}$
5. $\{1, 2, 3, \ldots, 100\} = \{1, 2, 3, 4, \ldots, 100\}$
6. $\{1, 3, 5, 7, \ldots, 101\}$ is a subset of $\{1, 3, 5, 7, 101\}$
7. $20 \in \{1, 3, 5, 7, \ldots, 99\}$
8. $\{2, 4, 6, 8\}$ is a subset of $\{2, 4, 6, \ldots, 20\}$
9. 4 is an element of $\{3, 2 + 2, 5\}$
10. $3 \in \{2 + 3, 5 + 4\}$
11. $\{2 + 1, 3 + 5\}$ is a subset of $\{3, 4 + 4, 9\}$
12. $\{2 + 2, 1 + 5\} = \{3 + 3, 1 + 3, 2 + 4\}$
13. $\{2\}$ is a subset of $\{2, 3, 4\}$
14. $\{3\}$ is a member of $\{2, 3, 4\}$
15. $\{4, 3, 2\}$ is a subset of $\{2, 3, 4\}$
16. $\{0\}$ is an element of $\{0, 2, 4\}$
17. $\emptyset$ is a subset of $\{0, 1, 2\}$
18. $\emptyset$ is a member of $\{0, 1, 2\}$
19. $\{\emptyset\}$ is a subset of $\{0, 1, 2\}$
20. $\emptyset$ is a member of $\{\emptyset, 1, 2\}$
21. $\{\emptyset\} = \{0\}$
22. $\{\emptyset\} = \emptyset$
23. $\emptyset = 0$
24. $\emptyset = \{0\}$

For Exercises 25–32 use braces to designate the indicated set.

25. $\{2, 3, 4\} \cup \{3, 4, 5\}$
26. $\{1, 2, 3, 5\} \cap \{2, 5, 6\}$
27. $\{1, 2, 3\} \cap \{\ \}$
28. $\{1, 2, 3\} \cup \emptyset$
29. $\{1, 2, 3, \ldots, 10\} \cap \{4, 5, 6, 7\}$
30. $\{1, 2, 3, \ldots, 10\} \cup \{7, 8, 9, \ldots, 20\}$
31. $\{1, 2, 3, \ldots, 10\} \cap \{7, 8, 9, \ldots, 20\}$
32. $\{3, 5, 7\} \cap \{2, 4, 6\}$

1.5 Infinite Sets

For Exercises 33–60 let the universe be $V = \{1, 2, 3, 4, 5\}$ and consider the sets $A = \{1, 2, 3\}$, $B = \{3, 4, 5\}$, $C = \{5, 6\}$. Use either braces or an appropriate upper-case letter to designate the indicated set, if possible.

33. $A \cap B$
34. $A \cup B$
35. $A \cap C$
36. $B \cup V$
37. $A \cap V$
38. $A \cup \emptyset$
39. $B \cap \emptyset$
40. $A \cup A$
41. $B \cap B$
42. $\overline{A}$
43. $\overline{B}$
44. $\overline{C}$
45. $\overline{V}$
46. $\overline{\emptyset}$
47. $\overline{A} \cup B$
48. $A \cap \overline{B}$
49. $\overline{A} \cup \overline{B}$
50. $\overline{A \cap B}$
51. $\overline{A} \cap \overline{B}$
52. $\overline{A \cup B}$
53. $\overline{B \cap C}$
54. $\overline{A \cap C}$
55. $A \cup \overline{A}$
56. $B \cap \overline{B}$
57. $A \cap \overline{\emptyset}$
58. $B \cup \overline{\emptyset}$
59. $\overline{V \cup \emptyset}$
60. $\overline{V \cap \emptyset}$

For Exercises 61–68, some of these have been given a meaning in this text, while some have not been defined (and never will be defined). Simplify those which have been defined, and indicate meaningless for the others.

61. $\{2\} + \{3\}$
62. $\{4\} \cup \{5\}$
63. $235 \cap 35$
64. $\{3\} \cap \{4\}$
65. $\{\emptyset\} \cup \{0\}$
66. $\emptyset + 0$
67. $\{\emptyset\} \cup \emptyset$
68. $\{\emptyset\} \cap \emptyset$

1.5 Infinite Sets

Each of the sets we have considered until now has had an obvious, unstated property. It has always been possible to list every one of its members. True, when there were many members we sometimes elected to make only a partial list or to use a verbal description of the members. But it has always been *possible* to list them all even when we chose not to do so. Sets which have this property we shall call **finite sets.** Thus, each of the following is a finite set: $\{1, 2, 3\}$, $\{4, 8, 12, \ldots, 40\}$, $\emptyset$, "the set of natural numbers less than 1000."

There are sets which are not finite. Such sets have so many members that no matter how long we might try, we could not list them all. We shall call such nonfinite sets **infinite sets.** It is important to realize that the reason a set will be called infinite is not our lack of time to list its members, or our lack of space to display the list, or our unwillingness to do so. Rather, it is because no listing can possibly be complete. Let us consider some examples of infinite sets we shall need as we continue our study of algebra.

We have referred to **natural numbers** several times. Without trying to define carefully what a natural number is, we shall simply say that the natural numbers are the numbers used to count. Indeed, "counting number" is a synonym for "natural number." We have agreed to use the symbols 1, 2, 3, etc. as the simplest forms of the natural numbers. When we consider the set of *all* natural numbers, we encounter an infinite set, which we represent by $\{1, 2, 3, \ldots\}$. Note that we do not end the list with some last natural number for there is no last natural number. Other examples of infinite sets and their representations are the set of odd natural numbers $\{1, 3, 5, 7, \ldots\}$, and the set of all multiples of 3 $\{3, 6, 9, 12, \ldots\}$. Frequently we shall use letters to name certain infinite sets. Thus, hereafter we shall always use the letter N as a symbol for the set of all natural numbers: $N = \{1, 2, 3, 4, \ldots\}$.

We shall now introduce some numbers not previously considered. Let us define a number called the **additive inverse** of 1, represent that number by the symbol -1, and define its sum with 1 to be 0. Then $1 + \text{-}1 = 0$. Similarly, we define the additive inverse of 2 to be -2 with $2 + \text{-}2 = 0$. In fact we can continue in this way, defining an additive inverse for each of the natural numbers. The additive inverse of 3 will be represented by -3, and $3 + \text{-}3 = 0$, and so on. Since the set of natural numbers is infinite, there will be infinitely many additive inverses of the sort we have been defining. We could express the set of all such additive inverses by writing $\{\text{-}1, \text{-}2, \text{-}3, \text{-}4, \ldots\}$ or $\{\ldots, \text{-}4, \text{-}3, \text{-}2, \text{-}1\}$, since these are equal sets.

Let us form a new set whose members are precisely the natural numbers, their additive inverses, and zero. One way to represent this set is to form the union of two sets, $\{0, 1, 2, 3, 4, \ldots\} \cup \{\text{-}1, \text{-}2, \text{-}3, \ldots\}$, but we shall often use the representation $\{0, 1, \text{-}1, 2, \text{-}2, 3, \text{-}3, \ldots\}$ or better still $\{\ldots, \text{-}3, \text{-}2, \text{-}1, 0, 1, 2, 3, \ldots\}$. We shall call this set the **integers** and represent it always with the letter I. Note that since every member of $N = \{1, 2, 3, 4, \ldots\}$ is also a member of $I = \{\ldots, \text{-}3, \text{-}2, \text{-}1, 0, 1, 2, 3, \ldots\}$, then N is a subset of I. Another subset of I is the set of the additive inverses of the natural numbers; that is, $\{\text{-}1, \text{-}2, \text{-}3, \text{-}4, \ldots\}$. These infinite subsets of I will sometimes be referred to by other names. We will call $N = \{1, 2, 3, \ldots\}$ the set of **positive integers;** the set $\{\text{-}1, \text{-}2, \text{-}3, \ldots\}$ we will call the set of **negative integers.** We observe that 0 is an integer, but it is not a member of either of these subsets. Hence 0 is neither a positive nor a negative integer.

Let us consider fractions such as $\frac{3}{4}$, -5/-16, 12/-7 and -5/-10, whose numerators and denominators are integer symbols. Any number which can be represented by such a fraction will be called a **rational number,** since it is the quotient or ratio of two integers. The letter F (for fraction) will be used to represent the set of all rational numbers. We shall see in Section 2.4 that a fraction whose *denominator* represents zero is not a number at all. But any fraction whose numerator and denominator are integer symbols, with the denominator not zero, represents a rational number. The numbers represented by the decimal expressions .75 and 2.5 can also be represented by the fractions $\frac{3}{4}$ and $\frac{5}{2}$, and hence they are also rational numbers. Decimal expressions with only a finite number of digits such as .75 and 2.5 shall be called terminating decimals or, more simply, decimals. It can be shown that every (terminating) decimal, positive or negative, can be represented by a fraction whose numerator and denominator are integer symbols and hence represents a rational number. We shall not at this time consider decimals with infinitely many digits. In Section 2.6 we shall see that such (nonterminating) decimals sometimes do represent rational numbers, but sometimes do not.

Example 1 Show that each of the following is a rational number by expressing each as a fraction whose numerator and denominator are (positive) integers.

(a) .125 (b) 32.2 (c) 3

Solution

(a) $.125 = \frac{125}{1000} = \frac{1}{8}$ (b) $32.2 = \frac{322}{10} = \frac{161}{5}$ (c) $3 = \frac{3}{1}$

The set F has some interesting subsets. Since $\frac{1}{1} = 1, \frac{2}{1} = 2, \frac{3}{1} = 3$, etc., we see that every natural number *can* be represented as a ratio of integers, and so every natural number is also a rational number. Then N is a subset of F. Similarly, we will show in the next chapter that $\frac{0}{1} = 0, -1/1 = -1, -2/1 = -2, -3/1 = -3$, etc., and so every integer *can* be represented as a ratio of integers. Then every integer is also a rational number and I is a subset of F. To summarize, we can say that N is a subset of I and both N and I are subsets of F.

Example 2 Each of these is a member of one or more of the special sets, $N, I,$ and F. Indicate which.

(a) -2 (b) -3/5 (c) $\frac{1}{2} + \frac{1}{2}$

Solution (a) I, F (b) F (c) N, I, F (since $\frac{1}{2} + \frac{1}{2} = 1$)

As we see, the set of rational numbers contains both positive and negative numbers. We shall call the set of all positive rational numbers the set of **quotient numbers** and represent it by the letter Q. Included in this set are all positive integers or natural numbers, and all numbers represented by fractions such as $\frac{2}{3}$ and $\frac{7}{4}$, whose numerators and denominators are natural number symbols, or by positive decimals. Since every natural number is also a quotient number, we see that N is a subset of Q. But since the negative integers are not quotient numbers, we see that I is not a subset of Q. Of course Q, the set of positive rational numbers, is a subset of F, the set of all rational numbers.

We may now summarize. There are four rather special infinite sets of numbers.

1. $N = \{1, 2, 3, \ldots\}$, the set of natural numbers.
2. $I = \{\ldots -2, -1, 0, 1, 2, \ldots\}$, the set of integers.
3. F = the set of rational numbers, any number which is the ratio of two integers. All fractions with numerator and denominator integer symbols (and denominator not zero), all integers, and all (terminating) decimals represent rational numbers.
4. Q = the set of all positive rational numbers. All fractions with numerator and denominator natural number symbols, all natural numbers, and all (terminating) positive decimals represent quotient numbers.

Example 3 Each of these is a member of one or more of the special sets N, I, Q, and F. Indicate which.

(a) 0 (b) $\frac{3}{2}$ (c) -2/5
(d) 4.3 (e) -.3 (f) $\frac{14}{7}$

Solution

(a) I, F (b) Q, F (c) F
(d) Q, F (e) F (f) N, I, Q, F (since $\frac{14}{7} = 2$)

Example 4 Which of the special sets have the given set as a subset?
(a) $A = \{-3, \frac{1}{2}\}$ (b) $B = \{1, 2\}$

Solution

(a) Both -3 and $\frac{1}{2}$ are rational numbers, so A is a subset of F. Since -3 is not a quotient number A is not a subset of Q. Since $\frac{1}{2}$ is not an integer, A is not a subset of I. Since -3 is not a natural number, A is not a subset of N.

(b) Both 1 and 2 are natural numbers, so B is a subset of N. Since 1 and 2 are also integers, also quotient numbers, and also rational numbers, B is also a subset of I, Q, and F.

1.5 Infinite Sets

Exercises 1.5

Each of the numbers in Exercises 1–12 is a member of one or more of the special sets of this section. Indicate which.

1. $\frac{2}{3}$
2. -3/4
3. -5
4. 1.5
5. $\frac{8}{-3}$
6. -3.7
7. 4
8. $\frac{7}{3}$
9. $5 - 2$
10. $\frac{0}{3}$
11. $2.6 + 1.5$
12. $4(3.5)$

For Exercises 13–16 consider the following numbers: 0, 3.6, -1, 2, -5/2. List those which are members of the special set indicated.

13. N
14. I
15. F
16. Q

For Exercises 17–24 indicate which of the special sets the given set is a subset of.

17. $\{1, -1\}$
18. $\{0, \frac{1}{2}\}$
19. $\{-2/5, 3.3\}$
20. $\{1.111, 3\}$
21. N
22. Q
23. F
24. I

For Exercises 25–36 use braces or a letter (if possible) to designate the indicated set.

25. $\{1, 2, 3\} \cup \{1, 2, 3, \ldots\}$
26. $\{0, 1, 2, 3, \ldots\} \cup \{-1, -2, -3, \ldots\}$
27. $\{-1, 0, 1\} \cap \{\ldots, -2, -1, 0, 1, 2, \ldots\}$
28. $\{0, 1, 2\} \cap N$
29. $\{-1, 0, 1\} \cup N$
30. $\{-1/2, 0, \frac{1}{2}\} \cap F$
31. $\{-.5, 0, .5\} \cap Q$
32. $\{.4, -3/2, 0\} \cup F$
33. $N \cap I$
34. $F \cup Q$
35. $I \cup F$
36. $Q \cap I$

For Exercises 37–44 an open sentence and a replacement set for the variable are given. Find the solution set.

37. $x + 2 = 5, N$
38. $3x = 6, F$
39. $x < 5, N$
40. $2x < 8, N$
41. $3x = 2, Q$
42. $x + 3 = 0, I$

28   Fundamental Concepts

43. $x + 5 = 1$, N **44.** $2x = 3$, F

For Exercises 45–52 two expressions and a domain are given. Decide whether or not the expressions are equivalent over the given domain.

45. $x + 2$ and $2 + x$, N
46. $3x$ and $x(1 + 2)$, N
47. $2(3x)$ and $(2 \cdot 3)x$, N
48. $2 + (3x)$ and $(2 + 3)x$, N
49. $2x + 3x$ and $(2 + 3)x$, N
50. $(2x)(3x)$ and $6x$, N
51. $2(x + 3)$ and $2x + 3$, N
52. $2(x + 3)$ and $2x + 6$, N

For Exercises 53–56 find the range of the function which is defined by an expression and a domain.

53. $f(x) = 2x$, N
54. $g(x) = x + 3$, N
55. $h(x) = x - 2$, $\{3, 4, 5, \ldots\}$
56. $k(x) = 2x - 1$, N

1.6 Sets of Points

All of the sets we have so far considered have been sets of numbers. Sometimes these sets were finite, sometimes infinite; one set was empty. We may also consider sets whose members are not numbers. For example, we could form the set whose members are people (such as the set of all current members of the United States Senate), nations (the set of members of the United Nations), or automobiles (the set of all automobiles licensed by the state of New York this year). We shall not want to investigate these sets in our study of algebra, but we will want to consider certain sets of points.

The concept of a plane is difficult to define. Let us agree that the flat page of a book, the top surface of a table, and the writing surface of a classroom blackboard are all examples of a plane surface. By a *point of a plane* we shall mean a particular position or location in the plane. A *set of points* in a plane is any collection of some points of the plane. We shall often describe a set of points by means of a picture. Thus in Figure

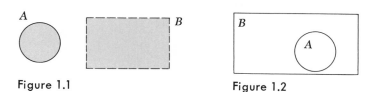

Figure 1.1   Figure 1.2

1.1, A is the set of all points on and within the drawn circle and B includes all points within, but not on, the drawn rectangle.

Sometimes the context of a problem will make it clear what the set of points is without any shading. In Figure 1.2, A is the set of points inside and on the circle, while B is the set of points inside and on the rectangle. Since every point of A is also a point of B, we see that A is a subset of B.

Example 1   Decide whether A is or is not a subset of B.

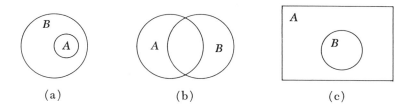

(a)   (b)   (c)

Solution

(a) A is a subset of B (b) A is not a subset of B
(c) A is not a subset of B (but B is a subset of A)

Figure 1.3 shows how to represent the union and intersection of sets of points. A is the set of points on and inside the left-hand circle, and B is the set of points on and inside the right-hand circle. Notice that $A \cup B$, represented by shading, contains all the points which are either in A or in B, while $A \cap B$ contains those points which are common to both A and B.

Figure 1.3

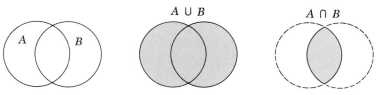

Example 2 Shade sets $A \cup B$ and $A \cap B$.

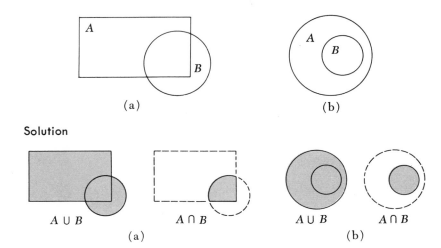

Solution

We may also represent the complement of a set of points with a picture as shown in Figure 1.4.

Remember that it is necessary to have some universe whenever the complement of a set is discussed. The universe here is the set of points inside and on the rectangle U. Note that $\bar{A}$ consists of all points which are outside of the circle but in the universe. Since the circle itself *is* a part of set A, then it is *not* a part of set $\bar{A}$, and this is indicated by making a boundary of $\bar{A}$ a dotted circle.

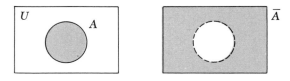

Figure 1.4

Example 3 In each case the universe is the rectangular set, and A is the set indicated by shading. Represent $\bar{A}$ by shading.

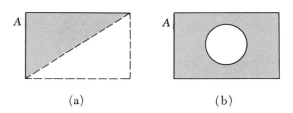

1.6 Sets of Points

Solution

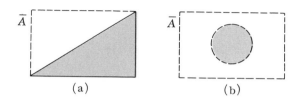

At times we shall want to consider sets of points each of which lies on some straight line. Figure 1.5 illustrates some sets of this sort. Set A represents *all* the points in a certain straight line. The arrowheads show that set A continues forever in both directions. We shall call set B a ray, a half-line which includes one end point and all points on a straight line which lie on only one side of that point. Set C will be called a line segment; note that it has two end points and does not continue forever in either direction. Set D is that subset of a line segment whose end points are not members of the set. We shall indicate this by using small empty circles at the ends.

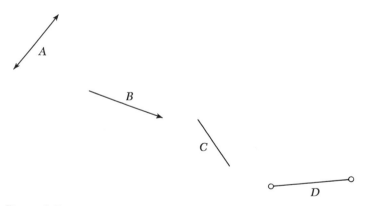

Figure 1.5

Example 4 Find the set indicated.

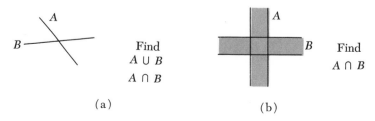

Solution

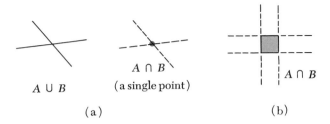

(a)

Pictorial representations of sets of points such as those we have been considering are often called Venn diagrams. Let us see how Venn diagrams can help us picture and understand the relationships that sets of numbers possess. We have seen that the set of natural *numbers*, N, is a subset of the set of integers, I. While neither N nor I is a set of *points*, it is possible to show the subset relation between them by a Venn diagram such as that in Figure 1.6. We use the Venn diagram here only to point out that N is a subset of I. Neither set is round, nor even a set of points.

Figure 1.6

Example 5 Make Venn diagrams to show the subset relation of these sets of numbers.

(a) N, Q, and F (b) I, Q, and F

Solution

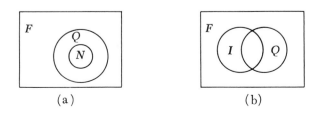

1.6 Sets of Points

Example 6 Indicate these sets by shading in the Venn diagram of Example 5(a).
(a) $N \cup Q$ (b) $N \cap Q$ (c) $\overline{Q}$ (use F as universe)

Solution

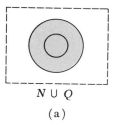

$N \cup Q$
(a)

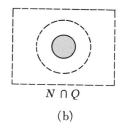

$N \cap Q$
(b)

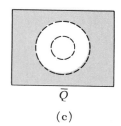

$\overline{Q}$
(c)

Example 7 Indicate these sets by shading in the Venn diagram for Example 5(b). Use F as the universe.
(a) $\overline{Q} \cap I$ (b) N

Solution

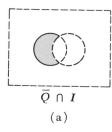

$\overline{Q} \cap I$
(a)

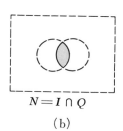

$N = I \cap Q$
(b)

Examples 5(b) and 7(b) can be used to show the subset relation for all the special sets N, I, Q, and F, as illustrated in Figure 1.7.

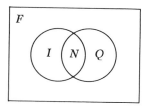

Figure 1.7

Example 8 Indicate where in Figure 1.7 the following numbers belong: 2, -3, 5/6, -2/3.

Solution

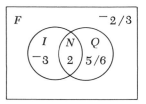

Sometimes a Venn diagram can be used to help solve a problem which seems at first to have little to do with sets of points. An example will illustrate.

Example 9 There are 100 boys in Delta Tau Delta fraternity. Of these, 75 take English, 60 take math, and 10 take neither English nor math. How many take:

(a) English, but not math?   (b) Math, but not English?
(c) Both math and English?

Solution We make a Venn diagram showing the sets E (those taking English) and M (those taking math) in the universe (the set of all fraternity men). Since 10 take neither English nor math, the size of $\overline{E \cup M}$ is 10 and this we show by writing 10 outside of both E and M. But then we must account for 90 students in the set $E \cup M$ and do so in such a way that 60 students are in M and 75 are in E. Since $90 - 60 = 30$ we must place 30 students in $E \cup M$ but *not* in M. And since $90 - 75 = 15$ we must place 15 students in $E \cup M$ but *not* in E. As shown in Figure 1.8 this requires that we place 45 students in $E \cap M$.

(a) 30   (b) 15   (c) 45

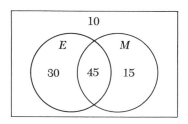

Figure 1.8

1.6 Sets of Points

Exercises 1.6

For Exercises 1–8 consider this Venn diagram which shows that A and B are both subsets of some universe V. Use shading to show each of these sets.

1. $\overline{A}$
2. $\overline{B}$
3. $A \cap V$
4. $B \cap V$
5. $\overline{A} \cap B$
6. $\overline{B} \cap A$
7. $\overline{A \cup B}$
8. $\overline{A \cap B}$

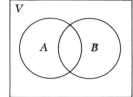

For Exercises 9–12 consider this Venn diagram and shade each of these sets.

9. $A \cup B$
10. $A \cap B$
11. $\overline{A} \cap B$
12. $\overline{B} \cup A$

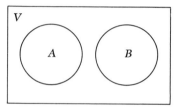

For Exercises 13–24 sets A, B, and C are the shaded triangles, while D is the diagonal of a rectangle. Each is a subset of the set of points on and inside the rectangle, which is the universe. Show each of these sets by shading.

13. $A \cup B$
14. $A \cap B$
15. $A \cup C$
16. $A \cap C$
17. $A \cup D$
18. $B \cap D$
19. $C \cap D$
20. $A \cap D$
21. $\overline{A} \cup D$
22. $\overline{A} \cap D$
23. $\overline{B} \cap D$
24. $\overline{B} \cup D$

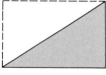
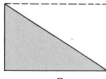

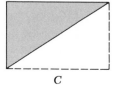
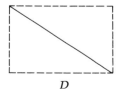

For Exercises 25–32 indicate these sets by shading the Venn diagram for the sets of numbers, N, I, and F. The universe is F.

25. $N \cap I$ 26. $N \cup I$
27. $\overline{N} \cap I$ 28. $\overline{I} \cup N$
29. $\overline{I} \cap N$ 30. $\overline{N} \cup N$
31. $\overline{I} \cap I$ 32. $\overline{N} \cap \overline{I}$

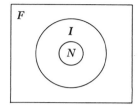

For Exercises 33–44 consider this Venn diagram which shows the special sets of numbers N, I, Q, and F, with F the universe. Use shading in this diagram to indicate these sets.

33. $N \cap Q$ 34. $N \cup I$
35. $\overline{I} \cap Q$ 36. $I \cap \overline{Q}$
37. $\overline{I} \cap N$ 38. $\overline{N} \cap I$
39. $\overline{N} \cap Q$ 40. $\overline{I} \cap \overline{Q}$

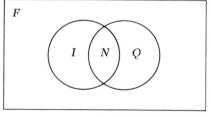

41. The set of integers which are not natural numbers.
42. The set of quotient numbers which are not integers.
43. The set of rational numbers which are not quotient numbers.
44. The set of rational numbers which are neither integers nor quotient numbers.

For Exercises 45–46 indicate where in the Venn diagram for the sets N, I, Q, and F each of these numbers belongs (as in Example 8).

45. 3, 0, $\frac{2}{3}$, -5/2   46. 1.6, -3.5, .222, -5.55

47. A market survey revealed that in a certain city 7% of the homes had no TV, 24% had a color TV, and 72% had a black and white TV. What percent had: (a) A color set only? (b) A black and white set only? (c) Both a color set and a black and white set?

48. There are twenty members in the Jet Set Club who are planning to travel together to either Hawaii or Mexico. Seven of them have never been to Hawaii, eleven of them have never been to Mexico, but five have been to both Mexico and Hawaii. How many have been to neither Mexico nor Hawaii?

The Real Numbers Chapter TWO

2.1 Five Basic Axioms

In the first chapter we considered some examples of numerical statements, some true and some false. $7 + 3 = 2 + 8$ is a true numerical statement, while $3 + 6 = 3 \cdot 6$ is a false statement. We have also considered open sentences such as $2x + 3 = 5x$ which are neither true nor false until a replacement is made for the variable.

Now let us consider some examples of **algebraic statements,** some true and some false. The algebraic statement "For every $x \in \{2, 3, 4\}$, $x + 1 = 1 + x$" is true since each replacement of x by a member of $\{2, 3, 4\}$ results in a true numerical statement. That is, $2 + 1 = 1 + 2$, $3 + 1 = 1 + 3$, and $4 + 1 = 1 + 4$ are *all* true numerical statements. On the other hand, the algebraic statement, "For every $x \in \{1, 2\}$, $2x + 3 = 5x$" is false. When x is replaced by 1 a true numerical statement results, but when x is replaced by 2, we obtain the false statement, $2 \cdot 2 + 3 = 5 \cdot 2$. Hence it is *false* that $2x + 3 = 5x$ for *every* $x \in \{1, 2\}$.

Let us next consider the algebraic statement, "For every $x \in N$, $x + 1 = 1 + x$." Since the replacement set for x is infinite, we cannot make every possible replacement for x. Our experience may lead us to believe that for every replacement we might make, a true numerical statement results. Therefore we will *assume* that this is a true algebraic statement. Similarly, we will assume that "For every $x \in N$, $x + 2 = 2 + x$," "For every $x \in N$, $x + 3 = 3 + x$," etc. are all true statements. We shall call this general assumption about such algebraic statements the **commutative axiom of addition** of natural numbers. We shall represent this axiom symbolically by writing, "For every $x \in N$ and $y \in N$, $x + y = y + x$." Thus we are assuming that the sum of any two natural numbers does not depend on which number we consider first.

We shall make the similar assumption that the product of any two natural numbers does not depend on which number we write first. We will call this assumption the **commutative axiom of multiplication** of natural numbers and symbolize it by writing, "For every $x \in N$ and $y \in N$, $xy = yx$."

Another algebraic statement which we may easily show to be true is, "For every $y \in \{1, 2, 3, \ldots, 10\}$, $(5 + y) + 2 = 5 + (y + 2)$." For example, if y is replaced by 7 we have $(5 + 7) + 2 = 5 + (7 + 2)$, a true numerical statement. Recall that the symbols of grouping indicate that in $5 + (7 + 2)$ we first add 7 and 2. Thus $5 + (7 + 2) = 5 + 9 = 14$. And also $(5 + 7) + 2 = 12 + 2 = 14$. We cannot verify that "For every $y \in N$, $(5 + y) + 2 = 5 + (y + 2)$," but our experience leads us to believe that it is a true statement and we will *assume* that it is true. More generally we will assume that "For every $x \in N$, $y \in N$ and $z \in N$, $(x + y) + z = x + (y + z)$" is a true statement and call this the **associative axiom of addition** of natural numbers. Similarly, we will assume the truth of "For every $x \in N$, $y \in N$ and $z \in N$, $(xy)z = x(yz)$," and we will call this assumption the **associative axiom of multiplication** of natural numbers.

Example 1 Decide whether or not the expressions are equivalent over N.
(a) $2 + 3a$ and $3a + 2$ (b) $5(a + 2)$ and $(a + 2)5$
(c) $2(a + 3)$ and $2a + 3$

Solution

(a) Equivalent. In the commutative axiom of addition, $x + y = y + x$, replace x by 2 and y by $3a$.
(b) Equivalent. In the commutative axiom of multiplication, $xy = yx$, replace x by 5 and y by $(a + 2)$.
(c) Not equivalent. Replace a by 1 in the given expressions.

In the last example we see that $2(a + 3)$ is not equivalent to $2a + 3$, since the respective values of the expressions are not the same for every

2.1 Five Basic Axioms

natural number replacement of a. If a is replaced by 2, for example, $2(a + 3)$ becomes $2(2 + 3) = 2 \cdot 5 = 10$, while $2a + 3$ becomes $2 \cdot 2 + 3 = 4 + 3 = 7$. Note that when finding the value of $2(a + 3)$, the *last* operation performed is multiplication, $2 \cdot 5$, while the *last* operation performed when finding the value of $2a + 3$ is addition, $4 + 3$. For this reason we shall call $2(a + 3)$ a **product** and $2a + 3$ a **sum**. Further, we shall say that $2(a + 3)$ is a product whose **factors** are 2 and $(a + 3)$, while $2a + 3$ is a sum whose **terms** are $2a$ and 3.

Example 2 Identify the terms of each sum and the factors of each product.
(a) $4x + 4 \cdot 2$ (b) $4(x + 2)$

Solution

(a) This is a sum whose terms are $4x$ and $4 \cdot 2$. Also 4 and x are factors of $4x$; 4 and 2 are factors of $4 \cdot 2$.
(b) This is a product whose factors are 4 and $(x + 2)$. Also x and 2 are terms of $x + 2$.

As in Example 2 the expression $3x + 6$ is a sum of two terms. The first term has factors 3 and x, while the second term can be thought of as having factors 2 and 3. Also the expression $3(x + 2)$ is a product of two factors. The second factor has x and 2 as terms. The words "term" and "factor" will be important in our study of algebra, and we must be careful to use them correctly. Notice that they are words which describe one expression as it relates to another expression. Thus, it makes no sense to say that $2x$ is either a term or a factor, but it does make sense to say that $2x$ is a term of the expression $2x + 3$, or that $2x$ is a factor of the expression $2x(x + 3)$.

Example 3 Identify all of the terms and factors in the expression $3x + 4(5x + 1)$.

Solution $3x$ and $4(5x + 1)$ are terms of $3x + 4(5x + 1)$.
3 and x are factors of $3x$.
4 and $(5x + 1)$ are factors of $4(5x + 1)$.
$5x$ and 1 are terms of $5x + 1$.
5 and x are factors of $5x$.

It would be time consuming, but possible, to show that $4(x + 2)$ and $4x + 4 \cdot 2$ are equivalent over $\{1, 2, 3, \ldots, 100\}$. For example, if x is replaced by 9, then $4(x + 2) = 4(9 + 2) = 4 \cdot 11 = 44$. Also $4x + 4 \cdot 2 = 4 \cdot 9 + 4 \cdot 2 = 36 + 8 = 44$. Let us assume that for every $x \in N$, $4(x + 2)$ and $4x + 4 \cdot 2$ are equivalent expressions. Similarly, we shall assume that the statement, "For every $x \in N$, $y \in N$ and $z \in N$,

$x(y + z) = xy + xz$," and call this assumption the **distributive axiom** of natural numbers. Notice that $x(y + z)$ is a *product* whose factors are x and $(y + z)$, while $xy + xz$ is a *sum* each of whose terms has the factor x. Thus the distributive axiom assumes that certain products are equivalent to certain sums. The sum, $xy + xz$, must *always* have a factor repeated in each term and that same factor, x, must also be one of the factors of the product, $x(y + z)$.

Example 4   Use the distributive axiom to find an equivalent expression over N.
(a) $3(x + 4)$ (b) $5x + 5 \cdot 3$ (c) $4x + 12$

Solution

(a) $3(x + 4) = 3x + 12$ (b) $5x + 5 \cdot 3 = 5(x + 3)$
(c) $4x + 12 = 4(x + 3)$

Example 5   Use the axiom named to find an equivalent expression over N.
(a) $3(x + 5)$, commutative axiom of addition
(b) $3(x + 5)$, commutative axiom of multiplication
(c) $3(x + 5)$, distributive axiom

Solution

(a) $3(x + 5) = 3(5 + x)$ (b) $3(x + 5) = (x + 5)3$
(c) $3(x + 5) = 3x + 15$

Example 6   Which axioms can be used to show that for every $x \in N$, $3(2x + 5) = 6x + 15$?

Solution

$3(2x + 5) = 3(2x) + 3 \cdot 5$ [Distributive axiom]
$= (3 \cdot 2)x + 3 \cdot 5$ [Associative axiom of multiplication]
$= 6x + 15$ [Simplifying]

Example 7   Which axioms can be used to show that for every $x \in N$, $4x + 3x = 7x$?

Solution

$4x + 3x = x \cdot 4 + x \cdot 3$ [Commutative axiom of multiplication]
$= x(4 + 3)$ [Distributive axiom]
$= x \cdot 7$ [Simplifying]
$= 7x$ [Commutative axiom of multiplication]

2.1 Five Basic Axioms

Example 7 suggests that the form of the distributive axiom can be modified in several ways. We shall not try to carefully establish the fact that the following true statement is a consequence of the distributive and commutative axioms. For every $x \in N$, $y \in N$, $z \in N$, $(x + y)z = xz + yz$. In fact, we shall often refer to this statement as "the distributive axiom" even though it is merely a similar statement which follows from other assumptions.

While they have caused us no trouble, expressions such as $3 + 2 + 7$ and $2 \cdot 3 \cdot 5$ are, in a way, meaningless. This is because both addition and multiplication are **binary operations**—they connect a *pair* of numbers. Thus, the sum of the two numbers 3 and 2 is a third number, 5. But $3 + 2 + 7$ indicates a sum of three numbers! Since $(3 + 2) + 7$ and $3 + (2 + 7)$ have the same value, let us agree that $3 + 2 + 7$ will mean either $(3 + 2) + 7$ or $3 + (2 + 7)$. And we will agree that a product of three numbers such as $2 \cdot 3 \cdot 5$ means either $(2 \cdot 3)5$ or $2(3 \cdot 5)$, since they have the same value. In general let us agree that $x + y + z$ means either $(x + y) + z$ or $x + (y + z)$, since the associative axiom of addition assumes that these latter two expressions have the same value for every natural number replacement of the variables. Similarly, we will agree that xyz means either $(xy)z$ or $x(yz)$ since the associative axiom of multiplication assumes that the latter two expressions have the same value for every natural number replacement of the variables.

Let us now summarize the five basic axioms we have assumed. Each of these is true for every natural number replacement of the variables.

1. Commutative axiom of addition $x + y = y + x$
2. Commutative axiom of multiplication $xy = yx$
3. Associative axiom of addition $(x + y) + z = x + (y + z)$
4. Associative axiom of multiplication $(xy)z = x(yz)$
5. Distributive axiom $\begin{cases} x(y + z) = xy + xz \\ (x + y)z = xz + yz \end{cases}$

Exercises 2.1

Each pair of expressions in Exercises 1–20 is equivalent over N. Name the axiom which can be used to prove this.

1. $x \cdot 4$ and $4x$
2. $3 + x$ and $x + 3$
3. $2(x \cdot 3)$ and $(2x)3$
4. $(3x)2$ and $3(x \cdot 2)$
5. $(5x)7$ and $7(5x)$
6. $(2 + x) + 7$ and $2 + (x + 7)$
7. $(3 + x) + 8$ and $8 + (3 + x)$

42 *The Real Numbers*

8. $(9 + x) + 4$ and $9 + (x + 4)$
9. $5(2x)$ and $(5 \cdot 2)x$
10. $5(2x)$ and $10x$
11. $2 + (3 + x)$ and $5 + x$
12. $(x + 4) + 3$ and $x + 7$
13. $2x + 3$ and $3 + 2x$
14. $2(x + 8)$ and $(x + 8)2$
15. $3(4 + x)$ and $3(x + 4)$
16. $(x + 6)5$ and $(6 + x)5$
17. $3x + 15$ and $3(x + 5)$
18. $2x + 3x$ and $(2 + 3)x$
19. $4x + 5x$ and $9x$
20. $3(x + 2)$ and $(x + 2)3$

For Exercises 21–40 an expression and an axiom are given. Use the axiom to write an expression which is equivalent to the given one over N. Also use arithmetic to simplify this new expression, if possible.

21. $(x + 3) + 9$, associative axiom of addition
22. $3(2x)$, associative axiom of multiplication
23. $3(x + 5)$, distributive axiom
24. $2 + (5 + x)$, associative axiom of addition
25. $7x + 28$, distributive axiom
26. $9 + 3x$, commutative axiom of addition
27. $x \cdot 7 + 2$, commutative axiom of multiplication
28. $5(x + 7)$, distributive axiom
29. $(x + 2)3$, commutative axiom of multiplication
30. $4(5 + x)$, commutative axiom of addition
31. $6(7 + x)$, distributive axiom
32. $(3 + x)4$, commutative axiom of addition
33. $2x + 5x$, distributive axiom
34. $9x + 4x$, distributive axiom
35. $5x + x$, distributive axiom
36. $x + 3x$, distributive axiom
37. $5 + (4 + 3x)$, associative axiom of addition
38. $(2x + 7) + 3$, associative axiom of addition
39. $3[x(x + 1)]$, associative axiom of multiplication
40. $4[3(2x + 1)]$, associative axiom of multiplication

For Exercises 41–60 decide which of the given pairs of expressions are equivalent over N. You need not do so, but you should be able to name the axiom or axioms which can be used to prove they are equivalent, or you should be able to give a replacement for the variable which will show they are not equivalent.

41. $(3x)2$ and $6x$
42. $2x + 5x$ and $7x$
43. $3x + 5$ and $8x$
44. $(2 + x) + 7$ and $x + 9$
45. $3(2x + 5)$ and $6x + 15$
46. $4x + 44$ and $(x + 11)4$
47. $2(5x)$ and $(25)x$
48. $(2 + 3)x$ and $2 + 3x$
49. $(3 + 2x) + 5$ and $2x + 8$
50. $(3x + 1) + 5$ and $3(x + 2)$
51. $2x + 5$ and $2(x + 5)$
52. $2(3x)$ and $5x + x$
53. $2(x + 3)$ and $2x + 23$
54. $2(x + 1) + 3$ and $2x + 5$
55. $3(x + 2) + 1$ and $3x + 4$
56. $5[4(x + 2)]$ and $20(x + 2)$
57. $3(x + 1)2$ and $6x + 6$
58. $2x + 5 + 3x$ and $5x + 5$
59. $2 + 3(x + 1)$ and $5(x + 1)$
60. $2 + 3(x + 1) + 1$ and $3(x + 2)$

2.2 Other Basic Axioms

Let us consider a special set, $\{0, 1\}$. This set has only two members, the numbers 0 and 1. We can find all possible products using these two numbers, $0 \cdot 0 = 0$, $0 \cdot 1 = 0$, $1 \cdot 0 = 0$, and $1 \cdot 1 = 1$. Note that in *every* case the product is again a member of the original set $\{0, 1\}$. Because this is so, we shall say that the set $\{0, 1\}$ is **closed** with respect to multiplication.

Consider again $\{0, 1\}$, but this time let us find all possible sums using the members of the set. $0 + 0 = 0$, $0 + 1 = 1$, $1 + 0 = 1$, and $1 + 1 = 2$. Here we see that the sum of two members of the set $\{0, 1\}$ is not *always* another member of that set. For this reason we shall say that $\{0, 1\}$ is **not closed** with respect to addition.

Let us next return to the set, N, of natural numbers and consider whether N is closed with respect to addition or with respect to multiplication. Since N is an infinite set, we cannot possibly test every sum or product to determine whether or not each is a member of N. But neither can we exhibit a pair of natural numbers whose sum or whose product is *not* a natural number. This suggests that we make two additional assumptions concerning the set N. We shall assume that the sum of any two natural numbers is, again, a natural number, that the set N is closed with respect to addition. We shall call this assumption the **axiom of closure under addition** of natural numbers and express it symbolically by, "For every $x \in N$ and $y \in N$, $(x + y) \in N$." We shall also assume that the product of any two natural numbers is another natural number, that the set N is closed with respect to multiplication. We shall call this assumption the **axiom of closure under multiplication** of natural numbers and express it symbolically by, "For every $x \in N$ and $y \in N$, $(xy) \in N$."

It is also interesting to see whether Q, the set of quotient numbers, is closed with respect to addition and multiplication.

Example 1 Find the simplest forms of these sums and products of quotient numbers.

(a) $\frac{2}{3} + \frac{4}{5}$ (b) $(\frac{2}{3})(\frac{4}{5})$

Solution

(a) $\dfrac{2}{3} + \dfrac{4}{5} = \dfrac{2 \cdot 5 + 3 \cdot 4}{3 \cdot 5} = \dfrac{22}{15}$ (b) $\dfrac{2}{3} \cdot \dfrac{4}{5} = \dfrac{2 \cdot 4}{3 \cdot 5} = \dfrac{8}{15}$

The example should help us recall the general definitions for the sum and product of any two quotient numbers.

$$\frac{a}{b} + \frac{c}{d} = \frac{ad + bc}{bd} \qquad \frac{a}{b} \cdot \frac{c}{d} = \frac{ac}{bd}$$

As the example suggests, it is possible to use these definitions and the assumptions we have made about closure for natural numbers to *prove* that the sum and product of any two quotient numbers is again a quotient number. That is, we can prove it is true that "For every $x \in Q$ and $y \in Q$, $(x + y) \in Q$ and $(xy) \in Q$." Thus the quotient numbers are closed with respect to both addition and multiplication.

Example 2 Find the simplest form of each sum or product of quotient numbers.

(a) $\frac{1}{2} + \frac{2}{3}$ (b) $\frac{2}{3} + \frac{1}{2}$
(c) $(\frac{1}{4})(\frac{3}{2})$ (d) $(\frac{3}{2})(\frac{1}{4})$

Solution (a) $\frac{7}{6}$ (b) $\frac{7}{6}$ (c) $\frac{3}{8}$ (d) $\frac{3}{8}$

From the example we see that $\frac{1}{2} + \frac{2}{3} = \frac{2}{3} + \frac{1}{2}$ and $(\frac{1}{4})(\frac{3}{2}) = (\frac{3}{2})(\frac{1}{4})$. This suggests that these operations with quotient numbers, as well as with natural numbers, are commutative. In fact it is possible to use the general definitions for addition and multiplication of quotient numbers

2.2 Other Basic Axioms

and the assumptions we have made about natural numbers to *prove* that "For every $x \in Q$ and $y \in Q$, $x + y = y + x$ and $xy = yx$." There are commutative properties of both addition and multiplication for the quotient numbers.

Example 3 Use the order indicated by the symbols of grouping and find the simplest form.

(a) $(\frac{1}{2} + \frac{2}{5}) + \frac{1}{3}$

(b) $\frac{1}{2} + (\frac{2}{5} + \frac{1}{3})$

Solution

(a) $(\frac{1}{2} + \frac{2}{5}) + \frac{1}{3} = \frac{9}{10} + \frac{1}{3} = \frac{37}{30}$

(b) $\frac{1}{2} + (\frac{2}{5} + \frac{1}{3}) = \frac{1}{2} + \frac{11}{15} = \frac{37}{30}$

The example shows that $(\frac{1}{2} + \frac{2}{5}) + \frac{1}{3} = \frac{1}{2} + (\frac{2}{5} + \frac{1}{3})$ and suggests that addition of quotient numbers may be associative. Actually it is possible, by using the general definitions and the axioms for natural numbers, to *prove* that there are associative properties of both addition and multiplication with quotient numbers. "For every $x \in Q$, $y \in Q$ and $z \in Q$, $(x + y) + z = x + (y + z)$ and $(xy)z = x(yz)$."

Example 4 Use $\frac{1}{2}$ and $\frac{2}{3}$ as terms and $\frac{3}{4}$ as a factor to illustrate the distributive property of quotient numbers.

Solution $(\frac{3}{4})(\frac{1}{2} + \frac{2}{3}) = (\frac{3}{4})(\frac{7}{6}) = \frac{7}{8}$ and $(\frac{3}{4})(\frac{1}{2}) + (\frac{3}{4})(\frac{2}{3}) = \frac{3}{8} + \frac{1}{2} = \frac{7}{8}$

Again it is possible to prove that "For every $x \in Q$, $y \in Q$ and $z \in Q$, $x(y + z) = xy + xz$." There is a general distributive property for quotient numbers.

The number 1 plays a unique role in multiplication. Let us assume that the product of any natural number and 1 is that same natural number again: "For every $x \in N$, $x \cdot 1 = x$." We shall describe this special multiplicative property which only the number 1 has by calling 1 the **multiplicative identity**. And we shall call our assumption the **multiplicative identity axiom**. We see at once that it is also true that "For every $x \in Q$, $x \cdot 1 = x$." For example, if $x = \frac{2}{3}$, we have

$$\frac{2}{3} \cdot 1 = \frac{2}{3} \cdot \frac{1}{1} = \frac{2 \cdot 1}{3 \cdot 1} = \frac{2}{3}$$

Hence 1 is also the multiplicative identity for the quotient numbers.

So far we have made a total of eight general assumptions about the natural numbers—two closure axioms, two commutative axioms, two associative axioms, a distributive axiom, and a multiplicative identity axiom. And we have shown, by example at least, that these axioms lead to eight analogous properties of addition and multiplication with quotient numbers. We are now ready to make an assumption about the quotient numbers which will *not* be true for the natural numbers.

It often happens that the product of two quotient numbers is 1, the multiplicative identity. For example, $(\frac{2}{3})(\frac{3}{2}) = 1$, $(6)(\frac{1}{6}) = 1$, etc. When the product of two numbers is 1, we will call each factor the **multiplicative inverse** of the other. Then since $(6)(\frac{1}{6}) = 1$, we call $\frac{1}{6}$ the multiplicative inverse of 6, and we call 6 the multiplicative inverse of $\frac{1}{6}$.

Example 5 Find the multiplicative inverse.

(a) $\frac{4}{9}$ (b) $2\frac{3}{4}$ (c) 1.25   (d) 1

Solution

(a) $\frac{9}{4}$ (b) $\frac{4}{11}$ (since $2\frac{3}{4} = \frac{11}{4}$)
(c) $\frac{4}{5}$ (since $1.25 = \frac{5}{4}$)   (d) 1

Let us assume that every quotient number has a multiplicative inverse which is also a quotient number. We shall call this assumption the **multiplicative inverse axiom** of quotient numbers and represent it symbolically by, "For every $x \in Q$, there is some $(1/x) \in Q$, such that $x \cdot (1/x) = 1$." Note that the natural number 3 has a multiplicative inverse, $\frac{1}{3}$, but this multiplicative inverse is not a natural number. It is *false* that "For every $x \in N$, there is some $(1/x) \in N$, such that $(x) \cdot (1/x) = 1$."

Let us now consider the set $I = \{\ldots, -2, -1, 0, 1, 2, \ldots\}$. In the next section we will make general definitions for the sum and product of any two integers, and we will deliberately choose those definitions in such a way that the eight general axioms we have assumed for the natural numbers will lead to eight analogous properties for the integers! Before doing so, however, we need to observe that I contains a special number, zero, which plays a unique role in adding. Let us assume that "$x + 0 = x$ for every $x \in N$, $x \in Q$, or $x \in I$." That is, we are assuming that the sum of any number (natural number, quotient number, or integer) with zero is that same number again. We shall call zero the **additive identity** and call our assumption the **additive identity axiom.**

Another assumption about the integers has already been made when we first discussed the integers in Section 1.5. We called -1 the additive inverse of 1, since $1 + -1 = 0$. Whenever the sum of two integers is zero, the additive identity, let us call each one the **additive inverse** of the other. We shall assume that the additive inverse of each integer is again an integer. "For every $x \in I$, there is some $-x \in I$, such that $x + -x = 0$." We shall call this assumption the **additive inverse axiom** of integers.

2.2 Other Basic Axioms

Example 6 For what replacement of a will the following be a true statement? "For every $x \in I$, $x = x + a$."

Solution Since $x = x + 0$ for every $x \in I$, we see that the only replacement for a is 0.

The example shows that whenever the sum of an integer and some other number is that same integer again, then the other number must be zero, the additive identity.

Let us assume that the integers have the same basic properties we have been discussing for the natural numbers and show how those properties can be used to prove a theorem about the integers

Theorem 1

> For every $x \in I$, $x \cdot 0 = 0$

Proof Each of the following is true for every $x \in I$.

$$\begin{aligned} x &= x \cdot 1 & &\text{[Multiplicative identity]} \\ &= x(1 + 0) & &[1 = 1 + 0] \\ &= x \cdot 1 + x \cdot 0 & &\text{[Distributive property]} \\ &= x + x \cdot 0 & &\text{[Multiplicative identity]} \end{aligned}$$

Then $x = x + x \cdot 0$. Hence we see that the sum of x and some other number, $x \cdot 0$, is x again. Example 6 shows that this other number, $x \cdot 0$, must be zero. Therefore, $x \cdot 0 = 0$.

Paired with every positive integer there is a negative integer which is its additive inverse. While 1 and -1, for example, are two distinct integers, we may emphasize the fact that they form a pair by saying that they both have the same **absolute value,** namely 1. Similarly, both 2 and -2 can be said to have absolute value 2, and the absolute value of both 3 and -3 is 3. In general we shall say that the absolute value of any positive integer is that integer, while the absolute value of a negative integer is the additive inverse of that integer. We shall indicate the absolute value of -4 in symbols by writing $|\text{-}4|$, and we see that $|\text{-}4| = 4$. Similarly, we have $|\text{-}5| = 5$ and $|6| = 6$. For the integer zero we shall agree that $|0| = 0$. As the next example shows, the symbols for absolute value are also considered symbols of grouping and indicate "first."

Example 7 Find the simplest form.
(a) $|4| + |\text{-}3|$
(b) $|4 - 3|$
Solution (a) $|4| + |\text{-}3| = 4 + 3 = 7$ (b) $|4 - 3| = |1| = 1$

48   The Real Numbers

Exercises 2.2

Decide whether each of the given sets in Exercises 1–8 is closed or not closed with respect to first addition and then multiplication.

1. $\{1\}$
2. $\{0\}$
3. $\{1, 2\}$
4. $\{1, 2, 3, \ldots, 100\}$
5. $\{2, 4, 6, 8, \ldots\}$
6. $\{1, 3, 5, 7, \ldots\}$
7. $\{5, 10, 15, 20, \ldots\}$
8. $\{10, 20, 30, \ldots, 1000\}$

Consider a new operation called "average" represented by the symbol $*$. Then by $4 * 10$ we mean the average of 4 and 10, so $4 * 10 = \dfrac{4 + 10}{2} = 7$. In general, $a * b = \dfrac{a + b}{2}$. Decide whether each of the sets in Exercises 9–16 is closed or not closed with respect to the operation, average.

9. $\{0\}$
10. $\{1\}$
11. $\{0\} \cup \{1\}$
12. $\{\frac{3}{2}\}$
13. $\{1, 3, 5, 7, \ldots\}$
14. $\{2, 4, 6, 8, \ldots\}$
15. N
16. Q

For Exercises 17–24 use the named property to write an expression which is equivalent to the given one over Q. Use arithmetic to simplify it, if possible.

17. $4(x + \frac{1}{2})$, distributive property
18. $(2x + \frac{1}{5}) + \frac{4}{5}$, associative property of addition
19. $(\frac{2}{3})(6x)$, associative property of multiplication
20. $(x)(\frac{3}{4}) + \frac{1}{2}$, commutative property of multiplication
21. $(x + \frac{1}{3})(3)$, commutative property of multiplication
22. $(\frac{3}{5})(\frac{3}{8} + x)$, commutative property of addition
23. $[(x + \frac{1}{2})(\frac{3}{8})](16)$, associative property of multiplication
24. $(\frac{2}{5})[15(x + \frac{1}{3})]$, associative property of multiplication

For Exercises 25–32 find the simplest form of the multiplicative inverse.

25. $\frac{3}{7}$
26. $\frac{1}{8}$
27. 4
28. $\frac{1}{2} + \frac{1}{3}$
29. $4\frac{2}{5}$
30. 3.4
31. $1.2 + \frac{1}{2}$
32. $(\frac{3}{5})(1\frac{2}{3})$

The simplest form of an integer is one of the symbols in the set $\{\ldots -2, -1, 0, 1, 2, \ldots\}$. Assume the integers have all of the basic properties discussed for the natural numbers, and find the simplest form for the expressions in Exercises 33–52.

33. $12 + {-12}$
34. $(-17)(0)$
35. $-11 + 0$
36. $0 + {-5}$
37. $(2)(5 + {-5})$
38. $(3 + {-3})(-7)$
39. $(-8)(-9)(0)$
40. $3 + (4 + {-4})$
41. $-6 + (-8 + 8)$
42. $4 + 3 + {-3}$
43. $8 + {-5}$ [Hint: $8 = 3 + 5$]
44. $9 + {-4}$
45. $-3 + {-4} + 4 + 3$
46. $-2 + {-6} + 8$ [Hint: $8 = 6 + 2$]
47. $(-2)(3 + {-3})$
48. $(-2)(4) + (-2)(-4)$
49. $|-5| + 3$
50. $|4 + {-4}|$
51. $|4| + |-4|$
52. $|6 + {-2}|$

2.3 Integers: Addition and Multiplication

In this section we shall develop the definitions for adding and multiplying integers in such a way that all of the basic properties of natural numbers are also valid for the integers. Let us summarize those properties now, so that it will be clear just what our objective is. We want each of these to be true for every integer replacement of the variables.

1. $x + y = y + x$
2. $xy = yx$ [Commutative properties]

3. $(x + y) + z = x + (y + z)$
4. $(xy)z = x(yz)$ [Associative properties]

5. $x(y + z) = xy + xz$
   $(x + y)z = xz + yz$ [Distributive property]

6. $x + 0 = x$
7. $x \cdot 1 = x$ [Identity elements]

In addition we shall want the following to be true.

8. For every $x \in I$, $y \in I$, $(x + y) \in I$
9. For every $x \in I$, $y \in I$, $(xy) \in I$ [Closure]

10. For every $x \in I$, there is $-x \in I$,   [Inverse elements]
    such that $x + {-x} = 0$

The Real Numbers

For the positive integers, these objectives may be readily met by making an obvious agreement. Since a positive integer is also a natural number, we shall agree that the sum or product of two positive integers is found by adding or multiplying these natural numbers according to the rules already established for them.

Now consider the sum of two negative integers. How shall we define -3 + -4? We may approach this problem indirectly by first considering the sum of 7 and (-3 + -4). The associative, inverse and identity properties will require that 7 + (-3 + -4) = (4 + 3) + (-3 + -4) = (4 + -4) + (3 + -3) = 0 + 0 = 0. But this shows that the sum of 7 and (-3 + -4) is 0, and hence that 7 and (-3 + -4) are additive inverses. Then -3 + -4 = -7, the additive inverse of 7.

Example 1 Use the above method to show that -5 + -8 = -13.

Solution 13 + (-5 + -8) = (8 + 5) + (-5 + -8) = (8 + -8) + (5 + -5) = 0 + 0 = 0. Since 13 + (-5 + -8) = 0, then -5 + -8 = -13, the additive inverse of 13.

Example 2 Find the simplest form of these sums.
(a) -10 + -2   (b) -5 + -5   (c) -73 + -215

Solution (a) -12   (b) -10   (c) -288

Next we shall consider several sums of positive and negative integers.

$$8 + {-5} = (3 + 5) + {-5} = 3 + (5 + {-5}) = 3 + 0 = 3$$

Example 3 Use the above method to find these sums.
(a) 8 + -6   (b) 8 + -7

Solution

(a) 8 + -6 = (2 + 6) + -6 = 2 + (6 + -6) = 2 + 0 = 2
(b) 8 + -7 = (1 + 7) + -7 = 1 + (7 + -7) = 1 + 0 = 1

To find the sum of 8 and -9 we use the fact that -8 + -1 = -9. Then

$$8 + {-9} = 8 + ({-8} + {-1}) = (8 + {-8}) + {-1} = 0 + {-1} = {-1}$$

Example 4 Use a similar method to find the simplest form.
(a) 8 + -10   (b) 8 + -11

Solution

(a) 8 + -10 = 8 + (-8 + -2) = (8 + -8) + -2 = 0 + -2 = -2
(b) 8 + -11 = 8 + (-8 + -3) = (8 + -8) + -3 = 0 + -3 = -3

2.3 Integers: Addition and Multiplication

Let us now take a close look at the sums we have so far considered.

$$-3 + {-4} = -7 \qquad 8 + {-5} = 3$$
$$-5 + {-8} = -13 \qquad 8 + {-6} = 2$$
$$-10 + {-2} = -12 \qquad 8 + {-7} = 1$$
$$-5 + {-5} = -10 \qquad 8 + {-8} = 0$$
$$-73 + {-215} = -288 \qquad 8 + {-9} = -1$$
$$8 + {-10} = -2$$
$$8 + {-11} = -3$$

We see that in every case the sum of two negative integers is a negative integer, while the sum of a positive integer and a negative integer is sometimes positive and sometimes negative. However, the absolute value of an integer can be used to state some simple rules which will make it easier to find sums of integers.

We first note that when adding two negative integers the absolute value of the sum is the sum of the absolute values of the terms. For example, $-3 + {-4} = -7$. The absolute values of the terms are 3 and 4, while the absolute value of the sum is 7. Then we note that when adding a positive and a negative integer, the absolute value of the sum is the *difference* of the absolute values of the terms! For example, $8 + {-5} = 3$. The absolute values of the terms are 8 and 5, while the absolute value of the sum is 3, the difference of 8 and 5. Again with $8 + {-10} = -2$, the absolute values of the terms are 8 and 10, and the absolute value of the sum is 2, the difference $10 - 8$. Finally we note that for the sum of a positive and a negative integer, whenever the positive term has the larger absolute value, the sum is positive, but whenever the negative term has the larger absolute value, the sum is negative. For example, $8 + {-7} = 1$; the positive term, 8, has the larger absolute value, and the sum is positive. But $8 + {-11} = -3$; the negative term, -11, has the larger absolute value, and the sum is negative.

These observations are actually quite general and can be collected into a simple two-part rule for adding integers:

1. To add two integers which are both positive or both negative add their absolute values. The sum is positive if both terms are positive; the sum is negative if both terms are negative.
2. To add one positive and one negative integer subtract their absolute values. If the term with larger absolute value is positive, the sum is positive; if the term with larger absolute value is negative, the sum is negative.

Example 5 For each sum indicate: (a) whether the absolute values of the terms should be added or subtracted; (b) whether the sum is a positive integer or a negative integer; (c) the simplest form of the integer.

(i) $-4 + 9$ (ii) $-8 + {-2}$ (iii) $3 + {-5}$

Solution

(i) (a) Subtract (b) Positive (c) 5
(ii) (a) Add (b) Negative (c) -10
(iii) (a) Subtract (b) Negative (c) -2

Now let us turn our attention to multiplication of integers. We shall first consider the product of a positive and a negative integer, $(3)(-4)$. We may approach this product indirectly by first finding the sum of 12 and $(3)(-4)$. $12 + (3)(-4) = (3)(4) + (3)(-4) = (3)(4 + {-4}) = 3 \cdot 0 = 0$. But since the sum of 12 and $(3)(-4)$ is zero, then $(3)(-4)$ must be the additive inverse of 12, or $(3)(-4) = -12$.

Example 6 Show that $(-5)(6) = -30$.

Solution $30 + (-5)(6) = (5)(6) + (-5)(6) = (5 + {-5})(6) = 0 \cdot 6 = 0$. Hence $(-5)(6) = -30$, the additive inverse of 30.

Similar treatment would give $(-7)(4) = -28$ and $(12)(-3) = -36$.

Next consider the product of two negative integers, $(-3)(-4)$. Again we will make an indirect approach by first considering the sum of this product and -12. We will use the above result that $(3)(-4) = -12$. Then $(-3)(-4) + -12 = (-3)(-4) + (3)(-4) = (-3 + 3)(-4) = (0)(-4) = 0$. But since the sum of $(-3)(-4)$ and (-12) is 0, then this product must be the additive inverse of -12. Hence $(-3)(-4) = 12$.

Example 7 Show that $(-5)(-6) = 30$.

Solution $(-5)(-6) + -30 = (-5)(-6) + (-5)(6) = (-5)(-6 + 6) = (-5)(0) = 0$. Therefore $(5)(-6) = 30$, the additive inverse of -30.

Similarly we have $(-12)(-3) = 36$ and $(-7)(-4) = 28$.
Let us examine the products we have found so far.

$(3)(-4) = -12$ $(-3)(-4) = 12$
$(-5)(6) = -30$ $(-5)(-6) = 30$
$(-7)(4) = -28$ $(-7)(-4) = 28$
$(12)(-3) = -36$ $(-12)(-3) = 36$

Note that whenever one factor is positive and the other factor is negative, the product is negative. On the other hand we see that when both factors are negative the product is positive. Also note that in *all* cases the absolute value of the product can be obtained by multiplying the absolute values of the factors.

2.3 Integers: Addition and Multiplication

These observations are also quite general and we collect them into another two-part rule for multiplying integers:

> 1. To multiply two integers which are both positive or both negative, simply multiply their absolute values. The product is this positive result.
> 2. To multiply one positive integer and one negative integer, multiply their absolute values. The product is the additive inverse of this result.

Neither the rules for adding integers nor the rules for multiplying them have considered zero as a term or as a factor. We do not need to do so, for our assumptions about the integers already yield. "For every $x \in I$, $x + 0 = x$, and $x \cdot 0 = 0$."

Example 8 Let $f(x) = 3x + 5$, $\{-1, -2, -3\}$. Find the range of f.

Solution

$$f(-1) = 3(-1) + 5 = -3 + 5 = 2$$
$$f(-2) = 3(-2) + 5 = -6 + 5 = -1$$
$$f(-3) = 3(-3) + 5 = -9 + 5 = -4$$

The range = $\{2, -1, -4\}$.

Example 9 Find the solution set. $x(x + 3) + 6 = -2x$, $\{-1, -2, -3\}$.

Solution If $x = -1$, the open sentence becomes

$$(-1)(-1 + 3) + 6 = (-2)(-1)$$
$$(-1)(2) + 6 = 2$$
$$-2 + 6 = 2$$
$$4 = 2 \quad \text{False}$$

If $x = -2$, the sentence becomes

$$(-2)(-2 + 3) + 6 = (-2)(-2)$$
$$(-2)(1) + 6 = 4$$
$$-2 + 6 = 4$$
$$4 = 4 \quad \text{True}$$

If $x = -3$, the sentence becomes

$$(-3)(-3 + 3) + 6 = (-2)(-3)$$
$$(-3)(0) + 6 = 6$$
$$0 + 6 = 6$$
$$6 = 6 \quad \text{True}$$

The solution set = $\{-2, -3\}$.

54   The Real Numbers

Exercises 2.3

For Exercises 1–8 indicate: (a) whether the absolute values of the terms should be added or subtracted; (b) whether the sum is positive or negative; (c) the simplest form of the sum.

1. -12 + -3
2. -11 + 5
3. 13 + -9
4. 4 + -8
5. -21 + -21
6. -6 + 12
7. -10 + -5
8. 17 + -3

For Exercises 9–32 find the simplest form.

9. (-5)(-8)
10. (-6)(9)
11. (4)(-4)
12. (-5)(-5)
13. (-8) + (-5)
14. (-4) + 10
15. (-3) + (3)
16. (-9) + (6)
17. (3)(-2) + 5
18. (-4)(-5) + (-3)
19. (-8)(4) + (-5)
20. (-2)(-7) + (3)(5)
21. (-5)(-2 + -2)
22. (-4 + -5)(6)
23. (8 + -2)(-4)
24. (-3)(-3 + -3)
25. (-2)(-2)(3)
26. (3)(-5)(2)
27. (-2)(-2)(-2)
28. (4)(-1)(-5)
29. -6 + -2 + -5
30. -8 + 4 + -9
31. 3 + -8 + -1
32. -6 + 12 + -2

For Exercises 33–40 find the range of the function.

33. $f(x) = 3x + 2$, {-2}
34. $f(x) = x(x + 3)$, {-5}
35. $f(x) = -4x + 1$, {-3}
36. $f(x) = -5 + -3x$, {-4}
37. $f(x) = -3x(x + 1)$, {-1, -2}
38. $f(x) = x(2x + 3) + -5$, {-1, -2}
39. $f(x) = (x + 3)(x + 4)$, {-3, -7}
40. $f(x) = x(x + 1)(x + 2)$, {-2, -3}

For Exercises 41–48 find the solution set.

41. $3(x + 2) = x$, {-1, -2, -3}

42. $2x + 5 = x + 3$, $\{-1, -2, -3\}$
43. $x(x + 3) = 2x$, $\{-1, -2, -3\}$
44. $(x + 3)(x + 2) = 0$, $\{-1, -2, -3\}$
45. $x + 4 = 3$, I
46. $2x + 8 = 0$, I
47. $(x + 1)(x + 2) = 0$, I
48. $2x + 3 = 1$, I

For Exercises 49–56 find the zeros of the function.

49. $f(x) = x(x + 3) + 2$, $\{-1, -2, -3\}$
50. $f(x) = 3(2x + 1) + 3$, $\{-1, -2, -3\}$
51. $f(x) = (x + 1)(x + 2)$, $\{-1, -2, -3\}$
52. $f(x) = x(x + 5) + 6$, $\{-1, -2, -3\}$
53. $f(x) = x + 5$, I
54. $f(x) = (x + 2)(x + 3)$, I
55. $f(x) = 2x + 10$, I
56. $f(x) = (2x + 6)(x + 1)$, I

2.4 Integers: Subtraction and Division

Recall that with natural numbers the operation of subtraction is defined as the inverse of addition. For example, we say that the *difference* of 17 and 9 is 8, since the *sum* of 8 and 9 is 17. In the statement $17 - 9 = 8$ we call 17 the minuend, 9 the subtrahend, and 8 the difference. Then, in general, the difference of two natural numbers is a number whose sum with the subtrahend equals the minuend. We will adopt exactly this same definition for subtraction with integers.

DEFINITION

The difference of two integers is a number whose sum with the subtrahend equals the minuend. In symbols, $a - b = d$, such that $d + b = a$.

Example 1 Use the definition to find the simplest form of these differences.
(a) $3 - {-8}$ (b) $-12 - 4$ (c) $-15 - {-7}$

Solution

(a) $3 - {-8} = 11$, since $11 + {-8} = 3$
(b) $-12 - 4 = -16$, since $-16 + 4 = -12$
(c) $-15 - {-7} = -8$, since $-8 + {-7} = -15$

56   The Real Numbers

As the example shows, the definition can be easily used to "check" the difference, but it is not too useful for actually finding the difference. Consider again Example 1(c), $^-15 - {}^-7 = {}^-8$. Let us form the sum of the minuend and the additive inverse of the subtrahend. Remember the minuend is $^-15$ and the subtrahend is $^-7$, so the sum we want is $^-15 + 7 = {}^-8$. The fact that the *difference* of two integers is also equal to the *sum* of the minuend and the additive inverse of the subtrahend is generally true as is shown in the following theorem.

Theorem 2

> If $a \in I$ and $b \in I$, then $a - b = a + {}^-b$

Proof We shall use the definition, which requires that the sum of the difference and the subtrahend equals the minuend:

$(a + {}^-b) + b = a + ({}^-b + b)$ [**Associative property of addition**]
$\phantom{(a + {}^-b) + b} = a + 0$ [**Additive inverse**]
$\phantom{(a + {}^-b) + b} = a$ [**Additive identity**]

Then we have shown that the sum of the difference $(a + {}^-b)$ and the subtrahend b, equals the minuend a.

Example 2   Use the theorem to find the simplest form.
(a) $14 - {}^-9$ (b) $^-8 - 7$ (c) $0 - {}^-19$

Solution

(a) $14 - {}^-9 = 14 + 9 = 23$ (b) $^-8 - 7 = {}^-8 + {}^-7 = {}^-15$
(c) $0 - {}^-19 = 0 + 19 = 19$

The expression $4 - {}^-3$ illustrates the two uses we have made of the "minus sign." The first minus sign indicates subtraction, while the second elevated minus sign indicates the additive inverse. Then $4 - {}^-3$ means the difference of 4 and the additive inverse of 3. And we have shown that $4 - {}^-3 = 4 + 3 = 7$. Often we will find it more convenient to indicate the additive inverse by using a minus sign which is not elevated. The context of its use will make it clear whether this symbol means to subtract or to find the additive inverse.

Example 3   Find the simplest form of each of the following:
(a) $4 - 2$ (b) $7 + (-2)$ (c) $(-5)(4)$ (d) $-15 - (-1)$

Solution

(a) Here $-$ indicates subtraction and $4 - 2 = 2$.
(b) Here $-$ indicates additive inverse, as the parentheses make clear, and $7 + (-2) = 5$.
(c) Again, the parentheses used with the minus sign indicate the first factor is the additive inverse of 5. Thus $(-5)(4) = -20$.
(d) The first $-$ indicates additive inverse since no minuend precedes 15. The second $-$ is a subtraction symbol. And the third $-$ again illustrates additive inverse. Thus $-15 - (-1) = -15 + 1 = -14$.

Example 4 Use the integers 7 and -4 to show that subtraction is *not* a commutative operation.

Solution $(7) - (-4) = 11; (-4) - (7) = -11$

Note that when the order of subtraction is reversed, the differences are additive inverses. The next theorem shows that this is generally true.

Theorem 3

If $a \in I$ and $b \in I$, then $a - b = -(b - a)$

Proof We form the sum of $a - b$ and $b - a$. Then

$(a - b) + (b - a)$
$= [a + (-b)] + [b + (-a)]$ [Theorem 2]
$= [a + (-b) + b] + (-a)$ [Associative property of addition]
$= (a + 0) + (-a)$ [Additive inverse]
$= a + (-a)$ [Additive identity]
$= 0$ [Additive inverse]

Since the sum of $a - b$ and $b - a$ is zero, we see that each is the additive inverse of the other. Hence $a - b = -(b - a)$.

Example 5 Use the integers -4, 7, and -8 to show that subtraction is not an associative operation.

Solution $[(-4) - (7)] - (-8) = (-11) - (-8) = -3;$
$(-4) - [(7) - (-8)] = (-4) - (15) = -19$

We have shown by example that subtraction of integers is neither commutative nor associative. Is the set of integers closed with respect to sub-

58   The Real Numbers

traction? Theorem 2 says that the difference of two integers is also equal to the sum of the minuend and the additive inverse of the subtrahend. We have assumed that the additive inverse of every integer is also an integer, and we have assumed that the sum of any two integers is also an integer. Hence, we see that the difference of two integers is always an integer, and I is closed with respect to subtraction.

Example 6   Are these subsets of I closed with respect to subtraction?
(a) $\{0\}$ (b) $\{2, -2, 3, -3, 4, -4, \ldots\}$ (c) N

Solution

(a) Closed $[0 - 0 = 0]$
(b) Not closed $[2 - 2 = 0,$ and 0 is not a member of the set$]$
(c) Not closed $[2 - 5 = -3,$ and -3 is not a member of $N]$

As with the natural numbers, we define division of integers to be the inverse of multiplication. Then $(-15) \div (3) = -5$, since $(-5)(3) = -15$. Similarly, $\dfrac{-44}{-4} = 11$, since $(11)(-4) = -44$. In the statement $\dfrac{-44}{-4} = 11$, we call -44 the dividend, -4 the divisor, and 11 the quotient. Note that the product of the quotient and the divisor equals the dividend.

DEFINITION

The quotient of two integers is a number whose product with the divisor equals the dividend. In symbols, $a/b = q$, such that $qb = a$.

Example 7   Find the simplest form of the quotient.
(a) $\dfrac{45}{-9}$ (b) $\dfrac{-15}{-5}$ (c) $(27) \div (-5)$

Solution

(a) $\dfrac{45}{-9} = -5$, since $(-5)(-9) = 45$

(b) $\dfrac{-15}{-5} = 3$, since $(3)(-5) = -15$

(c) There is no integer whose product with -5 equals 27. Hence $(27) \div (-5)$ is not an integer.

2.4 Integers: Subtraction and Division

As is seen in the example, when the quotient of two integers is an integer, there are rules analogous to those for multiplication. That is, the quotient of two positive or two negative integers is positive, but the quotient of one positive and one negative integer is negative. And of course the last example shows that the set of integers is not closed with respect to division.

There is a problem with division when we try to use zero as the divisor. Consider $3 \div 0$, which we may also write $\frac{3}{0}$. Our definition requires that the quotient must be a number whose product with 0, the divisor, equals 3, the dividend. Perhaps the first impulse is to say that $\frac{3}{0} = 0$. But then we must have $0 \cdot 0 = 3$, a false statement. If we try $\frac{3}{0} = 3$, we are lead to the equally false statement, $3 \cdot 0 = 3$. In fact there is no number whose product with zero is 3, and hence we must conclude that $3 \div 0$ is not defined. Thus $\frac{3}{0}$ is not a numeral since *it does not represent any number at all!* A similar argument will show that $\frac{1}{0}, \frac{2}{0}, \frac{4}{0}, \frac{5}{0}$, etc. and $-\frac{1}{0}, -\frac{2}{0}, -\frac{3}{0}$, etc. are not numerals and do not represent numbers.

The expression $\frac{0}{0}$ presents a unique difficulty. We might write $\frac{0}{0} = 0$ since $0 \cdot 0 = 0$. But we might also accept $\frac{0}{0} = 1$, since $1 \cdot 0 = 0$. Similarly, it is reasonable to conclude $\frac{0}{0} = 2$ since $2 \cdot 0 = 0$ and $\frac{0}{0} = 17$ since $17 \cdot 0 = 0$. We see that $\frac{0}{0}$ is hopelessly ambiguous so we say that $\frac{0}{0}$ is not a numeral since it may represent any number. In conclusion we shall agree that division by zero is undefined and if the denominator of a fraction is a symbol which represents zero, then that fraction does not represent any number at all, but is in fact undefined.

Example 8 Determine which of the following are defined, and find the simplest form for those which are.

(a) $\dfrac{-7}{2-2}$

(b) $\dfrac{(-3)-(-3)}{3+3}$

(c) $0 \div (-5)$

(d) $\dfrac{(-2)(-3)-6}{2(-3)+6}$

Solution

(a) $\dfrac{-7}{2-2} = \dfrac{-7}{0}$, not defined

(b) $\dfrac{(-3)-(-3)}{3+3} = \dfrac{-3+3}{6} = \dfrac{0}{6} = 0$. Note that the numerator of a fraction may be zero.

(c) $0 \div (-5) = \dfrac{0}{-5} = 0$, defined

(d) $\dfrac{(-2)(-3)-6}{2(-3)+6} = \dfrac{6-6}{-6+6} = \dfrac{0}{0}$, not defined

Example 9 Use the integers -4 and 2 to show that division is not commutative.

Solution

$(-4) \div (2) = -2$ but $(2) \div (-4)$ is not an integer, certainly not -2

Example 10 Use the integers -24, 6, and 2 to show that division is not associative.

Solution

$$[(-24) \div 6] \div 2 = (-4) \div 2 = -2$$
$$(-24) \div [6 \div 2] = (-24) \div 3 = -8$$

Example 11 Use the assumed properties to find an expression which is equivalent over I to the given one.
(a) $3(x - 5)$ (b) $-2x - 3x$

Solution

(a) $3(x - 5) = 3[x + (-5)] = 3x + 3(-5) = 3x + (-15) = 3x - 15$
(b) $-2x - 3x = (-2x) + (-3x) = [(-2) + (-3)]x = (-5)x = -5x$

Exercises 2.4

For Exercises 1–8: (a) identify the subtrahend, (b) indicate the sum of the minuend and the additive inverse of the subtrahend, and (c) find the simplest form of the integer.

1. $(-3) - (-8)$
2. $(-4) - (5)$
3. $(8) - (-2)$
4. $(3) - (10)$
5. $8 - 12$
6. $-5 - 2$
7. $-6 - (-4)$
8. $-9 - 4$

For Exercises 9–32 find the simplest form.

9. $-8 + 2$
10. $-15 + (-12)$
11. $8 - (-4)$
12. $4 + (-9)$
13. $(-3)(-5) + 2$
14. $(5)(-6) - 15$
15. $-12 - (3)(-2)$
16. $(-3)(-2) - (-4)(-2)$
17. $-5 + (-2)(6)$
18. $(4)(-9) - (-3)(2)$
19. $(-4) - (2)(-3)(-5)$
20. $-3 - 5 - 4(-2)$
21. $5 - 9 + 2(-5)$
22. $-4 - (-2 + 5)$

2.4 Integers: Subtraction and Division

23. $6 + (-2 - 9)$
24. $(-4)(-2) - (3 - 9)$
25. $-\dfrac{16}{8}$
26. $\dfrac{3 - 9}{-2}$
27. $\dfrac{-4 - 8}{4}$
28. $\dfrac{5 - 9}{2 - 3}$
29. $(-30) \div (-4 - 6)$
30. $\dfrac{(-2)(-2 + 5)}{-2 - (4 - 7)}$
31. $\dfrac{8 - (4 - 5)}{-2 + (3 - 4)}$
32. $\dfrac{(2)(-4) - (-2)(4)}{-3 - (3 - 3)}$

For Exercises 33–44 find the range of the function.

33. $f(x) = 2x - 7$, $\{-3\}$
34. $f(x) = 8 - 3x$, $\{-3\}$
35. $f(x) = -2(3x - 1)$, $\{-2\}$
36. $f(x) = -5x - 8$, $\{-1\}$
37. $f(x) = -4 - 3x$, $\{2, -2\}$
38. $f(x) = x(4 - 3x)$, $\{3, -3\}$
39. $f(x) = (-x)(-2 + 3x)$, $\{4, -4\}$
40. $f(x) = -x - (4 - x)$, $\{5, -5\}$
41. $f(x) = (2x - 3)(x - 1)$, $\{-1, -2, -3\}$
42. $f(x) = (x - 3)(2 - x)$, $\{-1, -2, -3\}$
43. $f(x) = (4 - x)(2 + x)$, $\{-1, -2, -3\}$
44. $f(x) = (3 - 2x)(x - 4)$, $\{-1, -2, -3\}$

For Exercises 45–52 find an expression which is equivalent over I to the given one. There is more than one possible answer.

45. $5x - 3x$
46. $-2x - 5x$
47. $3(x - 2) + 5$
48. $2(3x - 4) - 1$
49. $2x + 1 - 5x$
50. $-(3x - 2)$
51. $x - (4 - 3x)$
52. $2x - 3(x + 1)$

For Exercises 53–60 decide whether the given set is closed with respect to the given operation.

53. $\{0, 1\}$, subtraction
54. $\{0, 1\}$, division

55. $\{-1, 1\}$, multiplication
56. $\{-1, 1\}$, division
57. $\{-1, 0, 1\}$, addition
58. $\{-1, 0, 1\}$, subtraction
59. $\{-1, 0, 1\}$, multiplication
60. $\{-1, 0, 1\}$, division

2.5 The Rational Numbers

Let us now turn our attention to the set of rational numbers. We have already considered the four operations of arithmetic with the natural numbers, quotient numbers, and integers. Since these sets are all subsets of F, we will want the definitions and rules we develop for the rational numbers to be consistent with those already discussed for N, Q, and I.

Every *quotient* number can be represented by a fraction whose numerator and denominator are both natural number symbols. But many *different* fractions can represent the *same* quotient number. For example, $\frac{2}{3}$, $\frac{4}{6}$ and $\frac{10}{15}$ all represent the same number, so we call them equal fractions. Notice that $\frac{2}{3} = \frac{4}{6}$ and $2 \cdot 6 = 3 \cdot 4$. More generally we say that if a/b and c/d represent quotient numbers, then $a/b = c/d$ if and only if $ad = bc$. Every *rational* number can be represented by a fraction whose numerator and denominator are integer symbols. (But the denominator may not be 0.) Let us make an analogous assumption about the rational numbers.

If a/b and c/d represent rational numbers, then $a/b = c/d$ if and only if $ad = bc$.

Example 1 Decide whether or not these pairs of numerals are equal.

(a) $\dfrac{-8}{-3}$ and $\dfrac{8}{3}$ (b) $\dfrac{4}{-7}$ and $\dfrac{-7}{4}$ (c) $\dfrac{-3}{4}$ and $\dfrac{6}{-8}$

Solution

(a) $\dfrac{-8}{-3} = \dfrac{8}{3}$, since $(-8)(3) = (-3)(8)$

(b) $\dfrac{4}{-7} \neq \dfrac{-7}{4}$, since $(4)(4) \neq (-7)(-7)$

(c) $\dfrac{-3}{4} = \dfrac{6}{-8}$, since $(-3)(-8) = (4)(6)$

2.5 The Rational Numbers

Canceling is a useful technique when working with fractions which represent quotient numbers. If the numerator and denominator of a fraction have exactly the same factor, this factor may be canceled. For example,

$$\frac{27}{36} = \frac{3 \cdot 9}{4 \cdot 9} = \frac{3}{4}$$

This same canceling property holds in the set F, as the following theorem proves.

Theorem 4

If $a \in I$, $b \in I$, $x \in I$, $b \neq 0$ and $x \neq 0$, then $\dfrac{ax}{bx} = \dfrac{a}{b}$

Proof We will show that the cross products are equal.

$(ax)b = b(ax)$ [Commutative property of multiplication with integers]
$= b(xa)$ [Same as above]
$= (bx)a$ [Associative property of multiplication with integers]

Example 2 Use the canceling theorem to simplify these.

(a) $\dfrac{-24}{-30}$ (b) $\dfrac{-64}{24}$ (c) $\dfrac{18}{-45}$

Solution

(a) $\dfrac{-24}{-30} = \dfrac{(-6)(4)}{(-6)(5)} = \dfrac{4}{5}$ (b) $\dfrac{-64}{24} = \dfrac{(8)(-8)}{(8)(3)} = \dfrac{-8}{3}$

(c) $\dfrac{18}{-45} = \dfrac{(-9)(-2)}{(-9)(5)} = \dfrac{-2}{5}$

Let us adopt the same rules for adding and multiplying rational numbers that we use for quotient numbers.

If $a/b \in F$ and $c/d \in F$, then

$$\frac{a}{b} + \frac{c}{d} = \frac{ad + bc}{bd}, \qquad \frac{a}{b} \cdot \frac{c}{d} = \frac{ac}{bd}$$

64   The Real Numbers

Example 3   Add or multiply as indicated.

(a) $\dfrac{-4}{5} + \dfrac{-3}{-7}$

(b) $\dfrac{-2}{-5} \cdot \dfrac{7}{-9}$

Solution

(a) $\dfrac{-4}{5} + \dfrac{-3}{-7} = \dfrac{(-4)(-7) + (5)(-3)}{(5)(-7)} = \dfrac{28 + (-15)}{-35} = \dfrac{13}{-35}$

(b) $\dfrac{-2}{-5} \cdot \dfrac{7}{-9} = \dfrac{(-2)(7)}{(-5)(-9)} = \dfrac{-14}{45}$

There is an easier method for adding quotient numbers whose denominators are the same. For example,

$$\dfrac{4}{19} + \dfrac{12}{19} = \dfrac{4+12}{19} = \dfrac{16}{19}$$

This method can also be used with rational numbers, as the next theorem illustrates.

Theorem 5   If $a/b \in F$ and $c/b \in F$, then $\dfrac{a}{b} + \dfrac{c}{b} = \dfrac{a+c}{b}$

Proof $\dfrac{a}{b} + \dfrac{c}{b} = \dfrac{ab + bc}{b \cdot b} = \dfrac{ba + bc}{b \cdot b} = \dfrac{b(a+c)}{b \cdot b} = \dfrac{a+c}{b}$

Example 4   Add:

(a) $\dfrac{4}{-7} + \dfrac{36}{-7}$

(b) $\dfrac{-5}{12} + \dfrac{5}{6}$

Solution

(a) $\dfrac{4}{-7} + \dfrac{36}{-7} = \dfrac{4+36}{-7} = \dfrac{40}{-7}$

(b) $\dfrac{-5}{12} + \dfrac{5}{6} = \dfrac{-5}{12} + \dfrac{10}{12} = \dfrac{-5+10}{12} = \dfrac{5}{12}$

We shall represent the additive inverse of a rational number in a manner analogous to that we used for the integers. The additive inverse of $\tfrac{5}{7}$ is represented by $-\dfrac{5}{7}$, and the additive inverse of $\dfrac{-5}{7}$ is $-\dfrac{-5}{7}$. This nota-

2.5 The Rational Numbers

tion leads to some interesting results if we consider the fact that $\frac{5}{7} = \frac{-5}{-7}$ [since $(5)(-7) = (7)(-5)$]. Then $-\frac{5}{7} = -\frac{-5}{-7}$; that is, the additive inverse of $\frac{5}{7}$ equals the additive inverse of $\frac{-5}{-7}$. Also $\frac{-5}{7} = \frac{5}{-7}$ [since $(-5)(-7) = 7 \cdot 5$]. Hence $-\frac{-5}{7} = -\frac{5}{-7}$; the additive inverse of $\frac{-5}{7}$ equals the additive inverse of $\frac{5}{-7}$.

Example 5 Show that $\frac{-5}{7} = -\frac{5}{7}$.

Solution Since $\frac{5}{7} + \frac{-5}{7} = 0$, we see that $\frac{-5}{7}$ is the additive inverse of $\frac{5}{7}$, or in symbols $\frac{-5}{7} = -\frac{5}{7}$.

Of course the example also shows that $-\frac{-5}{7} = \frac{5}{7}$. (The additive inverse of $\frac{-5}{7}$ is $\frac{5}{7}$.) When we combine the results of the example and the notation discussed in the preceding paragraph, we see that $\frac{-5}{7}$, $\frac{5}{-7}$, $-\frac{-5}{-7}$, and $-\frac{5}{7}$ all represent the same rational number. Also, $-\frac{-5}{7} = -\frac{5}{-7} = \frac{-5}{-7} = \frac{5}{7}$.

Example 6 Determine which of these are equal:

$$\frac{-3}{-4}, \frac{3}{-4}, -\frac{3}{4}, -\frac{-3}{-4}, -\frac{3}{-4}, -\frac{-3}{4}, \frac{-3}{4}, \frac{3}{4}$$

Solution

$\frac{-3}{-4} = -\frac{3}{-4} = -\frac{-3}{4} = \frac{3}{4}$ [Note that in each there is an even number of negative signs]

$\frac{3}{-4} = -\frac{3}{4} = -\frac{-3}{-4} = \frac{-3}{4}$ [Note that each has an odd number of negative signs]

66   The Real Numbers

Since there are so many ways to represent any rational number, we shall want to define the simplest form of each. If the rational number is an integer, we have already agreed to represent it with one of the symbols, ..., $-2, -1, 0, 1, 2, \ldots$. Positive rational numbers are also quotient numbers, and we have agreed to represent those which are not integers with fractions whose numerators and denominators are the simplest forms of natural numbers and are relatively prime. That leaves only the negative rational numbers which are not integers; for them let us agree that the simplest form is a fraction whose numerator is the simplest form of a negative integer and whose denominator is the simplest form of a positive integer. Of course, we will want the numerator and denominator to be relatively prime (that is, have no common factors which will cancel).

Example 7   Find the simplest form.

(a) $\dfrac{12}{-4}$ (b) $\dfrac{-3}{-8}$ (c) $\dfrac{-8}{28}$ (d) $-\dfrac{6}{-48}$ (e) $-\dfrac{-5}{-15}$

Solution   (a) -3 (b) $\dfrac{3}{8}$ (c) $\dfrac{-2}{7}$ (d) $\dfrac{1}{8}$ (e) $\dfrac{-1}{3}$

Although we shall usually want to express a rational number in its simplest form, there are times when its decimal form will be helpful. We know that $\dfrac{3}{2} = 1.5$. Then since $\dfrac{-3}{2} = -\dfrac{3}{2}$, we see that $\dfrac{-3}{2} = -1.5$. When the decimal form is not obvious, we may always find it by division as the next example illustrates.

Example 8   Find the decimal form.

(a) $\dfrac{-5}{16}$ (b) $\dfrac{8}{3}$ (c) $\dfrac{-3}{11}$

Solution

(a) $16 \overline{)5.0000}$
$\,\,.3125$
$4\,8$
20
16
40
32
80
80
; $\dfrac{-5}{16} = -\dfrac{5}{16} = -.3125$

2.5 The Rational Numbers

(b) $\quad 3 \overline{\smash{)}8.000}\!\!\!\!\!\!\!\!\!\!\!\!\!\!\!\!\!\!\!\overset{2.666}{}$

$$\begin{array}{r} \underline{6} \\ 20 \\ \underline{18} \\ 20 \\ \underline{18} \\ 2 \end{array}$$

We see that the division never ends and the decimal form has infinitely many digits which repeat in a set pattern. Let us call such a decimal a repeating decimal. We shall write $\frac{8}{3} = 2.\overline{6}\ldots$. Note the use of dots to indicate "and so forth" and the bar over the 6 to indicate that the 6, but not the 2 is repeating.

(c) $\quad 11 \overline{\smash{)}3.0000}\!\overset{.27\overline{27}}{}; \quad \dfrac{-3}{11} = -.\overline{27}\ldots$

$$\begin{array}{r} \underline{2\,2} \\ 80 \\ \underline{77} \\ 30 \\ \underline{22} \\ 80 \\ \underline{77} \\ 3 \end{array}$$

Note the bar over both the 2 and 7 to indicate both these digits repeat.

It can be shown that every rational number has a decimal form which either terminates, such as .3125 or is repeating, such as $.\overline{27}\ldots$. It is also true that every decimal which either terminates or is repeating represents a rational number.

We are now ready to define subtraction for rational numbers. Example 5 suggests that the additive inverse of every rational number is also a rational number. In general if $\dfrac{a}{b} \in F$, then $-\dfrac{a}{b} = \dfrac{-a}{b}$ and so $-\dfrac{a}{b} \in F$.

This fact and Theorem 2 on subtraction of integers lead us to make the following definition.

DEFINITION

The difference of two rational numbers equals the sum of the minuend and the additive inverse of the subtrahend. In symbols, if $a/b \in F$ and $c/d \in F$, then

$$\frac{a}{b} - \frac{c}{d} = \frac{a}{b} + \frac{-c}{d}$$

Example 9 Find the simplest form.

(a) $\dfrac{-7}{12} - \dfrac{4}{12}$ (b) $\dfrac{-8}{9} - \dfrac{3}{-5}$

Solution

(a) $\dfrac{-7}{12} - \dfrac{4}{12} = \dfrac{-7}{12} + \dfrac{-4}{12} = \dfrac{-11}{12}$

(b) $\dfrac{-8}{9} - \dfrac{3}{-5} = \dfrac{-8}{9} + \dfrac{3}{5} = \dfrac{-40+27}{45} = \dfrac{-13}{45}$

Since $\dfrac{-7}{2} \cdot \dfrac{2}{-7} = \dfrac{-14}{-14} = 1$, we see that $\dfrac{-7}{2}$ and $\dfrac{2}{-7}$ are multiplicative inverses. Similarly, every rational number (except zero) has a multiplicative inverse. In general the multiplicative inverse of a/b is b/a, if $a \neq 0$ and $b \neq 0$.

Example 10 Find in simplest form the multiplicative inverse of each of these.

(a) $\dfrac{4}{-9}$ (b) $\dfrac{-11}{-13}$ (c) -3 (d) 0

Solution

(a) $\dfrac{-9}{4}$ (b) $\dfrac{13}{11}$ (c) $\dfrac{-1}{3}$ (d) There is none

We shall now define division with rational numbers in a manner analogous to that for quotient numbers.

DEFINITION

The quotient of two rational numbers (divisor not zero) equals the product of the dividend and the multiplicative inverse of the divisor. In symbols, if $a/b \in F$ and $c/d \in F$ and $c/d \neq 0$, then

$$\frac{a}{b} \div \frac{c}{d} = \frac{a}{b} \cdot \frac{d}{c}$$

2.5 The Rational Numbers

Example 11 Find the simplest form.

(a) $\dfrac{-3}{2} \div \dfrac{2}{5}$ \qquad (b) $\dfrac{-5}{-8} \div \dfrac{25}{-6}$

Solution

(a) $\dfrac{-3}{2} \div \dfrac{2}{5} = \dfrac{-3}{2} \cdot \dfrac{5}{2} = \dfrac{-15}{4}$

(b) $\dfrac{-5}{-8} \div \dfrac{25}{-6} = \dfrac{5}{8} \cdot \dfrac{-6}{25} = \dfrac{(5)(2)(-3)}{(4)(2)(5)(5)} = \dfrac{-3}{20}$

In Example 9(a) we see that $\dfrac{-7}{12} - \dfrac{4}{12} = \dfrac{-11}{12}$. It is worth noting that $\dfrac{-11}{12} + \dfrac{4}{12} = \dfrac{-7}{12}$. The sum of the difference and the subtrahend equals the minuend. Also in Example 11(a) we have $\dfrac{-3}{2} \div \dfrac{2}{5} = \dfrac{-15}{4}$. And $\dfrac{-15}{4} \cdot \dfrac{2}{5} = \dfrac{-3}{2}$. The product of the quotient and the divisor equals the dividend. It is interesting to observe that the definitions we have chosen for subtraction and division with rational numbers are consistent with the definitions we have chosen for these operations with the integers.

Exercises 2.5

For Exercises 1–32 find the simplest form.

1. $\dfrac{12}{-18}$ \qquad 2. $-\dfrac{-10}{15}$

3. $-\dfrac{45}{-15}$ \qquad 4. $-\dfrac{-36}{-15}$

5. $\dfrac{-2}{5} \cdot \dfrac{-3}{5}$ \qquad 6. $\dfrac{18}{-5} \cdot \dfrac{15}{27}$

7. $\dfrac{5}{12} \cdot \dfrac{-8}{15}$ \qquad 8. $-\dfrac{6}{-11} \cdot \dfrac{-22}{9}$

9. $\dfrac{-3}{17} + \dfrac{-22}{17}$ \qquad 10. $\dfrac{5}{12} + \dfrac{-11}{12}$

11. $\dfrac{-3}{-8} + \dfrac{-5}{16}$ \qquad 12. $\dfrac{3}{-10} + \dfrac{-2}{5}$

13. $\dfrac{-3}{5} + \dfrac{3}{4}$ \qquad 14. $\dfrac{-5}{8} + \dfrac{-2}{9}$

15. $\dfrac{7}{15} + \dfrac{-9}{10}$

16. $\dfrac{-4}{25} + \dfrac{-2}{35}$

17. $\dfrac{-3}{7} \div \dfrac{4}{9}$

18. $\dfrac{-15}{11} \div \dfrac{10}{3}$

19. $\dfrac{-5}{8} \div \dfrac{-3}{16}$

20. $\dfrac{20}{-9} \div \dfrac{-10}{3}$

21. $\dfrac{-5}{12} - \dfrac{-7}{12}$

22. $\dfrac{3}{11} - \dfrac{9}{11}$

23. $-\dfrac{4}{15} - \dfrac{2}{15}$

24. $-\dfrac{5}{21} - \dfrac{-8}{21}$

25. $\dfrac{-3}{4} - \dfrac{2}{5}$

26. $\dfrac{-8}{15} - \dfrac{-3}{10}$

27. $\dfrac{-5}{-9} - \dfrac{-2}{-3}$

28. $-\dfrac{-4}{9} - \dfrac{2}{15}$

29. $\dfrac{-3}{4} + 2$

30. $6 - \dfrac{5}{2}$

31. $5 \cdot \dfrac{-2}{3}$

32. $\dfrac{-3}{8} \div -2$

For Exercises 33–56 the domain of each function is F. Find the simplest form of the indicated number in the range.

33. $f(x) = x + 1,\ f(-\tfrac{5}{3})$
34. $g(x) = 3x,\ g(-\tfrac{1}{2})$
35. $h(x) = -x,\ h(-\tfrac{3}{4})$
36. $k(x) = -5x,\ k(-\tfrac{2}{3})$
37. $f(y) = 4 - y,\ f(\tfrac{3}{2})$
38. $g(y) = 1 - 3y,\ g(-\tfrac{3}{2})$
39. $h(y) = -2y + 1,\ h(\tfrac{1}{3})$
40. $k(y) = -2 - y,\ k(-\tfrac{1}{4})$
41. $f(z) = 3(z - 1),\ f(\tfrac{3}{4})$
42. $g(z) = (4 - z)2,\ g(-\tfrac{5}{2})$
43. $h(z) = -z(z + 2),\ h(-\tfrac{1}{3})$
44. $k(z) = (2z - 3) - z,\ k(-\tfrac{1}{2})$
45. $f(w) = (w - 1)(w + 1),\ f(\tfrac{1}{2})$
46. $g(w) = (2w - 1)(w - 3),\ g(-\tfrac{1}{3})$

2.5 The Rational Numbers

47. $h(w) = (3 - 2w)(w + 1)$, $h(-\frac{1}{4})$
48. $k(w) = (2 - w)(5 + 3w)$, $k(-\frac{4}{3})$
49. $f(x) = \dfrac{2x - 3}{x}$, $f(-3)$
50. $g(x) = \dfrac{4 - x}{x + 1}$, $g(-2)$
51. $h(x) = \dfrac{x + 3}{5 - x}$, $h(-6)$
52. $k(x) = \dfrac{3 - x}{2x - 5}$, $k(-2)$
53. $f(y) = \dfrac{y + 1}{y}$, $f\left(-\dfrac{2}{3}\right)$
54. $g(y) = \dfrac{1 - y}{2y}$, $g\left(-\dfrac{3}{4}\right)$
55. $h(y) = \dfrac{3 - 2y}{1 + y}$, $h\left(-\dfrac{5}{2}\right)$
56. $k(y) = \dfrac{-2 - 3y}{3y - 1}$, $k\left(-\dfrac{2}{5}\right)$

For Exercises 57–64 an open sentence and a domain are given. Find the solution set.

57. $3x - 5 = x + 4$, $\{-\frac{1}{2}, \frac{3}{2}, \frac{9}{2}\}$
58. $x + 3 = 4(x - 1)$, $\{-\frac{1}{3}, \frac{2}{3}, \frac{7}{3}\}$
59. $4 - 3x = 10 + 2x$, $\{-\frac{6}{5}, -\frac{1}{5}, \frac{2}{5}\}$
60. $2(x - 3) = 5x + 1$, $\{-\frac{10}{3}, -\frac{7}{3}, -\frac{1}{3}\}$
61. $(2x + 3)(3x - 1) = 0$, $\{-\frac{2}{3}, -\frac{1}{3}, \frac{1}{3}, \frac{2}{3}\}$
62. $(4x - 1)(2x + 1) = 0$, $\{-\frac{1}{2}, -\frac{1}{4}, \frac{1}{4}, \frac{1}{2}\}$
63. $(2 - 3x)(2 + 5x) = 0$, $\{-\frac{2}{3}, -\frac{2}{5}, \frac{2}{3}, \frac{2}{5}\}$
64. $(3x + 4)(2x + 1) = 0$, $\{-\frac{4}{3}, -\frac{3}{4}, -\frac{1}{2}, \frac{4}{3}\}$

For Exercises 65–72 decide whether the given set is closed with respect to the given operation.

65. $\{-\frac{1}{2}, \frac{1}{2}\}$, addition
66. $\{-\frac{1}{2}, \frac{1}{2}\}$, division
67. $\{-1, -\frac{1}{2}, \frac{1}{2}, 1\}$, multiplication

72   The Real Numbers

68. $\{-1, -\frac{1}{2}, \frac{1}{2}, 1\}$, division
69. F, addition
70. F, subtraction
71. F, multiplication
72. F, division

For Exercises 73–76 find the decimal form.

73. $\frac{2}{9}$ **74.** $-\frac{5}{32}$
75. $-\frac{5}{7}$ **76.** $\frac{73}{33}$

2.6 The Number Line

Figure 2.1 shows how to associate a different point on a line with each integer. The points associated with 0 and 1 are chosen first, quite arbitrarily but with 1 to the right of 0. Then the remaining points are chosen along the line so that the distance between successive points is the same as the distance between the points for 0 and 1. Of course we understand that the line continues forever in both directions, and that there is a point for each integer, although this is not explicitly shown in the figure. The number associated with a point is called the **coordinate** of that point. We will usually shorten such phrases as "the point whose coordinate is -3" to "the point -3." While points are not numbers, there will be no confusion, since each point has exactly one coordinate, and each integer corresponds to exactly one point on the line.

There is a well established order for the natural numbers, the order in which we count. We say that 6 is greater than 2, in symbols $6 > 2$, and it is clear what we mean. Similarly, we say that $3 < 5$. Now we wish to order all the rational numbers. Observe that $6 > 2$ and $6 - 2 = 4$, a *positive* number. But $3 < 5$ and $3 - 5 = -2$, a *negative* number. This suggests that we make the following definition for all rational numbers.

> **DEFINITION**
>
> $a > b$ means that $a - b$ is positive.
> $a < b$ means that $a - b$ is negative.

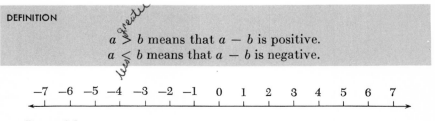

Figure 2.1

2.6 The Number Line

Example 1 Determine the order of these pairs of rational numbers.
(a) -7 and 2
(b) -3 and -5
(c) 0 and -10
(d) $-\frac{3}{4}$ and $-\frac{5}{7}$

Solution

(a) $(-7) - (2) = -9$. Then $-7 < 2$
(b) $(-3) - (-5) = 2$. Then $-3 > -5$
(c) $(0) - (-10) = 10$. Then $0 > -10$
(d) $-\frac{3}{4} - (-\frac{5}{7}) = -\frac{3}{4} + \frac{5}{7} = -\frac{1}{28}$. Then $-\frac{3}{4} < -\frac{5}{7}$

The statements $-7 < 2$ and $2 > -7$ are, of course, both true. More generally it is true that $a < b$ if and only if $b > a$. For if $a < b$, then $a - b$ is negative, so its additive inverse $b - a$ is positive, and $b > a$.

Notice that in Figure 2-1 all the points with positive coordinates are to the right of zero and all the points with negative coordinates are to the left of zero. Moreover the point 6 is to the right of the point 2, and $6 > 2$. The point -3 is to the right of the point -5, and $-3 > -5$. This is a general property of our line and the order of the integers. The coordinate of a point on the right is greater than the coordinate of any point on its left.

The *point* 3 on the number line is between the *points* 2 and 4. We also say that the *number* 3 is between the *numbers* 2 and 4, and write this in symbols $2 < 3 < 4$, since it is true that $2 < 3$ and also $3 < 4$. Similarly, we write $-6 < -4 < -1$, which means that -4 is between -6 and -1, and see that $-6 < -4$ and $-4 < -1$ are both true statements. As before, the point -4 is between the points -6 and -1 on the line.

Example 2 Find all the integers between -4 and 2.

Solution $-3, -2, -1, 0, 1$

We may also associate a different point on our line with each rational number as is suggested by Figure 2.2. The coordinates have been carefully chosen so that the order is increasing from left to right as was the case with the integers. And our observations about "between" for integers have been preserved for the rational numbers as well. It is easy to see by the figure that $\frac{7}{5} > -\frac{1}{3}$, and that $\frac{1}{2}$ is between -1 and $\frac{3}{4}$. But the figure is of little help in determining the order of $-\frac{3}{4}$ and $-\frac{5}{7}$, for example. For even if we carefully determine which points have these coordinates, they would be so close together as to make it difficult to decide which is on the right. However, the definition of "less than" can easily be used, and in fact in Example 1(d) we showed that $-\frac{3}{4} < -\frac{5}{7}$.

Now let us consider the question, "How many rational numbers are there between 0 and 1?" We see at once from the figure that $\frac{1}{4}$, $\frac{1}{2}$ and $\frac{3}{4}$ are all between 0 and 1. But by no means are these the only numbers between 0 and 1. The arithmetic mean or average of $\frac{1}{4}$ and $\frac{1}{2}$ is $\frac{\frac{1}{4}+\frac{1}{2}}{2}=$ $\frac{3}{8}$, and $\frac{1}{4} < \frac{3}{8} < \frac{1}{2}$. Then $\frac{3}{8}$ is between $\frac{1}{4}$ and $\frac{1}{2}$, and hence also between 0 and 1. Similarly, the average of $\frac{1}{4}$ and $\frac{3}{8}$ is $\frac{\frac{1}{4}+\frac{3}{8}}{2} = \frac{5}{16}$, and $\frac{5}{16}$ is between $\frac{1}{4}$ and $\frac{3}{8}$ and also between 0 and 1. The average of any two rational numbers is a *rational* number between the original pair. By continuing this averaging process we can show that there are infinitely many rational numbers between 0 and 1. We can similarly show that there are infinitely many rational numbers between $-\frac{3}{4}$ and $-\frac{5}{7}$, yet these *points* were already so close together on the line as to be difficult to distinguish. In fact we can show that between any two rational numbers, no matter how close together their points may be, there are infinitely many other rational numbers!

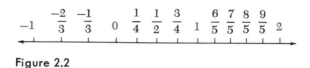

Figure 2.2

Surely the points on the number line with rational number coordinates are close together. Yet a remarkable fact about this line is that there are still other points to which no rational number coordinate has been assigned! Figure 2.3 shows how to locate one such point. The square is one unit long on each side. In geometry we can prove that the coordinate of point A must be a number whose product with itself equals 2. Later in this

2.6 The Number Line

text we will prove that there is no such *rational* number. (Try to find a fraction whose product with itself equals *exactly* 2!) Since numbers like the coordinate of point A are not rational numbers, we shall call them **irrational** numbers. And it is a fact that between any two rational numbers there are not only infinitely many other rational numbers, but also infinitely many irrational numbers!

Irrational numbers, unlike the rationals, cannot be represented exactly with fractions. The symbol $\sqrt{2}$ is often used to represent the irrational number which is the coordinate of point A in Figure 2.3. There are many decimals which we can use to *approximate* $\sqrt{2}$. When we write $\sqrt{2} = 1.4142\ldots$, we mean that the *irrational* number $\sqrt{2}$ cannot be represented exactly by any decimal, either terminating or repeating, but that 1.4142, a *rational* number, is the best four-place decimal approximation for $\sqrt{2}$. Recall that every rational number has either an exact (terminating) decimal form, such as $\frac{3}{2} = 1.5$, or may be represented by a repeating decimal, such as $\frac{1}{3} = .\overline{3}\ldots$. The irrational numbers cannot be represented exactly with either terminating or repeating decimals. And every nonrepeating, nonterminating decimal represents an irrational number.

Example 3 Which are rational and which are irrational?
(a) $-\frac{3}{7}$ (b) $5.\overline{23}\ldots$ (c) 2.561 (d) $2.561\ldots$

Solution

(a) Rational   [A fraction with numerator and denominator integer symbols]

(b) Rational   [Repeating decimal]

(c) Rational   [Terminating decimal]

(d) Irrational   [Nonterminating, nonrepeating decimal]

We will not consider the arithmetic operations with irrational numbers at this time, but we will call the union of the set of rational numbers and the set of irrational numbers the set of **real** numbers, use R to designate that set, and make the assumption that corresponding to each point on a line there is a different real number. We will call the line the real number line. Then every real number corresponds to exactly one point on the line, is either rational or irrational, and both the set of rational numbers and the set of irrational numbers are subsets of R.

76   The Real Numbers

Example 4   Make a Venn diagram to show the subset relation of N, I, F, and R.

Solution

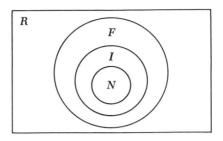

Example 5   Indicate by shading in the diagram of Example 4 the set of irrational numbers.

Solution

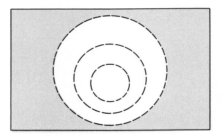

Example 6   Locate each of these in the Venn diagram of Example 4:
$$-\tfrac{2}{3},\ -3,\ \sqrt{2},\ 1,\ 1.5$$

Solution

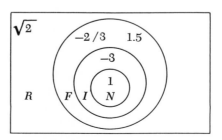

2.6 *The Number Line*

Example 7 Which of the numbers 0, 1, −1, $\frac{1}{2}$, −$\frac{1}{2}$, 2.5, 1.$\overline{2}$..., .3427... are
(a) Natural numbers?   (b) Integers?   (c) Rational numbers?
(d) Irrational numbers? (e) Real numbers?

Solution

(a) 1   (b) 0, 1, −1   (c) All except .3427...
(d) .3427...   (e) All are real numbers

Let us return to the rational numbers and the number line, Figure 2.4. The point 0 is a rather special point, and we call it the **origin** of the number line. It is interesting to note that the distance between the origin and either 3 or −3 is 3, and that 3 is the absolute value of both 3 and −3.

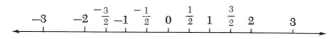

Figure 2.4

Similarly, $|-2| = |2| = 2$, and 2 is the distance between the origin and either 2 or −2. In general the absolute value of any integer is the distance between the origin and the point whose coordinate is that integer. Let us define the absolute value of every real number in a way that this geometric observation about points with integer coordinates is preserved. The absolute value of a positive real number (to the right of the origin) is that same real number, but the absolute value of a negative real number (to the left of the origin) is the additive inverse of that real number.

Example 8 Find the simplest form.
(a) $|-\frac{3}{2}|$ (b) $|\frac{1}{2}|$

Solution

(a) $\frac{3}{2}$ (b) $\frac{1}{2}$

By definition,

$$\frac{-3}{2} + \frac{1}{3} = \frac{-9 + 2}{6} = \frac{-7}{6}$$

It is interesting to observe that this is the sum of one positive and one negative rational number, that the negative term has the larger absolute value, that the sum is also negative, and that the absolute value of the sum can be found by subtracting the absolute values of the terms.

$$\tfrac{3}{2} - \tfrac{1}{3} = \tfrac{7}{6} \quad \text{and} \quad \tfrac{7}{6} = |-\tfrac{7}{6}|$$

That is, the same rule we used for adding *integers* may also be used to add *rational* numbers. Likewise $-\tfrac{1}{2} \cdot \tfrac{3}{4} = -\tfrac{3}{8}$; the product of a positive and a negative rational number is negative, and the absolute value of the product can be found by multiplying the absolute values of the factors. In general the rules for adding and multiplying *integers* can also be used for rational numbers. And these rules are actually easier to use when the rational numbers are written in decimal form, as the next example illustrates.

Example 9 Find the decimal indicated.
(a) $-3.6 - 1.5$ (b) $(-2.1)(5.6)$

Solution

(a) $-3.6 - 1.5 = (-3.6) + (-1.5) = -5.1$
(b) $(-2.1)(5.6) = -11.76$

It is the algebra of real numbers with which we will be concerned in this text. While we will not study the irrationals until much later, we will now assume that the various properties we have observed for natural numbers, integers, and rational numbers are preserved as well in the set of real numbers. We shall assume that each of the following is a true statement for every real number replacement of the variables.

1. If $a \in R$ and $b \in R$, then $(a + b) \in R$ and $(ab) \in R$ [Closure axioms]
2. $a + b = b + a$
   $ab = ba$ [Commutative axioms]
3. $(a + b) + c = a + (b + c)$
   $(ab)c = a(bc)$ [Associative axioms]
4. $a(b + c) = ab + ac$
   $(a + b)c = ac + bc$ [Distributive axiom]
5. $a + 0 = a$
   $a \cdot 1 = a$ [Identity axioms]
6. If $a \in R$, there is $-a \in R$, such that $a + (-a) = 0$. If $a \in R$ and $a \neq 0$, there is $1/a \in R$, such that $a(1/a) = 1$.
       [Inverse axioms]

7. If $a \in R$, then exactly *one* of these is true:
   (i) $a > 0$
   (ii) $a = 0$ [Trichotomy axiom]
   (iii) $a < 0$
8. If $a > 0$ and $b > 0$, then $a + b > 0$ and $ab > 0$ [Axiom of positives]
9. The coordinate of every point on the number line is a member of R, and each member of R corresponds to a different point on the line.   [Completeness axiom]

We will not attempt to do so, but these axioms can be used to prove that Theorems 1–5 in this chapter are also true for real numbers. We will assume that they are. And hereafter, unless we explicitly indicate otherwise, the replacement set for each variable will be the set of all real numbers for which the expression in which the variable appears has a real value. Expressions which are equivalent over R will be described as merely equivalent.

Example 10

(a) What is the domain of $f(x) = 3x + 5$?

(b) Are $2(x + 3)$ and $2x + 6$ equivalent?

(c) What is the replacement set for the variable in the expression $\dfrac{5}{x - 2}$?

(d) What is $0 \cdot \sqrt{2}$?

Solution

(a) R, since $3x + 5$ has a real value for every real replacement of the variable.

(b) Yes, by the distributive axiom.

(c) The set of all real numbers except 2. (The expression has no value when $x = 2$.)

(d) 0. (We are assuming Theorem 1 is true for all reals.)

Exercises 2.6

For Exercises 1–12 determine the order as in Example 1.

1. -6 and -10
2. 3 and -5
3. -4 and 0
4. -5 and -4

80   The Real Numbers

5. $-\frac{3}{8}$ and $-\frac{5}{13}$
6. $-\frac{4}{7}$ and $-\frac{3}{5}$
7. $\frac{9}{5}$ and $\frac{11}{6}$
8. $-\frac{5}{8}$ and $-\frac{2}{3}$
9. -2.65 and -2.64
10. $.37$ and -4.16
11. $-\frac{5}{8}$ and $-.63$
12. $-\frac{2}{3}$ and $-.66$

For Exercises 13–16 arrange in increasing order.

13. $\frac{9}{2}, 4, \frac{14}{3}$
14. $-1, -\frac{5}{3}, -2$
15. $-\frac{7}{2}, -3, -\frac{10}{3}$
16. $-2.1, -2.5, -\frac{9}{4}$

For Exercises 17–20 find two consecutive integers such that the given number is between them.

17. $\frac{7}{2}$ 18. $-\frac{5}{3}$ 19. $-12\frac{1}{3}$ 20. -6.3

For Exercises 21–24 list all the integers which are between the two given numbers.

21. -8 and -3
22. $-\frac{7}{2}$ and $\frac{1}{3}$
23. $-\frac{1}{2}$ and $-\frac{1}{4}$
24. $-\frac{12}{7}$ and $\frac{8}{3}$

For Exercises 25–28 find a single decimal numeral which is equal to the given expression.

25. $-7.2 + 11.5$
26. $-18.3 - 7.14$
27. $\dfrac{-.63}{.7}$
28. $(-1.5)(.9)$

Each of the numbers in Exercises 29–36 is a member of one or more of the sets N, I, F, R. Indicate which.

29. $-\frac{3}{8}$
30. -5
31. 7.6
32. -3.7
33. $.7645\ldots$
34. $-.\overline{35}\ldots$
35. $-1.4 + 3.4$
36. $\dfrac{0}{-5}$

For Exercises 37–44 consider the following numbers.

(a) -7 (b) 2.4 (c) $32.41\ldots$ (d) $-\frac{6}{5}$ (e) $.\overline{54}\ldots$

2.6 The Number Line

Indicate which of these are members of the given set.

37. Natural numbers
38. Integers
39. Rational numbers
40. Real numbers
41. Quotient numbers
42. Irrational numbers
43. $F \cap R$
44. $I \cup R$

For Exercises 45–60 exactly one of the following sets is described in each question. Indicate which.

(a) Natural numbers
(b) Integers
(c) Quotient numbers
(d) Rational numbers
(e) Irrational numbers
(f) Real numbers
(g) Empty set

45. A set which contains $\frac{2}{3}$ but not -1.
46. A set which contains -2.3 but not $\sqrt{2}$.
47. A set which contains $\sqrt{2}$ but not 1.
48. A set which contains both $\sqrt{2}$ and 1.
49. A set which contains neither $\sqrt{2}$ nor 1.
50. The union of the rational and irrational numbers.
51. The intersection of the rational and irrational numbers.
52. The intersection of the irrational and real numbers.
53. A subset of the integers which contains 0.
54. The union of the irrational and real numbers.
55. A subset of the quotient numbers which contains 1 but not $\frac{1}{2}$.
56. A set which is closed with respect to division, but is not $\emptyset$.
57. A subset of the rational numbers which contains -1, but not $-\frac{1}{2}$.
58. A subset of the irrational numbers which does not contain $\sqrt{2}$.
59. A subset of the real numbers which does not contain the multiplicative identity but which is not the empty set.
60. A set which contains the additive identity and the multiplicative inverse of -2, but does not contain $\sqrt{2}$.

First Degree Sentences   Chapter THREE

3.1 Polynomial Expressions

When a product involves a repetition of the same factor a new notation will be convenient. We shall write $7 \cdot 7 \cdot 7$ as 7^3 and call this the third **power** of 7. The sixth power of 2 will be written 2^6 and $2^6 = 2 \cdot 2 \cdot 2 \cdot 2 \cdot 2 = 64$. Similarly, $625 = 5 \cdot 5 \cdot 5 \cdot 5 = 5^4$. In the expression 5^4 we shall call 5 the **base** and note that 5 is the repeated factor; we shall call 4 the **exponent** and observe that there are 4 factors. We may also write a first power such as $3^1 = 3$, but we will seldom use an exponent when it is 1. To find the simplest form of an expression such as $2 + 3^4$, we must, as usual, multiply before adding. Then $2 + 3^4 = 2 + 3 \cdot 3 \cdot 3 \cdot 3 = 2 + 81 = 83$. But the symbols of grouping in $(2 + 3)^3$ indicate that addition is performed first, so $(2 + 3)^3 = 5^3 = 125$. When the base is a negative number, symbols of grouping must *always* be used. Thus the second power or square of -3 is written $(-3)^2 = 9$, while -3^2 indicates the additive inverse of 3^2 and $-3^2 = -9$.

Example 1   Find the simplest form.

(a) $3 \cdot 2^4$ (b) $(2 \cdot 5)^3$ (c) -5^2
(d) $(-3)^4$ (e) $(2^2)^3$ (f) 0^4

3.1 *Polynomial Expressions*

Solution

(a) $3 \cdot 2^4 = 3 \cdot 16 = 48$ (b) $(2 \cdot 5)^3 = 10^3 = 1000$
(c) $-5^2 = -25$ (d) $(-3)^4 = (-3)(-3)(-3)(-3) = 81$
(e) $(2^2)^3 = 4^3 = 64$ (f) $0^4 = 0 \cdot 0 \cdot 0 \cdot 0 = 0$

Example 2 If $f(x) = 2 + x^3$, find:
(a) $f(3)$ (b) $f(-3)$

Solution

(a) $f(3) = 2 + 3^3 = 2 + 27 = 29$
(b) $f(-3) = 2 + (-3)^3 = 2 + (-27) = -25$

Example 3 If $g(x) = -4 - x^2$, find:
(a) $g(3)$ (b) $g(-4)$

Solution

(a) $g(3) = -4 - 3^2 = -4 - 9 = -13$
(b) $g(-4) = -4 - (-4)^2 = -4 - 16 = -20$

Example 4 Let $h(x) = x^3 + x^2$, $\{0, -1, -3\}$. Find the range of h.

Solution $h(0) = 0^3 + 0^2 = 0$; $h(-1) = (-1)^3 + (-1)^2 = -1 + 1 = 0$; $h(-3) = (-3)^3 + (-3)^2 = -27 + 9 = -18$. Range $= \{0, -18\}$.

Example 5 If $k(x) = -x - 2x^3$, find:
(a) $k(-1)$ (b) $k(-2)$

Solution

(a) $k(-1) = -(-1) - 2(-1)^3 = 1 - 2(-1) = 1 + 2 = 3$
(b) $k(-2) = -(-2) - 2(-2)^3 = 2 - 2(-8) = 2 + 16 = 18$

The expressions in Examples 2–5 have all been what we shall call **polynomial** expressions in the variable x. The polynomial $3x^4 + 2x^3 + 4$ has three terms. In the first term, $3x^4$, we shall call 3 the numerical **coefficient** and 4 the **degree.** Similarly, the degree of the second term, $2x^3$, is 3 and the coefficient is 2. The last term, 4, is called a constant term, and it is said to have degree zero.

Example 6 Identify the coefficient and degree of each term in the polynomial $4x^3 + x^2 + 3x + 2$.

Solution In $4x^3$ the coefficient is 4 and the degree is 3. In x^2 the coefficient is 1 and the degree is 2. In $3x$ the coefficient is 3 and the degree is 1. In 2 the coefficient is 2 and the degree is 0.

84   *First Degree Sentences*

The polynomial $-3x^2$ has just one term,* and we say that its coefficient is -3 since $-3x^2$ and $(-3)x^2$ are equivalent, as the next theorem shows.

Theorem 1
$$(-a)b = -ab$$

Proof We will find the sum of $(-a)b$ and ab. Then $(-a)b + ab = (-a + a)b = 0 \cdot b = 0$. Hence $(-a)b$ is the additive inverse of ab, or in symbols $(-a)b = -ab$.

By Theorem 2 of Chapter 2 the polynomial $2x^2 - 3x - 1$ is equivalent to $(2x^2) + (-3x) + (-1)$, and we say it has three terms whose coefficients are 2, -3, and -1 respectively, and whose degrees are respectively 2, 1 and 0. The degree of a polynomial is the largest degree of any of its terms. Then the degree of $2x^2 - 3x - 1$ is 2, but the degree of $x^2 - 5x^4$ is 4.

Example 7 Identify the degree of this polynomial and the coefficients of its terms. $4x^3 - 3x^2 + x - 1$.

Solution The degree is 3 and the coefficients are 4, -3, 1, and -1.

The polynomial $3x^2 + 2x^2$ has two terms with the same degree. We shall call such terms **similar** terms, and we may always use the distributive axiom to find a single term which is equivalent to the sum of any two similar terms. Thus $3x^2 + 2x^2 = (3 + 2)x^2 = 5x^2$. Also $-4x^3 + x^3 = (-4 + 1)x^3 = -3x^3$.

Example 8 Find a polynomial with a single term, which is equivalent to the given one.
(a) $-3x + 4x$ (b) $5x^2 - 2x^2 - 7x^2$ (c) $3x^2 + 2x$

Solution
(a) x (b) $-4x^2$ (c) Impossible. The terms are not similar.

Example 9 Find an equivalent polynomial in which no two terms are similar.
(a) $3x^2 + x - x^2$ (b) $x^2 - 3x + 2 + 2x^2 - x - 4$

---

* This does not contradict our agreement in Section 2.1 that there must be a sum in order to have terms. For in a trivial sense $-3x^2$ is a term of the sum $-3x^2 + 0$.

3.1 Polynomial Expressions

Solution

(a) $3x^2 + x - x^2 = (3x^2 - x^2) + x = 2x^2 + x$
(b) $x^2 - 3x + 2 + 2x^2 - x - 4 = (x^2 + 2x^2) + (-3x - x) + (2 - 4) = 3x^2 - 4x - 2$

We have been discussing polynomials without saying precisely what a polynomial is. Let us do so now.

DEFINITION

A polynomial in x is any expression such as:

or   (i) a with $a \in R$
or   (ii) ax^n with $a \in R$ and $n \in N$
or   (iii) the sum of two or more expressions of the form (i) and (ii)
      (iv) Any expression which is equivalent to one of the form (i), (ii), or (iii).

Of course we may replace x by y or z, etc. in the above definition and have polynomials in y or in z, etc. Each term, except the constant term, in a polynomial must be the product of a real number and a natural number power of one variable.

Example 10 Which of these are polynomials?

(a) $2y^2 - 3y + 4$ (b) $5x^2$ (c) 7
(d) $\frac{2}{3}z + 5$ (e) $4x^{-2}$ (f) $3/x^2$

Solution

(a), (b), (c), and (d) are polynomials.
(e) is not a polynomial, since -2 is not a member of N.
(f) is not, since $3/x^2$ is not the *product* of a real number, and a natural number power of a variable.

We shall say that a polynomial is in **standard form** if no two terms have the same degree and the degree of each term is less than the degree of the term which it follows. Thus $3x^2 + 5x + 2$ is in standard form, but the standard form of $3x + x^3 + 4$ is $x^3 + 3x + 4$. We shall also agree that the standard form of $3x^2 + (-2x) + 5$ is $3x^2 - 2x + 5$. The sum of two polynomials can be written in standard form by adding similar terms. Thus $(3x + 2) + (4x - 5) = 3x + 2 + 4x - 5 = (3x + 4x) + (2 - 5) = 7x - 3$.

Example 11 Find the standard form of each of these sums.
  (a) $(x^2 - 3x + 1) + (3x^2 + 4x - 9)$
  (b) $(x^2 - 3) + (2x - 3x^2 + 4)$
  (c) $(x^2 - x + 3) + (-x^2 + x - 3)$

Solution

(a) $(x^2 - 3x + 1) + (3x^2 + 4x - 9) = x^2 - 3x + 1 + 3x^2 + 4x - 9 = (x^2 + 3x^2) + (-3x + 4x) + (1 - 9) = 4x^2 + x - 8$

(b) $(x^2 - 3) + (2x - 3x^2 + 4) = (x^2 - 3x^2) + 2x + (-3 + 4) = -2x^2 + 2x + 1$

(c) $(x^2 - x + 3) + (-x^2 + x - 3) = (x^2 - x^2) + (-x + x) + (3 - 3) = 0$

When the sum of two numbers is 0 we call each the additive inverse of the other. Similarly, when the sum of two polynomials is 0, we can call them additive inverses. From Example 11(c) we see that the additive inverse of $x^2 - x + 3$ is $-x^2 + x - 3$. The next theorem can be used to show that the additive inverse of any polynomial is obtained by replacing each coefficient by its additive inverse.

Theorem 2

$$-(a + b) = -a - b$$

Proof $(a + b) + (-a - b) = [a + (-a)] + [b + (-b)] = 0 + 0 = 0$, and so the additive inverse of $a + b$ is $-a - b$, or in symbols $-(a + b) = -a - b$.

Example 12 Find the standard form.
  (a) $-(3 - x^2 + 4x)$ (b) $-(2x - 3x^2 - 5x^3)$

Solution

(a) $-(3 - x^2 + 4x) = -3 + x^2 - 4x = x^2 - 4x - 3$
(b) $-(2x - 3x^2 - 5x^3) = -2x + 3x^2 + 5x^3 = 5x^3 + 3x^2 - 2x$

Subtraction of polynomials uses the fact that the difference of two numbers is also the sum of the minuend and the additive inverse of the subtrahend.

Example 13 Find the standard form.
  (a) $(4x + 3) - (x - 2)$ (b) $(2 + 3x - 4x^2) - (x + x^2 - 3)$

3.1 Polynomial Expressions

Solution

(a) $(4x + 3) - (x - 2) = (4x + 3) + (-x + 2)$
$= 4x + 3 - x + 2 = 3x + 5$

(b) $(2 + 3x - 4x^2) - (x + x^2 - 3) = 2 + 3x - 4x^2 - x - x^2 + 3$
$= -5x^2 + 2x + 5$

Exercises 3.1

For Exercises 1–32 find the simplest form.

1. $2 + 3^3$
2. $(1 + 3)^2$
3. $2 \cdot 5^2$
4. $(3 \cdot 4)^2$
5. $(-2)^2$
6. $(-3)^3$
7. -2^2
8. -3^3
9. 1^3
10. 1^4
11. 1^{10}
12. 1^{55}
13. $(-1)^3$
14. $(-1)^4$
15. $(-1)^{10}$
16. $(-1)^{55}$
17. $(-2)^5$
18. -2^6
19. $-(-2)^4$
20. $-(-2)^5$
21. $5 - 4^2$
22. $-3 - 2^3$
23. $-9 - 3^2$
24. $-9 - (-3)^2$
25. $(-2 - 3)^2$
26. $(5 - 9)^3$
27. $(2 \cdot 3 - 8)^3$
28. $2(-4 + 1)^3$
29. $(2^2)^2$
30. $(3^2)^2$
31. $(-2 + 2^2)^4$
32. $(-2^2)^3$

For Exercises 33–40 find the value of the expression when $x = 2$.

33. x^7
34. $-x^4$
35. $(-x)^4$
36. $-(-x)^5$
37. $-3x^3$
38. $5x - x^2$
39. $3x^2 - 4x - 1$
40. $-x^3 + x^2 - 3x$

For Exercises 41–48 find the value of the expression when $x = -2$.

41. $-x$
42. x^2
43. $(-x)^3$
44. $-x^4$

First Degree Sentences

45. $x^2 + 3x$
46. $x^3 - 2x^2 + x$
47. $-4x^2 - x + 1$
48. $2x^2 - 2x - 1$

For Exercises 49–56 find the indicated value of the function.

49. $f(x) = x^2 + 2x - 1$, $f(0)$
50. $g(x) = 2x^2 - x + 5$, $g(3)$
51. $h(x) = -x^2 + 5x - 9$, $h(4)$
52. $k(x) = x^2 - 3x + 7$, $k(6)$
53. $f(y) = y^2 + y + 1$, $f(-1)$
54. $g(y) = y^2 - y$, $g(-4)$
55. $h(y) = -y^2 - 5$, $h(-5)$
56. $k(y) = -2y^2 - y + 10$, $k(-3)$

For Exercises 57–64 an open sentence and a domain are given. Find the solution set.

57. $3x + 2 = 5x - 4$, $\{-2, 0, 3\}$
58. $x^2 = x + 2$, $\{0, 1, 2\}$
59. $x^2 + 3x = 2x$, $\{-2, -1, 0\}$
60. $x^2 - 2x = x^2 - 6$, $\{-1, 2, 3\}$
61. $x^2 + 7 = x + 9$, $\{-1, 1, 2\}$
62. $x^2 - 2x + 3 = 2x^2 + x + 5$, $\{-2, -1, 2\}$
63. $2x^2 - 3x + 1 = x^2 - 3x + 5$, $\{-2, 0, 2\}$
64. $3x^2 - 2x + 1 = 2x^2 - 6x - 2$, $\{-3, -2, -1\}$

cannot combine unlike terms

For Exercises 65–100 find the standard form of the polynomial.

use distributive axiom to find like terms

65. $5x + 4 - 2x + 1$
66. $4 - 3x + 8x$
67. $-2x - 3x + 4$
68. $-3 + x - 7x + 9$
69. $(2x + 5) + (3x - 9)$
70. $(x - 6) + (5 - 4x)$
71. $(-4 - 3x) + (3x + 1)$
72. $(-3x + 1) + (5 - 3x)$
73. $(2x + 6) - (x + 2)$
74. $(x + 7) - (4x - 1)$
75. $(4 - 3x) - (5x - 3)$
76. $(2x - 3) - (-3 + x)$
77. $(x^2 - 3x + 1) + (5x + 3x^2)$
78. $(5x - x^2 + 9) + (x^2 - 3x)$
79. $(4x^2 - 3x + 1) - (2x^2 - x + 3)$
80. $(5x - x^2 + 3) - (x^2 + 5x + 3)$
81. $3(x - 4)$
82. $-2(x + 5)$

83. $(x + 4) + 2(x - 3)$
84. $-3(x + 1) + 4(x - 1)$
85. $2(4 - x) - 3(x + 2)$
86. $-5(x + 1) - 2(x - 1)$
87. $5 + 2(x + 2)$
88. $2 - 7(2x - 1)$
89. $(\frac{1}{3})(3x)$
90. $(\frac{1}{5})(5x)$
91. $(-\frac{1}{2})(2x)$
92. $(-\frac{1}{4})(4x)$
93. $(-\frac{1}{2})(-2x)$
94. $(-\frac{1}{6})(-6x)$
95. $(\frac{2}{3})(12x)$
96. $(-\frac{3}{4})(8x)$
97. $-4 + 4x$
98. $-5 + 6x$
99. $6 - 6x$
100. $8 - 7x$

3.2 First Degree Equations

The open sentence $3x - 1 = 5$ is neither true nor false until a replacement is made for the variable. If the replacement set for the variable is $\{-2, 0, 2\}$, we can make the three replacements and find that $3x - 1 = 5$ is true when $x = 2$, but false when $x = -2$ or $x = 0$. Hence its solution set is $\{2\}$. An open sentence or statement which contains the symbol $=$ will be called an **equation**. We shall call the expression written to the left of $=$ the **left expression** and that which appears to the right, the **right expression**. Then $3x - 1$ is the left expression and 5 is the right expression in the equation $3x - 1 = 5$.

Example 1 Using $\{-2, 0, 2\}$ as the domain, find the solution set of these equations.
(a) $3x = -6$ (b) $2x = -4$ (c) $x + 2 = 0$

Solution (a) $\{-2\}$ (b) $\{-2\}$ (c) $\{-2\}$

Notice that the open sentences in Example 1, although they are all different equations, have the same solution set. Whenever two or more open sentences have exactly the same domain and exactly the same solution set, they will be called **equivalent** over that domain. Then the equations of Example 1 are equivalent over $\{-2, 0, 2\}$.

Example 2 Determine whether the following pairs of equations are equivalent over $\{-2, 0, 2\}$.
(a) $2(x + 9) = 24 - x$ and $x - 2 = 0$
(b) $3x - 2 = 4x$ and $2x + 7 = x + 7$

Solution (a) Equivalent (b) Not equivalent

When two equations both have the set R as a domain and are equivalent over R, we shall merely say that they are equivalent. Thus, $2x = 4$ and $x = 2$ are equivalent, since each has R as a domain and each has the solution set $\{2\}$. Recall that two algebraic expressions may also be equivalent, but in this latter context the word "equivalent" has another meaning. For example, $4(x - 3)$ and $4x - 12$ are equivalent since for any real number replacement of the variable the two expressions have the same value. Of course, neither of these expressions has a solution set; only open sentences have solution sets.

Example 3 Are these equations equivalent?
(a) $x = 3$ and (b) $3x - 1 + 2x - 10 = -3 + 7$.

Solution It is obvious that the solution set for (a) is $\{3\}$. And by making the replacement $x = 3$ in Equation (b) it is found that a true statement results. Then 3 is a member of the solution set for (b). But is 3 the only member? This is a question which we cannot answer without making some assumptions about equivalent equations. (We will answer it later in this section.)

Observe that $3 + 1 = 2 + 2$ is a true statement, since both the left and right expressions represent 4. If we add the same number, for example -7, to each expression we obtain $(3 + 1) + (-7) = (2 + 2) + (-7)$, which is also a true statement, for now each expression represents -3. Similarly, if we multiply each expression by the same number, for example -2, we have $(3 + 1)(-2) = (2 + 2)(-2)$, another true statement, where each expression represents -8. Also observe that $2 = 3$ is a false statement, and if we add -3 to both expressions we have $2 + (-3) = 3 + (-3)$, a false statement. If we multiply each by -3 we have the false statement $2(-3) = 3(-3)$. But if we multiply by 0 the false statement $2 = 3$ is replaced by the true statement $2 \cdot 0 = 3 \cdot 0$.

Let us make three assumptions about equations which are equivalent over R.

> 1. If in any equation we replace one *expression* by another which is equivalent to it, then the resulting equation is equivalent to the original. [Substitution axiom]
> 2. If we add the same expression to the left and right expressions of an equation, the resulting equation is equivalent to the original. [Addition axiom]
> 3. If we multiply the left and right expressions in an equation by the same expression (whose value is not zero), the resulting equation is equivalent to the original. [Multiplication axiom]

3.2 First Degree Equations

Example 4   Which axiom shows that these pairs of equations are equivalent?
(a) $5 = x + 7 - 19$,
$5 = x - 12$
(b) $2x - 7 = 13$,
$(2x - 7) + 7 = 13 + 7$
(c) $5x = 21$,
$\left(\frac{1}{5}\right)(5x) = \left(\frac{1}{5}\right)(21)$

Solution

(a) Substitution, since $x + 7 - 19$ is equivalent to $x - 12$.
(b) Addition, since 7 has been added to each of the original expressions.
(c) Multiplication, since each of the original expressions has been multiplied by $\frac{1}{5}$.

Sometimes the substitution axiom can be used to replace the given equation by an equivalent one whose solution set is obvious.

Example 5   Find the solution set.
(a) $x = 19 - 5 - 12$
(b) $3x - 2x = 4 - 15$

Solution

(a) $x = 19 - 5 - 12$
$x = 2$
$\{2\}$
$[19 - 5 - 12$ is equivalent to 2]

(b) $3x - 2x = 4 - 15$
$x = -11$
$\{-11\}$
$[3x - 2x$ is equivalent to x and $4 - 15 = -11]$

Often we may use both the substitution and addition axioms together to replace a given equation by an equivalent one whose solution set is obvious.

Example 6   Find the solution set of each of the following.
(a) $x + 3 = 15$
(b) $x - 9 = -20$
(c) $-4 = 10 + x$

Solution

(a) $x + 3 = 15$
$(x + 3) + (-3) = 15 + (-3)$ [Addition axiom]
$x = 12$ [Substitution axiom]
$\{12\}$

(b) $x - 9 = -20$
$(x - 9) + 9 = -20 + 9$ [Addition axiom]
$x = -11$ [Substitution axiom]
$\{-11\}$

92   First Degree Sentences

(c) $\quad -4 = 10 + x$
$-4 + (-10) = (10 + x) + (-10)$ [Addition axiom]
$-14 = x$ [Substitution axiom]
$\{-14\}$

Example 6 shows that often we will want to add to the left and right expressions the additive inverse of the constant term of the first degree polynomial.

Example 7 Find the solution set.
(a) $4x = -20$ (b) $-9 = -3x$

Solution

(a) $\quad 4x = -20$
$\left(\frac{1}{4}\right)(4x) = \left(\frac{1}{4}\right)(-20)$ [Multiplication axiom]
$x = -5$ [Substitution axiom]
$\{-5\}$

(b) $\quad -9 = -3x$
$\left(-\frac{1}{3}\right)(-9) = \left(-\frac{1}{3}\right)(-3x)$ [Multiplication axiom]
$3 = x$ [Substitution axiom]
$\{3\}$

This example shows that we may sometimes find the solution set if we multiply both the left and right expressions by the multiplicative inverse of the coefficient of the variable. By using all of the techniques illustrated in the previous examples, the solution sets of many equations may be determined.

Example 8 Find the solution set of each of the following
(a) $4x - 1 = 3$ (b) $-2 = 2x + 5$
(c) $3x - 1 + 2x - 10 = -3 + 7$

Solution

(a) $\quad 4x - 1 = 3$
$(4x - 1) + 1 = 3 + 1$
$4x = 4$
$\left(\frac{1}{4}\right)(4x) = \left(\frac{1}{4}\right)(4)$
$x = 1$
$\{1\}$

(b) $\quad -2 = 2x + 5$
$(-2) + (-5) = (2x + 5) + (-5)$
$-7 = 2x$
$\left(\frac{1}{2}\right)(-7) = \left(\frac{1}{2}\right)(2x)$
$-\frac{7}{2} = x$
$\left\{-\frac{7}{2}\right\}$

3.2 First Degree Equations

(c) $3x - 1 + 2x - 10 = -3 + 7$
$5x - 11 = 4$
$(5x - 11) + 11 = 4 + 11$
$5x = 15$
$\left(\frac{1}{5}\right)(5x) = \left(\frac{1}{5}\right)(15)$
$x = 3$
$\{3\}$

In Example 8(c) we have used the axioms to show that the equation $3x - 1 + 2x - 10 = -3 + 7$ [of Example 3(b)] is equivalent to $x = 3$—that these equations have the same solution set. But the only member of the solution set of $x = 3$ is 3 and hence there is only this single member in the solution set of $3x - 1 + 2x - 10 = -3 + 7$.

It is often easy to verify whether a particular real number is a member of the solution set of a given equation. When there are only one or two members in the solution set we would be wise to "check" our solutions even though the proper use of Axioms 1–3 makes such verification logically unnecessary.

Example 9 Find the solution set and check it. $3x - 4 + 15 - 8x = 2 - 21$.

Solution $3x - 4 + 15 - 8x = 2 - 21$
$-5x + 11 = -19$ [Substitution axiom]
$(-5x + 11) + (-11) = -19 + (-11)$ [Addition axiom]
$-5x = -30$ [Substitution axiom]
$\left(-\frac{1}{5}\right)(-5x) = \left(-\frac{1}{5}\right)(-30)$ [Multiplication axiom]
$x = 6$ [Substitution axiom]
$\{6\}$

Check: Replace x by 6 in the original equation.
$3 \cdot 6 - 4 + 15 - 8 \cdot 6 = 2 - 21$
$18 - 4 + 15 - 48 = -19$
$-19 = -19$ [A true statement]

The techniques we have used to find the solution sets of certain equations will enable us to solve some related problems. The following examples will illustrate.

Example 10 Three more than what number is nineteen?

Solution Let us use the placeholder x to represent the desired number. Then the problem suggests the open sentence $x + 3 = 19$. Taking R as the domain we see that the solution set of this equation is $\{16\}$. We then verify that "Three more than 16 is (indeed) nineteen." Hence the solution is 16.

Example 11 If 27 is subtracted from twice a number, the difference is 45. Find the number.

Solution With x as the placeholder for the desired unknown number we have the equation $2x - 27 = 45$. Finding its solution set, we have $2x = 72$ and $x = 36$, $\{36\}$. We then verify that "If 27 is subtracted from twice 36 the result is (indeed) 45." Hence the required number is 36.

Exercises 3.2

For Exercises 1–40 find the solution set and check your answer.

1. $x + 4 = 7$
2. $x - 3 = -2$
3. $1 = 9 + x$
4. $-3 + x = -8$
5. $3x = 5$
6. $8x = -3$
7. $-4x = 7$
8. $-2 = -3x$
9. $2x + 1 = 5$
10. $5x - 3 = 2$
11. $8 + 4x = 0$
12. $7 = -5 + 6x$
13. $-x = -5 + 3$
14. $-x = (-2)^4$
15. $3 - x = 1$
16. $-2 - x = 5$
17. $1 = -3x + 4$
18. $-2x - 3 = 5$
19. $6 - 5x = 16$
20. $-3 - 2x = -9$
21. $3x - 4 = -2$
22. $-2x + 1 = 6$
23. $4 - 3x = 0$
24. $-2 - 7x = 1$
25. $2x - 5 = -1 - 3$
26. $x - 7 = -3 - 4$
27. $4 + x = -8 + 12$
28. $-x + 9 = (-3)^2$
29. $0 = 3x - 2^2$
30. $-6 - 3x = -3^2 + 3$
31. $2x + 5 + x = 2$
32. $-x + 1 + 5x = -3$
33. $-3x - 2x + 4 = 1$
34. $(2x + 3) + (x - 1) = 6$
35. $(1 - 4x) + (x - 2) = -2$
36. $(3x - 2) - (x - 5) = 1$
37. $(1 - 5x) - (1 + 5x) = 3$
38. $4(x + 2) = 5$
39. $-3(5 - x) = 8$
40. $2(x - 5) - 3(x + 1) = 1$

For Exercises 41–48 first write an open sentence in the variable x as described and then find the solution set for this open sentence.

3.3 *First Degree Equations, Continued*

41. Three more than x is negative five.
42. Two less than x is six.
43. The product of 3 and 2 more than x is 11. $(3 \cdot 2) + x = 11$
44. The sum of 4 and 3 times x is 8.
45. Mary has x dollars, and John has four times as much money as Mary. If John had $5 more, he would have $33.
46. Bill is x years old. His brother Tom is twice as old as Bill. In three years Tom will be 23.
47. A TV repairman charges $15 for a house call plus $8 per hour for the time worked. The charge for a call that lasted x hours was $27.
48. To promote Christmas sales a magazine offers one-year subscriptions for $7 with each additional gift subscription only $5. The total charge for x subscriptions was $27.

3.3 First Degree Equations, Continued

When both the left and right expressions of an equation are polynomials, we shall call it a polynomial equation. The degree of a polynomial equation is the degree of its polynomial expression with larger degree. Hence $3x - 4 = 5$ and $2x + 3 = x + 4$ are both first degree polynomial equations, and $x^2 + 1 = 2x - 3$ is a second degree polynomial equation. In the last section we considered only first degree equations with one expression a first degree polynomial and the other expression a constant (polynomial with zero degree). We learned that the axioms may be used to find the solution set for such an equation.

Example 1 Find the solution set. $-4 + 7 = 5(2x - 1) + 3x - 1$.

Solution

$$
\begin{aligned}
-4 + 7 &= 5(2x - 1) + 3x - 1 \\
3 &= 10x - 5 + 3x - 1 & &\text{[Substitution axiom]} \\
3 &= 13x - 6 & &\text{[Substitution axiom]} \\
3 + 6 &= (13x - 6) + 6 & &\text{[Addition axiom]} \\
9 &= 13x & &\text{[Substitution axiom]} \\
\left(\tfrac{1}{13}\right)(9) &= \left(\tfrac{1}{13}\right)(13x) & &\text{[Multiplication axiom]} \\
\tfrac{9}{13} &= x & &\text{[Substitution axiom]}
\end{aligned}
$$

$\left\{\tfrac{9}{13}\right\}$

Now let us consider a first degree equation whose expressions are *both* first degree polynomials.

Example 2 Find the solution set. $23x = 8x + 45$.

Solution

$$23x = 8x + 45$$
$$23x + (-8x) = (8x + 45) + (-8x) \quad \text{[Addition axiom]}$$
$$15x = 45 \quad \text{[Substitution axiom]}$$
$$\left(\tfrac{1}{15}\right)(15x) = \left(\tfrac{1}{15}\right)(45) \quad \text{[Multiplication axiom]}$$
$$x = 3 \quad \text{[Substitution axiom]}$$
$$\{3\}$$

Note that the addition of $-8x$ in Example 2 enabled us to obtain an equivalent equation in which one expression was a constant. Then we proceeded as in the examples of the last section. This is a general technique which is also effective with more complicated equations.

Example 3 Find the solution set. $3 - 2x = 17(x - 2) - (x - 1)$.

Solution

$$3 - 2x = 17(x - 2) - (x - 1) \quad \text{[Copy equation]}$$
$$3 - 2x = 17x - 34 - x + 1$$
$$-2x + 3 = 16x - 33 \quad \text{[Standard form]}$$
$$(-2x + 3) + 2x = (16x - 33) + 2x$$
$$3 = 18x - 33 \quad \text{[One expression is constant]}$$
$$3 + 33 = (18x - 33) + 33$$
$$36 = 18x$$
$$\left(\tfrac{1}{18}\right)(36) = \left(\tfrac{1}{18}\right)(18x)$$
$$2 = x$$
$$\{2\}$$

The example shows that we will first write each expression in standard form; use the addition axiom to reduce one expression to a constant; then proceed as in the last section. One of our major goals will be to develop the ability to find the solution set of a variety of open sentences. We have learned a technique for finding the solution set for *any* polynomial equation of the first degree. Let us summarize that technique now.

1. Copy the original equation.
2. Find the standard form of the left and right expressions.
3. Use the addition axiom to find an equivalent equation with at least one of its expressions a constant.
4. Use the addition axiom to find an equivalent equation whose left and right expressions are both polynomials with a single term.
5. Use the multiplication axiom to find an equivalent equation whose solution set is obvious.

3.3 *First Degree Equations, Continued*

Step 3 above may always be accomplished by adding the additive inverse of one of the first degree terms to both expressions. Step 4 requires that the additive inverse of the constant term of the first degree polynomial be added to both expressions. For step 5, multiply both expressions by the multiplicative inverse of the coefficient of the first degree term. An example will illustrate the technique.

Example 4   Find the solution set of $4x + 3(4x - 1) = 5(2x + 3)$.

Solution

$$4x + 3(4x - 1) = 5(2x + 3) \quad \text{[Step 1]}$$
$$4x + 12x - 3 = 10x + 15$$
$$16x - 3 = 10x + 15 \quad \text{[Step 2]}$$
$$(16x - 3) + (-10x) = (10x + 15) + (-10x)$$
$$6x - 3 = 15 \quad \text{[Step 3]}$$
$$(6x - 3) + 3 = 15 + 3$$
$$6x = 18 \quad \text{[Step 4]}$$
$$\left(\tfrac{1}{6}\right)(6x) = \left(\tfrac{1}{6}\right)(18)$$
$$x = 3 \quad \text{[Step 5]}$$
$$\{3\}$$

The next two examples consider two special types of equations.

Example 5   Find the solution set of $(3x - 2) + (9x - 7) = -3(3 - 4x)$.

Solution

$$(3x - 2) + (9x - 7) = -3(3 - 4x)$$
$$3x - 2 + 9x - 7 = -9 + 12x$$
$$12x - 9 = 12x - 9$$

We see that the left and right expressions are identical and hence equivalent over R. Then *any* replacement for x by a real number results in a true statement for both the final equation and the original equation and the solution set is R, the set of *all* real numbers. Whenever an equation is true for every possible replacement of the variable, it is called an **identical equation** or an identity. Identical equations play an important part in certain areas of mathematics.

At the other extreme consider the following.

Example 6   Find the solution set of $4(x - 3) = 4x + 3$.

Solution

$$4(x - 3) = 4x + 3$$
$$4x - 12 = 4x + 3$$

$$(4x - 12) + (-4x) = (4x + 3) + (-4x)$$
$$-12 = 3$$
$$\varnothing$$

No replacement of the variable can make the *numerical* statement $-12 = 3$ true. Yet our original equation is equivalent to $-12 = 3$. Therefore there can be no replacement for x in $4(x - 3) = 4x + 3$ which will result in a true statement. Hence the solution set is the empty set, $\varnothing$. This conclusion may be disappointing, but it would be even more upsetting if we were to conclude the solution set has any members when, in fact, every replacement of the variable in $4(x - 3) = 4x + 3$ results in a false statement. Whenever an equation has no solution we must be sure to recognize this fact.

There is a useful procedure which can sometimes shorten our work. When the left and right expressions of a polynomial equation contain exactly the same term, the addition axiom can be used to delete this term from each expression and obtain an equivalent equation. Some examples will illustrate.

Example 7   Find the solution set for each of these:

(a) $3x^2 - 5x + 7 = 3(x^2 + 3x)$
(b) $2(2x - 1) + 5x = (3 - 2x) + (5x - 11)$
(c) $10x - 3(x + 1) = 3 + 7x$

Solution

(a) $3x^2 - 5x + 7 = 3(x^2 + 3x)$
$3x^2 - 5x + 7 = 3x^2 + 9x$
$\quad\quad\; -5x + 7 = 9x$ [Adding $-3x^2$ will delete $3x^2$]
$\quad\quad\quad\quad\;\; 7 = 14x$
$\quad\quad\quad\quad\; \frac{1}{2} = x$
$\quad\quad\quad\quad\; \{\frac{1}{2}\}$

(b) $2(2x - 1) + 5x = (3 - 2x) + (5x - 11)$
$\quad\; 4x - 2 + 5x = 3 - 2x + 5x - 11$
$\quad\quad\quad\; 4x - 2 = -2x - 8$ [We deleted $5x$]
$\quad\quad\quad\; 6x - 2 = -8$
$\quad\quad\quad\quad\;\; 6x = -6$
$\quad\quad\quad\quad\quad\; x = -1$
$\quad\quad\quad\quad\; \{-1\}$

3.3 First Degree Equations, Continued

(c) $10x - 3(x+1) = 3 + 7x$
     $10x - 3x - 3 = 3 + 7x$
     $7x - 3 = 7x + 3$
     $-3 = 3$ [$7x$ was deleted]
     $\emptyset$ [The last statement is always false]

Exercises 3.3

For Exercises 1–28 find the solution set.

1. $3x = 2x + 5$
2. $7x = 4 - x$
3. $-2x = x + 6$
4. $5x = 3 - 2x$
5. $4x + 1 = 3x - 1$
6. $1 - 3x = -3 - 4x$
7. $3x - 6 = 5 - 2x$
8. $-3 - 5x = 1 + 2x$
9. $2(x + 3) = 5x - 1$
10. $2x - 3(x + 1) = 3x + 5$
11. $3x - 5 - 7x = 2x - 9$
12. $2(x - 5) - 3x = x - 2$
13. $5(x + 1) = 4x + 5$
14. $4(2 - x) = 3(x + 2) + 2$
15. $2(x - 3) = 5(x + 1) - 11$
16. $2(x + 3) - (5x + 6) = 0$
17. $3(x + 1) - 3x = 4$
18. $5(1 - x) + 5x = 7$
19. $2x + 3 = 2(x + 1)$
20. $1 - 3x = 3(4 - x)$
21. $2(x + 3) - 2x = 6$
22. $(1 - 4x) + (4x + 3) = 4$
23. $3(x - 2) + 9 = 3(x + 1)$
24. $2(3 - 4x) = -3x - 5x$
25. $(x^2 + 3x - 1) + (x - x^2) = 7$
26. $(2 + 3x^2 - x) - (3x^2 + x - 1) = 0$
27. $2x^2 + 3(x - 2) = (x^2 - 1) + x^2 + x$
28. $5x^2 - 3x = 5(2x + x^2 - 1)$

For Exercises 29–32 first write an open sentence and then find its solution set to answer the stated question.

29. If 3 more than twice a certain number is 7 less than that number, what is it?
30. If the product of 5 and 2 less than a certain number is 6 more than that number, what is it?
31. If a certain number is increased by 6 it will be 2 less than 3 times as large as before. What is it?
32. If a number is added to 5 less than twice that number, the result is 6 times the original number. What is it?

3.4 First Degree Inequalities

The open sentence $2x - 9 > -2x$ is neither true nor false until a replacement is made for the variable. For example if x is replaced by 2, we have the numerical statement $-5 > -4$ which is false; if $x = 3$, we have the true statement $-3 > -6$. Let us call an open sentence or statement which includes the connective $>$ or $<$ an **inequality,** and define the degree of an inequality as we did for an equation. Then $2x - 9 > -2x$ is a first degree polynomial inequality whose left expression is $2x - 9$ and whose right expression is $-2x$. We shall want to find the solution sets for such inequalities.

Example 1 If the domain is $\{-5, -4, -3, -2, \ldots, 5\}$ find the solution sets for these inequalities:
(a) $x > 3$ (b) $x < 2$

Solution (a) $\{4, 5\}$ (b) $\{-5, -4, -3, -2, -1, 0, 1\}$.

Example 2 If the domain is I, find the solution set of:
(a) $x > 3$ (b) $x < 2$

Solution (a) $\{4, 5, 6, 7, \ldots\}$ (b) $\{\ldots -2, -1, 0, 1\}$

The examples illustrate that when sets contain only integers we may sometimes list only a few of their members and use dots to suggest the remaining members. Often, however, we will want to consider sets which contain not only some of the integers, but also all those rational and irrational numbers which are between these integers. We will need a new notation for such sets. We will represent the set of *all* real numbers which are between 2 and 7 by writing $\langle 2, 7 \rangle$. Similarly, $\langle -\frac{5}{2}, 1 \rangle$ is the set of all real numbers between $-\frac{5}{2}$ and 1. Sets of this kind are called **intervals** of numbers. Note that the interval $\langle 2, 7 \rangle$ contains not only the integers 3, 4, 5 and 6, but also all such rational numbers as 2.1, 3.5 and 6.9999 which are between 2 and 7, as well as all the irrational numbers between 2 and 7. But neither 2 nor 7 itself is a member of $\langle 2, 7 \rangle$. And of course such numbers as -3, 1.99 and 7.5 are not members of the interval $\langle 2, 7 \rangle$.

The set of all real numbers which are greater than 3 is an interval, and we may represent it by $\langle 3, \infty \rangle$. And the interval $\langle -\infty, 2 \rangle$ is the set of all real numbers which are less than 2.* Note that $\{\ldots, -2, -1, 0, 1\}$ is

---

* The symbols ∞ and $-\infty$ will not be given any meaning when used alone or in a context other than the above. We shall agree in general that $\langle a, \infty \rangle$ represents the set of all real numbers greater than a, while $\langle -\infty, b \rangle$ represents the set of all real numbers less than b.

3.4 First Degree Inequalities

the set of all *integers* which are less than 2, and this is not the same as $\langle -\infty, 2 \rangle$. The interval $\langle -\infty, 2 \rangle$ contains such numbers as $1.5, -3.7$, etc., which are not in $\{\ldots, -2, -1, 0, 1\}$.

Example 3 Find the solution set of the following:
(a) $x > 3$ (b) $x < 2$ (c) $x > -5$
(d) $-6 > x$ (e) $0 < x$

Solution The domain is understood to be R.
(a) $\langle 3, \infty \rangle$ (b) $\langle -\infty, 2 \rangle$ (c) $\langle -5, \infty \rangle$
(d) $\langle -\infty, -6 \rangle$ since $-6 > x$ means $x < -6$ (e) $\langle 0, \infty \rangle$

As was the case with equations, we shall call two inequalities equivalent over a domain if they have exactly the same solution set over that domain, and we will often want to know when two inequalities are equivalent over R. Let us make an assumption about equivalent inequalities which is analogous to one which we made for equations.

> If any expression in an inequality is replaced by an equivalent expression, the resulting inequality is equivalent to the original.
> [Substitution axiom for inequalities]

Example 4 Find the solution set.
(a) $x < 4(2 - 7) + 1$ (b) $3x - 2x > 5$

Solution

(a) $x < 4(2 - 7) + 1$ (b) $3x - 2x > 5$
$x < 4(-5) + 1$ $x > 5$
$x < -20 + 1$ $\langle 5, \infty \rangle$
$x < -19$
$\langle -\infty, -19 \rangle$

Notice that $-2 < 3$ is a true statement. If we add 4 to both the left and right expressions, we have $-2 + 4 < 3 + 4$, also a true statement since $2 < 7$ is true. And if we add -3 to both expressions, we have $-2 + (-3) < 3 + (-3)$, true since it is true that $-5 < 0$. On the other hand if we begin with the false statement $2 > 5$, adding 4 to both expressions gives $2 + 4 > 5 + 4$ or $6 > 9$, another false statement. This suggests that we make the following assumption about inequalities.

> If the same expression is added to the left and right expressions of an inequality, the resulting inequality is equivalent to the original.
> [Addition axiom for inequalities]

Example 5 Find the solution set for each of these:
(a) $5 < 8 + x$
(b) $8x - 1 > 7x + 4 - 6$

Solution

(a)
$$5 < 8 + x$$
$$5 + (-8) < (8 + x) + (-8)$$
$$-3 < x \quad \text{which we can write as } x > -3$$
$$\langle -3, \infty \rangle$$

(b)
$$8x - 1 > 7x + 4 - 6$$
$$8x - 1 > 7x - 2$$
$$(8x - 1) + (-7x) > (7x - 2) + (-7x)$$
$$x - 1 > -2$$
$$(x - 1) + 1 > -2 + 1$$
$$x > -1$$
$$\langle -1, \infty \rangle$$

It is tempting to make an assumption about equivalent inequalities which is analogous to the multiplication axiom for equations. However, such an assumption would lead to contradictions. Consider the true statement $-3 < 2$. If we multiply both the left and right expressions by 3 we obtain $(-3)(3) < (2)(3)$ which is also a true statement since $-9 < 6$. But if we multiply by -4 we have $(-3)(-4) < 2(-4)$ which is *false* since $12 < -8$ is false. Note, however, that while $(-3)(-4) < 2(-4)$ is false, it is true that $(-3)(-4) > 2(-4)$, or $12 > -8$. Similarly, $-2 > -4$ is true, and if we multiply each expression by -3 and also replace $>$ by $<$ we obtain $(-2)(-3) < (-4)(-3)$ which is true since $6 < 12$. These observations suggest that we make the following assumption.

> If the left and right expressions of an inequality are multiplied by the same *positive* number, the resulting inequality is equivalent to the original. If the two expressions are multiplied by a *negative* number, and if also the symbols $>$ and $<$ are interchanged, the resulting inequality is equivalent to the original.
>
> [Multiplication axiom for inequalities]

We may use the three assumptions about equivalent inequalities and the techniques described in the last section for equations to find the solution set for a variety of first degree inequalities.

Example 6 Find the solution set of $3x - 4 > 11 - 2x$.

Solution
$$3x - 4 > 11 - 2x$$
$$3x - 4 > -2x + 11$$

3.4 First Degree Inequalities

$$(3x - 4) + 2x > (-2x + 11) + 2x$$
$$5x - 4 > 11$$
$$(5x - 4) + 4 > 11 + 4$$
$$5x > 15$$
$$\left(\tfrac{1}{5}\right)(5x) > \left(\tfrac{1}{5}\right)(15) \qquad \text{[Multiplication axiom with}$$
$$x > 3 \qquad \qquad \text{a positive multiplier]}$$
$$\langle 3, \infty \rangle$$

Example 7 Find the solution set of $3(2 - x) > 12$.

Solution
$$3(2 - x) > 12$$
$$6 - 3x > 12$$
$$-3x + 6 > 12$$
$$(-3x + 6) + (-6) > 12 + (-6)$$
$$-3x > 6 \qquad \text{[Multiplication axiom with a}$$
$$\left(-\tfrac{1}{3}\right)(-3x) < \left(-\tfrac{1}{3}\right)(6) \qquad \text{negative multiplier, } -\tfrac{1}{3}.$$
$$x < -2 \qquad \text{Note that } > \text{ is replaced by } <]$$
$$\langle -\infty, -2 \rangle$$

Example 8 Find the solution set of $4x + 5 < 4(x + 5)$.

Solution
$$4x + 5 < 4(x + 5)$$
$$4x + 5 < 4x + 20$$
$$(4x + 5) + (-4x) < (4x + 20) + (-4x)$$
$$5 < 20$$
$$R$$

Since the last inequality is a true statement and is equivalent to the original, we conclude that every replacement for the variable results in a true statement.

Example 9 Find the solution set of $4x - 5 > (2x - 3) + (3 + 2x)$.

Solution
$$4x - 5 > (2x - 3) + (3 + 2x)$$
$$4x - 5 > 4x$$
$$(4x - 5) + (-4x) > 4x + (-4x)$$
$$-5 > 0$$
$$\emptyset$$

Since the inequality $-5 > 0$ is a false statement and is equivalent to the original, we conclude that every replacement of x by a real number results in a false statement.

While we have called the three rules given for equivalent inequalities assumptions, in fact they are theorems which can be proved from the definitions for > and < and the previously assumed properties of real numbers and equations. The proofs are given in more advanced algebra texts.

Exercises 3.4

For Exercises 1–16 answer true or false.

1. $-1 \in \langle -4, 10 \rangle$
2. $\frac{1}{2} \in \langle -4, 10 \rangle$
3. $\sqrt{2} \in \langle -4, 10 \rangle$
4. $-3.99 \in \langle -4, 10 \rangle$
5. $10 \in \langle -4, 10 \rangle$
6. $-4 \in \langle -4, 10 \rangle$
7. $-4.1 \in \langle -4, 10 \rangle$
8. $10.1 \in \langle -4, 10 \rangle$
9. $3 \in \{-4, 10\}$
10. $3 \in \{-4, -3, -2, \ldots, 10\}$
11. $\frac{3}{2} \in \{-4, -3, -2, \ldots, 10\}$
12. $9.99 \in \{-4, -3, -2, \ldots, 10\}$
13. $\langle -4, 10 \rangle = \{-4, 10\}$
14. $\langle -4, 10 \rangle = \{-4, -3, -2, \ldots, 10\}$
15. $\{-3, -2, -1, \ldots, 9\}$ is a subset of $\langle -4, 10 \rangle$
16. $\{-2, -1, 0, \ldots, 6\}$ is a subset of $\langle -2, 6 \rangle$

For Exercises 17–44 find the solution set.

17. $x + 3 < 1$
18. $-2 + x > -1$
19. $4 < x + 9$
20. $-2 < 4 + x$
21. $3x < 9$
22. $5 > 2x$
23. $-2x < 8$
24. $-3x > 12$
25. $1 - 5x < 6$
26. $8 - 3x > 2$
27. $2(1 + x) < 5 - 8$
28. $-4 - 5 < 3(2 - x)$
29. $4x + 3 < 2x - 1$
30. $2x - 5 > 5x + 1$
31. $3(x + 4) < 5(x - 3)$
32. $-2(3 - x) > 3x + 1$
33. $2(x - 5) > x - 10$
34. $5x - 3 < 2(x - 1) - 1$
35. $3x - 7 > 0$
36. $0 > 5x - 1$
37. $2x + 7 < 2(1 + x)$
38. $-2 - 3 < 5(1 - x) + 2x$
39. $4x - 1 - 3 > 2x - (5 - 2x)$
40. $-2x - 4 - x > 3(1 - x)$
41. $3(x + 2) - 1 < 3x + 5$
42. $-2x - 3x + 1 > 5(1 - x) - 4$
43. $2x^2 - 3x < 2(x^2 - x)$

44. $3 - 4x^2 < -4(x^2 + x)$

For Exercises 45–48 write an open sentence and find its solution set.

45. Twice a certain number plus 6 is more than three times that number.
46. The product of 2 and 3 less than a number is more than the sum of that number and 5.
47. If a number is increased by 6 it will still be less than the product of 4 and that number.
48. If twice a certain number is decreased by 5 it will be less than 1.

3.5 Graphs of Sets

Associated with each real number there is exactly one point on the number line. Then given any set of *numbers* there corresponds a set of *points*, and we shall call this set of points the **graph** of the number set. Figure 3.1 illustrates how to graph certain sets. Note that the empty circle indicates that the end point of the interval is not included in the set.

Often we will want to find the union of two sets and graph it. Let us introduce the notation $[-2, 3]$ to indicate the union of $\langle -2, 3 \rangle$ and $\{-2, 3\}$, which is shown in Figure 3.2. Then $[-2, 3]$ is an interval of numbers with both end points included. We shall call this a **closed** interval to distinguish it from the **open** interval $\langle -2, 3 \rangle$ which does not include the end points -2 and 3.

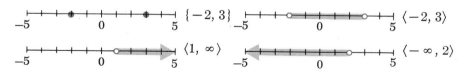

Figure 3.1

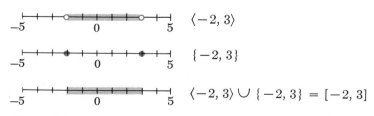

Figure 3.2

106 First Degree Sentences

When an interval contains one of its end points but not the other, it is said to be half closed (and half open). Then $\langle -1, 4]$ is half closed since it contains 4 but not -1. This set is the union of $\langle -1, 4\rangle$ and $\{4\}$, and its graph is shown in Figure 3.3.

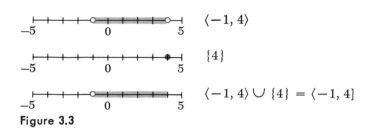

Figure 3.3

Example 1 Graph each of these sets:
(a) $[-3, 1]$ (b) $[-4, -1\rangle$ (c) $\{2\} \cup \langle -1, 2\rangle$ (d) $[1, \infty\rangle$

Solution

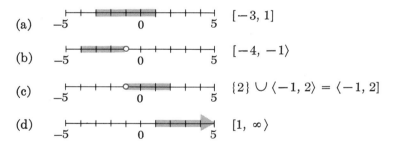

Let us introduce a new set notation. By $\{x: x > 3\}$, which we read "the set of all x such that x is greater than 3," we shall mean the solution set of the open sentence $x > 3$. Then $\{x: x > 3\} = \langle 3, \infty\rangle$. Similarly, $\{x: x < -1\}$ indicates the solution set of $x < -1$, which is $\langle -\infty, -1\rangle$.

Example 2 Find the indicated set and graph it.
(a) $\{x: 2x + 1 < 5\}$ (b) $\{x: 2x + 1 = x - 3\}$

Solution We will first write the open sentence and then find its solution set.

3.5 Graphs of Sets

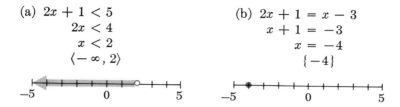

The symbol $\leq$ is often used as a connective in open sentences, and it means "is less than *or* equal to." Then $\{x: x \leq 2\}$ is the set of all x such that either x is less than 2 or x is equal to 2. But this is exactly the *union* of $\{x: x < 2\}$ and $\{x: x = 2\}$, which is shown in Figure 3.4. Then we see that $\{x: x \leq 2\} = \langle-\infty, 2]$. Similarly, $\{x: x \geq -1\}$ is the set of all x such that x is either greater than -1 or x is equal to -1 and $\{x: x \geq -1\} = [-1, \infty\rangle$ as is shown in Figure 3.4. In general when the word "or" is used in this context, it implies the union of two sets.

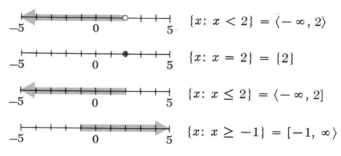

Figure 3.4

Example 3 Find the graph of this set and represent it with interval notation: $\{x: x \geq 1 \text{ or } x > -1\}$.

Solution $\{x: x \geq 1 \text{ or } x > -1\} = \{x: x \geq 1\} \cup \{x: x > -1\} = [1, \infty\rangle \cup \langle-1, \infty\rangle = \langle-1, \infty\rangle$.

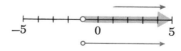

In Section 2.6 we introduced the idea of a number being between two other numbers. We say that 3 is between -1 and 5 and write $-1 < 3 < 5$, which means $-1 < 3$ and $3 < 5$. Similarly, $\{x: -1 < x < 5\}$ is the set of all x such that x is between -1 and 5, and the graph of this set is shown in Figure 3.5. Of course $-1 < x < 5$ also means that $-1 < x$ and $x < 5$, and hence $\{x: -1 < x < 5\}$ is also the *intersection* of $\{x: -1 < x\}$ and $\{x: x < 5\}$ as is shown in the figure. Generally the word "and" when used in this context implies the intersection of two sets.

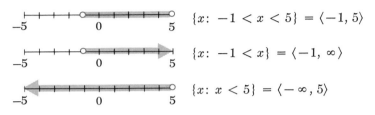

Figure 3.5

Example 4 Find the graph of each set and represent it with interval notation.
(a) $\{x: -3 \leq x \leq 2\}$ (b) $\{x: x \leq -1 \text{ and } x < 1\}$

Solution

(a) $\{x: -3 \leq x \leq 2\} = [-3, 2]$

(b) $\{x: x \leq -1 \text{ and } x < 1\} = \{x: x \leq -1\} \cap \{x: x < 1\}$
$= \langle -\infty, -1] \cap \langle -\infty, 1\rangle = \langle -\infty, -1]$

Note that while $-3 \leq x \leq 2$ means $x \geq -3$ and $x \leq 2$, we shall *not* write $-1 \geq x < 1$ to mean $x \leq -1$ and $x < 1$. We shall agree that sentences such as $2 < x > 3$ or $-2 > x < 3$ are meaningless and not write them at all. Some examples will help to summarize the notation which we have introduced in this section.

Example 5 Find the graph and represent the set with interval notation.
(a) $\{x: x > -1 \text{ and } x > -4\}$ (b) $\{x: x \leq 2 \text{ or } x < -1\}$
(c) $\{x: -1 \leq x < 3\}$ (d) $\{x: x > 1 \text{ or } x \leq -1\}$
(e) $\{x: x \leq -1 \text{ and } x \geq 2\}$ (f) $\{x: x > -1 \text{ or } x < 2\}$

3.5 Graphs of Sets

Solution

(a) $\langle -1, \infty \rangle \cap \langle -4, \infty \rangle = \langle -1, \infty \rangle$

(b) $\langle -\infty, 2] \cup \langle -\infty, -1 \rangle = \langle -\infty, 2]$

(c) $[-1, 3\rangle$

(d) $\langle 1, \infty \rangle \cup \langle -\infty, -1]$

(e) $\langle -\infty, -1] \cap [2, \infty \rangle = \emptyset$

(f) $\langle -1, \infty \rangle \cup \langle -\infty, 2 \rangle = \langle -\infty, \infty \rangle = R$

Exercises 3.5

For Exercises 1–20 find the solution set and graph it.

1. $2x + 3 < 5$
2. $3x - 1 > -7$
3. $1 + x \geq 3$
4. $5 \leq 2x + 1$
5. $x + 3 < 2x + 1$
6. $3x - 1 < x + 1$
7. $2x + 5 \leq 4x + 1$
8. $3(x - 1) \geq 5x - 1$
9. $2x + 6 = 5x - 3$
10. $5(x - 1) = 2x + 1$
11. $4x - 6 = 3(x - 2)$
12. $2(x + 1) - 1 = 3x - 2$
13. $4 - 3x < 1$
14. $2(1 - x) \geq 4$
15. $-5 - 2x \leq 3(2 - 5)$
16. $-3 - 4x > -7$
17. $2x + 4 < 2(x + 3) - 2$
18. $3x + 5 = 3(x + 1) + 2$
19. $1 - 5x \geq 5(1 - x) - 4$
20. $2(3 - 4x) \leq 4(1 - 2x)$

For Exercises 21–32 find the graph of the indicated set and represent it as a single interval, if possible.

21. $\langle 1, 4 \rangle \cup \{1, 4\}$
22. $[-3, 2 \rangle \cup \{2\}$
23. $[-3, 3] \cap \langle -3, 3 \rangle$
24. $[-1, 4] \cap \{-1, 4\}$
25. $\langle 2, 4 \rangle \cap \{2, 4\}$
26. $\langle -2, 3 \rangle \cup \langle -1, 4 \rangle$
27. $\langle -5, 2 \rangle \cap \langle -1, 3 \rangle$
28. $[-3, 2 \rangle \cap \langle -2, 2 \rangle$
29. $\langle -\infty, -1 \rangle \cap \langle -3, \infty \rangle$
30. $\langle 1, \infty \rangle \cap \langle -\infty, 2]$
31. $\langle -\infty, -2] \cup [2, \infty \rangle$
32. $\langle -\infty, 1 \rangle \cup \langle -1, \infty \rangle$

For Exercises 33–40 answer true or false.

33. $[-1, 3] = \{-1, 3\}$
34. $\langle -1, 3 \rangle = \{-1, 3\}$
35. $\langle -1, 3 \rangle$ is a subset of $[-1, 3]$
36. $\langle -1, 3 \rangle$ is a subset of $\{-1, 3\}$
37. $\{-1, 3\}$ is a subset of $\langle -1, 3 \rangle$
38. $\{-1, 3\}$ is a subset of $[-1, 3]$
39. $\{-1, 3\}$ contains infinitely many members
40. $[-1, 3]$ contains exactly two members

For Exercises 41–60 find the graph of the given set and represent it with interval notation.

41. $\{x \colon x > 1\}$
42. $\{x \colon x < 2\}$
43. $\{x \colon x \leq -1\}$
44. $\{x \colon x \geq -1\}$
45. $\{x \colon -1 < x < 3\}$
46. $\{x \colon -2 \leq x \leq 2\}$
47. $\{x \colon 0 < x \leq 5\}$
48. $\{x \colon -3 \leq x < 0\}$
49. $\{x \colon x > 1 \text{ or } x > -1\}$
50. $\{x \colon x \leq 2 \text{ or } x \leq 3\}$
51. $\{x \colon x > 4 \text{ or } x < -2\}$
52. $\{x \colon x \geq -1 \text{ or } x \leq 2\}$
53. $\{x \colon x < 3 \text{ and } x < 2\}$
54. $\{x \colon x \geq -1 \text{ and } x \geq -2\}$
55. $\{x \colon x \leq -2 \text{ and } x < 3\}$
56. $\{x \colon x > 1 \text{ and } x \geq 0\}$
57. $\{x \colon -1 < x < 2 \text{ or } x > 1\}$
58. $\{x \colon x \leq 3 \text{ and } -4 < x < 4\}$
59. $\{x \colon x \geq 2 \text{ and } -3 \leq x \leq 2\}$
60. $\{x \colon -3 < x < 2 \text{ or } 2 < x \leq 3\}$

3.6 Applications

In this chapter we have considered a number of equations and inequalities and we have found and graphed their solution sets. Let us now look rather

111   3.6 Applications

carefully at some problems which can be solved by first writing down an appropriate open sentence and then showing that certain members of its solution set are, in fact, solutions of the original problem.

While some of the problems we shall encounter will not have obvious solutions, others may be very easy to solve once we understand them. But consider the following problem: "John is now 7 years older than Mary. In 5 years John will be twice as old as Mary is now. How old is John at the present time?" It may well be that the most difficult part of solving this problem is understanding exactly what it states. Or consider the equally bewildering: "If the cost of socks is increased just 4 cents per pair, the number of pairs that can be bought for $20.24 is just one less than could be bought for that same amount at the original price. What was the original price for a pair of socks?" Here the problem is not an easy one to solve, but, again, a great deal of difficulty centers around merely understanding *what the problem states*. It would seem that a first step in solving any problem would be to *carefully read the problem*.

A careful reading implies a very clear understanding of exactly what constitutes a solution to the problem. Consider the problem: "If 3 times as many cars were sold this year as last year, and if this year 480 more cars were sold than last year, how many cars were sold last year?" It would be a pity to "solve" this problem and find the number of cars sold this year instead of the number sold last year. It would also be incorrect to find John's age 5 years from now when a problem asks for his age at the present time. Hence, a part of any reading is a clear *understanding of what solution is sought*.

Of course, every problem differs at least slightly from every other. But if there is one technique that is universal in solving problems algebraically, it is the idea of choosing a variable (and an associated replacement set) to represent a number used to measure the solution. The following examples illustrate this first step.

Example 1   The sum of a natural number and five times that number is 66. Find the number.

   Solution   We seek a number. Let us choose the placeholder x to represent this number. Since the number we seek is a natural number, we choose N as the replacement set, $x \in N$.

Example 2   A rectangle is three times as long as it is wide. If 5 inches are removed from each of the two longer sides and 6 inches are added to each of the two shorter sides, the resulting figure is a square. Find the width of the rectangle.

   Solution   Again, we shall use the placeholder x to represent the *number* of inches in the width of the rectangle. But the width of the rectangle need *not* be a natural number although it must be a positive number. Hence, the replacement set chosen will be the set of positive real numbers, $x \in \langle 0, \infty \rangle$.

Some problems require that two or more numbers be found. It is often advisable to attack such problems by choosing a placeholder to represent the smallest of the numbers sought and representing the other numbers by algebraic expressions in this variable.

Example 3 The sum of three consecutive integers is -33. Find the integers.

Solution We let x represent the smallest of the three integers. Clearly, $x \in I$. The next consecutive integer is represented by $x + 1$ and the integer following that is represented by $(x + 1) + 1 = x + 2$.

Example 4 The length of a rectangle is 2 inches less than three times its width. The perimeter of the rectangle is 98 inches. Find the dimensions of the rectangle.

Solution Since we seek the length and width of a rectangle, the replacement set of the variable will be the set of positive real numbers. We choose x to represent the *number* of inches in the width and then the expression $3x - 2$ will represent the number of inches in the length.

The most difficult single step in solving any problem is formulating an open sentence whose solution set may supply the answer to the problem. There is no general instruction which can be used to write such a sentence, but the words of the problem often suggest the operations which must appear in it. Used appropriately, the phrase "more than" should suggest addition, "less than," subtraction, and "is," equals. The following examples may help illustrate the technique.

Example 5 The largest of three consecutive integers is 7 less than twice the smallest. Find the three integers.

Solution We shall let x represent the smallest, $x + 1$ the next, and $x + 2$ the largest of the integers, with $x \in I$. Then our problem suggests the open sentence:

$$x + 2 = 2x - 7$$

The largest is 7 less than twice the smallest.

Example 6 At a school play production student tickets sold for $1.00 and regular adult admission was $1.50. For one performance, when four times as many students attended as adults, only $44 was taken in. How many students and how many adults attended this performance?

Solution We shall let x represent the *number* of adults and then $4x$

will represent the number of students who attended. Note that $x \in N$. Clearly, the money received at the box office from the sale of adult tickets is the product of the number of adult tickets sold and the cost of each ticket. To avoid decimal fractions we will express this in cents. Then $150x$ cents is the revenue from the sale of adult tickets. And $100(4x)$ cents represents the money received from the sale of student tickets. An open sentence we could use to solve this problem is:

$$150x \quad + \quad 100(4x) \quad = \quad 4400$$

Adult sales + student sales = total sales.

Example 7 A boy starts downstream on a raft. Four hours later he has traveled 6 miles, but he had actually stopped for one hour to repair the raft. How fast is the stream flowing?

Solution We shall let x represent the *number* of miles per hour in the rate of the current, and we see that x is a positive real number. To solve this problem we use the idea that the distance traveled is the rate of speed multiplied by the time of travel. But since only 3 hours were spent traveling, we write the equation

$$6 = 3x$$

We have not shown a complete solution to any of the problems we have considered so far. For the next five examples we shall not only select a placeholder and replacement set and formulate an open sentence, but we shall also find the solution set of the sentence and determine whether any of its members are solutions to the original problem.

Example 8 Thirteen less than 4 times a number is 25 more than twice the number. What number is it?

Solution We let $x \in R$ represent the number. Then a suitable open sentence is

$$4x - 13 = 25 + 2x$$

Thirteen less than 4 times a number is 25 more than twice that number.

$$4x - 13 = 25 + 2x$$
$$2x - 13 = 25$$
$$2x = 38$$
$$x = 19$$

Then the required number is 19 since, in fact, 4 times 19 is 76 and 13 less is 63, while twice 19 is 38 and 25 more is 63.

Example 9 The perimeter of a rectangular lot is 264 feet. If the length of the lot is 2 feet less than twice its width, what are its dimensions?

Solution We shall let x represent the *number* of feet in the width with $x \in R$ and $x > 0$. The number of feet in the length is represented by $2x - 2$. The 264 feet perimeter (distance around) suggests the open sentence:

$$x + x + (2x - 2) + (2x - 2) = 264$$
$$6x - 4 = 264$$
$$6x = 268$$
$$x = 44\tfrac{2}{3}$$

Then the width is $44\tfrac{2}{3}$ feet. The length is $2x - 2$ which is $87\tfrac{1}{3}$ when $x = 44\tfrac{2}{3}$. We may check by finding the perimeter. $44\tfrac{2}{3} + 44\tfrac{2}{3} + 87\tfrac{1}{3} + 87\tfrac{1}{3} = 264$.

Example 10 Find three consecutive odd integers such that the second integer is 11 more than the sum of the first and the third.

Solution Let x represent the first odd integer. Then $x + 2$ represents the next odd integer (since the next integer $x + 1$ would be *even*). And $x + 4$ represents the third odd integer. The open sentence we could use is $x + 2 = 11 + [x + (x + 4)]$. The solution is found by

$$x + 2 = 2x + 15$$
$$2 = x + 15$$
$$-13 = x$$

Then the first integer is -13. The next odd integer is $-13 + 2 = -11$, and the third odd integer is -9. Note that 11 more than the sum of the first and third is $11 + (-13) + (-9)$ which equals -11, the second integer.

Example 11 A man rows down a stream with a current whose rate is 3 miles per hour. He covers 11 miles in 2 hours time. How fast could he row in still water?

Solution Let x represent the number of miles per hour in his still-water speed. Then going down stream the man's speed is $(3 + x)$ miles per hour, and a suitable equation is

3.6 Applications

$$(3 + x)2 = 11$$
$$6 + 2x = 11$$
$$2x = 5$$
$$x = \tfrac{5}{2}$$

The man's rowing rate is $2\tfrac{1}{2}$ miles per hour.

Sometimes a problem may not have any solution. The next example illustrates one sort of difficulty we may encounter.

Example 12 A man has twice as many nickels as dimes and three times as many pennies as dimes. He adds up the value of the coins and announces that he has $3.87. Is he correct?

Solution Let x represent the number of dimes. Then $2x$ represents the number of nickels and $3x$ the number of pennies. Clearly the value of his nickels in cents is the product of 5 and the number of nickels, $5(2x)$. And the value of his dimes is $10x$ cents. This suggests the open sentence

$$10x + 5(2x) + 1(3x) = 387$$

The value of his dimes plus the value of his nickels plus the value of his pennies equals the total value of his coins.

$$10x + 5(2x) + 1(3x) = 387$$
$$10x + 10x + 3x = 387$$
$$23x = 387 \quad \cdot \tfrac{1}{23}$$
$$x = 16\tfrac{19}{23}$$

But since the variable x represents the *number* of dimes, it is clear that $x \in N$. Since $16\tfrac{19}{23}$ is *not* a natural number, we conclude that there is no solution, and the man made an error.

In conclusion, let us return to the three problems we discussed briefly at the beginning of the section and apply the techniques we have been illustrating to solve them.

Example 13 If three times as many cars were sold this year as last year, and if this year 480 more cars were sold than last year, how many cars were sold last year?

Solution Let x represent the number of cars sold last year. Then $3x$

represents the number of cars sold this year. A suitable open sentence would be

$$x + 480 = 3x \qquad \text{where } x \in N$$
$$480 = 2x$$
$$240 = x$$

Then 240 cars were sold last year.

Example 14 John is 7 years older than Mary. In 5 years John will be twice as old as Mary is now. How old is John at the present time?

Solution Let x represent John's age now. Then in 5 years John's age can be represented by $x + 5$. Since John is now 7 years older than Mary, she is now 7 years younger than John, and Mary's age could be represented by $x - 7$. Since John's age in 5 years is twice Mary's age now, we could write

$$x + 5 = 2(x - 7)$$
$$x + 5 = 2x - 14$$
$$5 = x - 14$$
$$19 = x$$

John is 19 years old now.

Example 15 If the cost of socks is increased by just 4 cents per pair, the number of pairs that can be bought for $20.24 is just one less than could be bought for that same amount at the original price. What was the original price of a pair of socks?

Solution We shall let x represent the *number* of cents in the original price of a pair of socks. Then the number of pairs that could be bought for 2024 cents is represented by $2024/x$. The increased price per pair of socks we can represent by $x + 4$, and the number of pairs that can be bought for this price is $\dfrac{2024}{x+4}$. But this number of pairs is one less than the original number of pairs. This suggests the open sentence

$$\frac{2024}{x+4} = \frac{2024}{x} - 1$$

Unfortunately this is not a polynomial equation of the first degree, which is the only type we have learned to solve. However, in Chapter 5 we shall be able to complete our solution of this problem.

Exercises 3.6

1. Find the complete solution to Example 7.
2. Find the complete solution to Example 6.
3. Find the complete solution to Example 5.
4. Find the complete solution to Example 1.
5. What number added to 85 gives a sum which is 5 less than 6 times the original number?
6. The product of 3 and 2 less than a number equals the sum of that number and 4 more than the number. What is it?
7. If the area of a plot of ground is tripled, the resulting area is just 42 acres more than it originally was. What was the original area?
8. John now has twice as much money as Fred, but after paying Fred $15, Fred will then have five times as much as John will have. How much does Fred have before John pays him?
9. One number is 3 more than another. If their sum is 1 less than 3 times the smaller, find the numbers.
10. Find two numbers such that the larger is 4 more than the smaller and such that 3 times the larger less twice the smaller is twice the larger plus 3 times the smaller.
11. There are 29 students in an algebra class. The number of boys is 5 more than twice the number of girls. How many girls and how many boys are in the class?
12. In one season the Giants played 160 games. Twice the number of games that they won is 10 less than 3 times the number of games that they lost. How many games did they win and how many did they lose?
13. Expenses for a ski trip are estimated to be $15 for transportation, $11 per day for room and board, and $7 per day for the ski lift. If a trip cost $87, how many days did it last?
14. Golf balls sell for $2 a package. After a certain purchase $8 is returned from a $20 bill. How many packages were purchased?
15. On one Saturday the Cinema 007 theater took in 3 times as much money at its snack bar as it did in ticket sales. If the total money received that day was $1620, how much money was obtained from the sale of tickets?
16. To impress his algebra teacher, John explained that one hour less than 4 times the number of hours he spent each day studying mathematics equaled the difference between the number of hours in a

day and the number of hours he studied algebra. How many hours did John study mathematics each day?

17. Find the complete solution to Example 3.
18. There are three consecutive integers such that the sum of the first and third is 3 less than the second. What are they?
19. The largest of three consecutive even integers is 6 less than twice the smallest. What are the integers?
20. The sum of three consecutive odd integers is 5 times the largest. What are they?
21. Find the complete solution to Example 2.
22. Find the complete solution to Example 4.
23. A rectangle is 5 inches longer than it is wide. If the width is tripled and the length is reduced by 2 inches, the figure will be a square. What are the dimensions of the original rectangle?
24. If one inch were removed from the length of a rectangle, its length would be 3 times its width. The perimeter of the original rectangle is 12 times its width. What are its dimensions?
25. When the price of pens went up 5 cents each, only 20 pens could be bought for the amount of money which would have purchased 22 pens at the old price. What was the original price of pens?
26. Fred has 3 times as many dimes as quarters and 4 times as many nickels as quarters. Altogether his coins are worth 75 cents. How many of each type of coin does he have?
27. Sue has as many quarters as she has dimes and nickels together. If she had one more dime, she would have the same number of dimes as nickels. Altogether she has $1.60 in coins. How many of each type does she have?
28. I have twice as many $5 bills as $10 bills. If I had one fewer $5 bill and one more $10 bill, I would have $105. How many of each type of bill do I have?
29. Mr. Williams can drive to the mountains in the family car in 5 hours. His son Jim makes the same trip in his sports car in only 4 hours. If Jim drives 10 miles per hour faster than his father, what is the average speed of his father's car?
30. When hiking downhill a man can walk at 3 times the rate he can when hiking uphill. In 5 hours of uphill walking he travels only 3 miles farther than he does in one hour of downhill walking. How fast does he walk uphill?
31. A man drove his car a total distance of 295 miles in 5 hours. But during the last 2 hours he had to slow down by 15 miles per hour because of a rain storm. How fast did he drive during the storm?

3.6 *Applications*

32. Two planes start from the same airport at the same time, one flying east, the other west. The ground speed of one plane is 70 miles per hour faster than the other and after 2 hours they are 1740 miles apart. What is the speed of each plane?

33. Bill is twice as old as his brother John. Six years ago Bill was 4 times as old as John. How old is each now?

34. Mrs. Gabor will not tell her age, since she is 4 years older than her husband. In only 2 years she will be twice as old as he was when they were married 19 years ago. How old is she?

35. Said the man to the youth, "I am, of course, older. I'm three times your age, and wiser and bolder." "Give me time" said the youth, "in another eight years I'll be half your age then and one of your peers." How old is each?

36. If Ann were 6 years older, she would be twice as old as Mary was when Ann was born. If Mary is 18, how old is Ann?

37. The current in a stream is 3 miles per hour. A motor boat travels upstream for 5 hours and returns downstream to its starting point in only 2 hours. What is the speed of the boat in still water?

38. If the sum of four consecutive odd integers is 32, what are they?

39. There are three consecutive integers such that the sum of the first and third is twice the second. What are they?

40. If the length of a rectangle is one inch more than twice the width, and its perimeter is five times its width, what are its dimensions?

Polynomials   Chapter FOUR

4.1   Natural Number Exponents

In Section 3.1 we defined a polynomial and agreed that when a polynomial expression is written in standard form, each term except the constant term is in the form ax^n, where a is *any* real number but n is a *natural* number. In this section we will investigate more closely expressions, both numerical and algebraic, which have natural number exponents.

Recall that the numerical expression 2^4 means $2 \cdot 2 \cdot 2 \cdot 2$. The exponent, 4, indicates how many times the base, 2, is used as a factor. Now consider the product $2^4 \cdot 2^3$. This is $(2 \cdot 2 \cdot 2 \cdot 2)(2 \cdot 2 \cdot 2)$, the product of 7 factors, each of which is 2. Then $2^4 \cdot 2^3 = 2^7$. Similarly, $3^2 \cdot 3^6 = (3 \cdot 3)(3 \cdot 3 \cdot 3 \cdot 3 \cdot 3 \cdot 3) = 3^8$. And $5^5 \cdot 5^4 = 5^9$. Note that in each example the exponents are *added*, while the *base remains the same.*

Example 1   If possible, write each expression using only one base and one exponent.

(a) $7^3 \cdot 7^5$ (b) $(-3)^5(-3)^2$ (c) $(\tfrac{2}{3})^2(\tfrac{2}{3})^4$ (d) $2^3 \cdot 3^4$

Solution

(a) $7^{3+5} = 7^8$ (b) $(-3)^{5+2} = (-3)^7$ (c) $(\tfrac{2}{3})^{2+4} = (\tfrac{2}{3})^6$

(d) Impossible, since the bases are not the same. But $2^3 \cdot 3^4 = 8 \cdot 81 = 648$.

Now consider the algebraic expression $x^3 \cdot x^4$. Since $x^3 \cdot x^4 = (x \cdot x \cdot x)(x \cdot x \cdot x \cdot x)$, we see that $x^3 \cdot x^4$ is equivalent to x^7. Similarly, $x^2 \cdot x^{10} = x^{12}$. More generally, we see that the *product* of any two expressions which have the *same base* and natural number exponents is equivalent to an expression whose base is unchanged and whose exponent is found by *adding* the original exponents. We may express this fact symbolically in the next theorem.

Theorem 1

If $a \in N$, and $b \in N$, then $x^a \cdot x^b = x^{a+b}$

We may use this theorem and our previous axioms about the set of real numbers to simplify many algebraic expressions as the following examples illustrate.

Example 2 Find the standard form.
(a) $x^5 \cdot x$ (b) $(3x^2)(x^5)$ (c) $(2x^3)(-3x^2)$

Solution

(a) $x^5 \cdot x = x^{5+1} = x^6$ (b) $(3x^2)(x^5) = 3(x^2 \cdot x^5) = 3x^7$
(c) $(2x^3)(-3x^2) = 2(-3)(x^3 \cdot x^2) = -6x^5$

Often the distributive axiom may be used with Theorem 1.

Example 3 Find the standard form.
(a) $3x^2(2x + 5)$ (b) $(-3x)(x^3 - 4x^2)$ (c) $3x^2 + 2x$

Solution

(a) $3x^2(2x + 5) = (3x^2)(2x) + (3x^2)(5) = 6x^3 + 15x^2$
(b) $(-3x)(x^3 - 4x^2) = (-3x)(x^3) + (-3x)(-4x^2) = -3x^4 + 12x^3$
(c) $3x^2 + 2x$ is already in standard form. Theorem 1 cannot be used to simplify a sum.

Now consider the numerical expression $2^3 \cdot 5^3$. We see that $2^3 \cdot 5^3 = (2 \cdot 2 \cdot 2)(5 \cdot 5 \cdot 5) = (2 \cdot 5)(2 \cdot 5)(2 \cdot 5) = (2 \cdot 5)^3 = 10^3$. Similarly, $3^4 \cdot 7^4 = (3 \cdot 7)^4 = 21^4$. Notice that when both factors in a *product* have the same exponent, we may *multiply* the bases, but the exponent remains unchanged.

Example 4 If possible, write each expression using only one base and one exponent.
  (a) $5^7 \cdot 3^7$ (b) $(\tfrac{2}{3})^4 \cdot 3^4$ (c) $(-3)^5(-2)^5$
  (d) $5^2 \cdot 5^5$ (e) $5^3 + 2^3$

Solution

(a) $(5 \cdot 3)^7 = 15^7$ (b) $(\tfrac{2}{3} \cdot 3)^4 = 2^4$
(c) $[(-3)(-2)]^5 = 6^5$ (d) $5^{2+5} = 5^7$
(e) $5^3 + 2^3 \neq 7^3$. But $5^3 + 2^3 = 125 + 8 = 133$

The algebraic expression $2^3 \cdot x^3$ is equivalent to $(2 \cdot 2 \cdot 2)(x \cdot x \cdot x) = (2x)(2x)(2x) = (2x)^3$. Similarly, $5^4 \cdot x^4 = (5x)^4$ and $(3x)^7 = 3^7 \cdot x^7$. The *product* of any two expressions which have the same natural number exponent is equivalent to an expression whose exponent remains unchanged and whose base is found by *multiplying* the original bases. This fact is represented symbolically in the next theorem.

Theorem 2   If $a \in N$, then $x^a \cdot y^a = (x \cdot y)^a$

The following examples show how to use Theorem 2 to find equivalent algebraic expressions.

Example 5 Find an equivalent expression which uses only one exponent.
  (a) $4^5 x^5$ (b) $(-3)^2 (2x)^2$

Solution

(a) $(4x)^5$ (b) $(-6x)^2$

Often we will want to use Theorem 2 to find the standard form of a polynomial.

Example 6 Find the standard form.
  (a) $(2x)^3$ (b) $(-2x)^2$ (c) $(-x)^2$ (d) $(-x)^3$

Solution

(a) $(2x)^3 = 2^3 x^3 = 8x^3$
(b) $(-2x)^2 = (-2)^2 x^2 = 4x^2$
(c) $(-x)^2 = (-1)^2 x^2 = 1 \cdot x^2 = x^2$

4.1 Natural Number Exponents

(d) $(-x)^3 = (-1)^3 x^3 = (-1)x^3 = -x^3$

We may use both Theorems 1 and 2 to simplify polynomials.

Example 7   Find the standard form.

(a) $(3x)^2(2x^3)$ (b) $(-2x)^3(5x^2 - x + 3)$

Solution

(a) $(3x)^2(2x^3) = (9x^2)(2x^3) = 18x^5$

(b) $(-2x)^3(5x^2 - x + 3) = (-8x^3)(5x^2 - x + 3) = -40x^5 + 8x^4 - 24x^3$

Now consider the numerical expression $(2^3)^2$. We see that $(2^3)^2 = (2^3)(2^3) = 2^6$. Similarly, $(5^2)^4 = (5^2)(5^2)(5^2)(5^2) = 5^8$. Note that in each case the expression was a power of an expression which was itself a power, and that we *multiplied* the exponents, but did not change the base of the original power.

Example 8   If possible, write each expression using a single base and a single exponent.

(a) $(5^3)^4$ (b) $[(\frac{2}{3})^4]^5$ (c) $[(-2)^3]^5$

Solution

(a) $5^{3 \cdot 4} = 5^{12}$ (b) $(\frac{2}{3})^{4 \cdot 5} = (\frac{2}{3})^{20}$

(c) $(-2)^{3 \cdot 5} = (-2)^{15}$

Again we may generalize this property of exponents. $(x^3)^7 = x^{21}$, $[(2x)^2]^5 = (2x)^{10}$. Whenever an expression is a natural number power of an expression which is itself a natural number power, it is equivalent to an expression whose *exponent* is the *product* of the two original exponents, and whose base is the base of the original expression.

Theorem 3

> If $a \in N$ and $b \in N$, then $(x^a)^b = x^{ab}$

We shall not try to prove any of them, but we may use the three theorems of this section to find many equivalent expressions.

Example 9   Find the standard form.

(a) $(x^3)^5$ (b) $[(2x)^3]^2$ (c) $(3x^2)^3$

(d) $(-x^3)^2$ (e) $(-2x^2)^3$

Solution

(a) $x^{3 \cdot 5} = x^{15}$
(b) $(2x)^{3 \cdot 2} = (2x)^6 = 2^6 x^6 = 64 x^6$
(c) $(3x^2)^3 = 3^3 (x^2)^3 = 27 x^6$
(d) $(-x^3)^2 = (-1)^2 (x^3)^2 = x^6$
(e) $(-2x^2)^3 = (-2)^3 (x^2)^3 = -8 x^6$

Exercises 4.1

For Exercises 1–16 use Theorem 1 to find the standard form.

1. $x^7 \cdot x^3$
2. $(3x)(2x^3)$
3. $(-4x)(2x^2)$
4. $(3x^2)(-2x^3)$
5. $(-x)(2x^5)$
6. $(-2x^2)(-x)$
7. $(2x^2)(x^3)(5x)$
8. $(x^3)(-2x)(4x^2)$
9. $x^2(3x + 2)$
10. $2x(5x^2 - 3x + 1)$
11. $-3x(4x + 5)$
12. $-4x^2(x^2 - 2x + 3)$
13. $(4x^2 - 9)(-2x)$
14. $(x^2 + x - 2)(3x^3)$
15. $(2x)(x^2)(5x - 1)$
16. $(2x^2)(3x + 4)(-x^3)$

For Exercises 17–24 use Theorem 2 to find the standard form.

17. $(2x)^3$
18. $(3x)^2$
19. $(-2x)^2$
20. $(-2x)^5$
21. $(-x)^4$
22. $(-x)^7$
23. $(-x)^{150}$
24. $(-x)^{301}$

For Exercises 25–28 use Theorem 3 to find the standard form.

25. $(x^5)^6$
26. $(x^2)^4$
27. $(x^4)^{15}$
28. $(x^{25})^3$

For Exercises 29–52 use any theorem to find the standard form.

29. $(2x^2)^3$
30. $(-3x^3)^2$
31. $(-3x^2)^3$
32. $(-x^3)^4$
33. $(2x)^2(x^3)$
34. $(3x)^2(2x)$
35. $(-3x)^3(3x)$
36. $(-2x)^3(x^3)$
37. $(4x)^2(-2x)^3$
38. $(-x)^4(3x)^2$
39. $(5x)^2(-x)^5$
40. $(-2x)^4(-x)^5$
41. $(x^2)^4(3x^2)$
42. $(x^3)^5(-2x)^2$

43. $(-3x)^3(x^2)^4$
44. $(-x)^7(2x)^3$
45. $(2x)^2(3x-5)$
46. $(-3x)^2(x+6)$
47. $(-2x^2)^3(5x^2-1)$
48. $(3x^2+4x+1)(3x^2)^2$
49. $(2x)(3x)^2(-x)$
50. $(-x)^4(2x^2)^3(x)$
51. $(x)(-x)^3(3x+1)$
52. $(-2x)^2(5x-1)(4x)$

For Exercises 53–64 write the expression using a single base and a single exponent.

53. $7^8 \cdot 7^5$
54. $3^2 \cdot 5^2$
55. $(2^3)^5$
56. $2^5(2^2)^3$
57. $2^3 \cdot 5^3 \cdot 10^2$
58. $(3^2)^3 \cdot 5^6$
59. $(7^3)^4 \cdot (3^6)^2$
60. $(-3)^2(2^2)^4$
61. $4 \cdot 2^5$ [Hint: $4 = 2^2$]
62. $8 \cdot 2^6$
63. $7^3 \cdot 8$
64. $(27)^2 \cdot 5^6$

4.2 Multiplication of Polynomials

Often we shall want to find the standard form for expressions which contain the product of two polynomials. The distributive axiom, which we can express symbolically by $(a+b)c = ac + bc$ or $a(b+c) = ab + ac$, and the other assumptions we have made about the set of real numbers can be used to accomplish this as the following example illustrates.

Example 1 Find the standard form.
(a) $(3x-4)(2x)$
(b) $2x(x+5) + 3(x+5)$
(c) $x(2x-1) - 2(2x-1)$

Solution

(a) $(3x-4)(2x) = (3x)(2x) + (-4)(2x) = 6x^2 - 8x$
(b) $2x(x+5) + 3(x+5) = 2x^2 + 10x + 3x + 15 = 2x^2 + 13x + 15$
(c) $x(2x-1) - 2(2x-1) = x(2x-1) + (-2)(2x-1)$
$= 2x^2 - x - 4x + 2 = 2x^2 - 5x + 2$

When each factor of a product has two or more terms, we may still use the distributive axiom to find the standard form, but the problem is considerably more complicated. For example, we can treat the product $(2x+3)(3x+4)$ as if $(2x+3)$ is considered as a sum $(a+b)$ and $(3x+4)$ is considered as a single factor, c. Then $(a+b)c = ac + bc$ becomes $(2x+3)(3x+4) = (2x)(3x+4) + (3)(3x+4)$. Each of these products can be simplified again using the distributive axiom. $2x(3x + 4) = 6x^2 + 8x$ and $3(3x + 4) = 9x + 12$. The standard form is then obtained by adding similar terms. $6x^2 + 8x + 9x + 12 = 6x^2 + 17x + 12$. That is, $(2x + 3)(3x + 4) = 2x(3x + 4) + 3(3x + 4) = 6x^2 + 8x + 9x + 12 = 6x^2 + 17x + 12$. In fact every product of polynomials may be written in standard form by repeated use of the distributive axiom. Some examples will further illustrate the technique of doing so.

Example 2 Find the standard form.

(a) $(2x^2 + 3)(x - 2)$
(b) $(4x - 5)(3x + 2)$
(c) $(2x - 3)(x^2 + 4x - 1)$
(d) $(x^2 + 2x - 1)(2x + 5)$

Solution

(a) $(2x^2 + 3)(x - 2) = 2x^2(x - 2) + 3(x - 2) = 2x^3 - 4x^2 + 3x - 6$
(b) $(4x - 5)(3x + 2) = 4x(3x + 2) + (-5)(3x + 2)$
$= 12x^2 + 8x - 15x - 10 = 12x^2 - 7x - 10$
(c) $(2x - 3)(x^2 + 4x - 1) = 2x(x^2 + 4x - 1) + (-3)(x^2 + 4x - 1)$
$= 2x^3 + 8x^2 - 2x - 3x^2 - 12x + 3 = 2x^3 + 5x^2 - 14x + 3$
(d) $(x^2 + 2x - 1)(2x + 5) = (2x + 5)(x^2 + 2x - 1)$
$= 2x(x^2 + 2x - 1) + 5(x^2 + 2x - 1)$
$= 2x^3 + 4x^2 - 2x + 5x^2 + 10x - 5 = 2x^3 + 9x^2 + 8x - 5$

To allow us to refer to polynomials in a convenient way we shall occasionally use the expressions trinomial, binomial, or monomial. A polynomial such as $3x^2 - 7x + 5$ or $4x^5 - 2x^2 + 3x$ which has three terms we shall call a **trinomial**. A polynomial such as $2x + 3$, $x^2 - 7$, and $3x^2 + 4x$ which has exactly two terms we shall call a **binomial**. Polynomials with fewer than two terms such as $4x^3$, $-2x$, and 17 we shall sometimes call **monomials**. Polynomials with more than three terms

$$(2x + 5)(3x + 4) = 2x(3x + 4) + 5(3x + 4)$$
$$= 6x^2 + 8x + 15x + 20$$
$$= 6x^2 + 23x + 20$$

$\overbrace{6x^2}$ $\overbrace{8x}$
$(2x + 5)(3x + 4)$ $(2x + 5)(3x + 4)$ $8x + 15x = 23x$
$\underbrace{}_{20}$ $\underbrace{}_{15x}$

Figure 4.1

4.2 Multiplication of Polynomials

will not be given any special names. Then $(2x + 5)(3x + 4)$ is the product of two first degree binomials. We shall want to find the standard form for such a product so often that it will be useful for us to develop a schematic method which will somewhat shorten the process by allowing us to do much of the work mentally.

Let us make several observations about the standard form of the polynomial in Figure 4.1. Notice that the product of the two *first* degree binomials $2x + 5$ and $3x + 4$ is equivalent to the *second* degree trinomial $6x^2 + 23x + 20$. The second degree *term* $6x^2$ is merely the product of $2x$ and $3x$, the first degree *terms* of the factors. And the constant term 20 is the product of the constant terms of the factors. But the first degree term $23x$ is the *sum* of two other first degree terms, one of which is the product of 5 and $3x$, the other of which is the product of $2x$ and 4. These observations are quite generally true for the product of two first degree binomials. Let us summarize them now and illustrate them with some examples.

> The product of two first degree binomials is always a second degree polynomial whose:
>
> 1. Second degree term is the product of the two first degree terms of the factors.
> 2. First degree term is found by adding two first degree terms each of which is the product of the constant term from one factor and the first degree term from the other factor.
> 3. Constant term is the product of the two constant terms of the factors.

Example 3 Find the standard form and show schematically how the first degree term is found.

(a) $(x + 3)(2x + 5)$ (b) $(2x - 1)(5x + 2)$
(c) $(4x - 3)(2x - 5)$

Solution

(a) $(x + 3)(2x + 5) = 2x^2 + 11x + 15$

$(x + 3)(2x + 5)$ with $5x$ above and $6x$ below

(b) $(2x - 1)(5x + 2) = 10x^2 - x - 2$

$(2x - 1)(5x + 2)$ with $4x$ above and $-5x$ below

(c) $(4x - 3)(2x - 5) = 8x^2 - 26x + 15$

$$(4x - 3)(2x - 5)$$
with $-20x$ from outer and $-6x$ from inner.

It will usually not be necessary to show schematically how the product is obtained. When all computations are done mentally, we shall say the result was obtained by inspection.

Example 4 Find the standard form by inspection.
(a) $(4x - 5)(3x + 1)$ (b) $(2x - 5)(x + 4)$
(c) $(x + 4)(x - 4)$

Solution

(a) $12x^2 - 11x - 5$ (b) $2x^2 + 3x - 20$ (c) $x^2 - 16$

In the last example we have $(x + 4)(x - 4) = x^2 - 16$. Notice that the first degree term of the standard form is the sum of $4x$ and $-4x$, which is 0. It is generally true that one of the terms will be zero when we multiply two binomials with one factor the sum of two terms and the other factor the difference of *those same two terms*. The next theorem expresses this fact symbolically.

Theorem 4

$$(a + b)(a - b) = a^2 - b^2$$

Proof $(a + b)(a - b) = a(a - b) + b(a - b)$
$= a^2 - ab + ba - b^2 = a^2 - b^2$

We may express Theorem 4 verbally by saying that the product of the *sum* of two terms and the *difference* of those same two terms is equivalent to the difference of the square of the first term and the square of the second term. And we need not restrict ourselves to products of first degree binomials, although we often shall.

Example 5 Use Theorem 4, if possible, to find the standard form.
(a) $(x + 5)(x - 5)$ (b) $(2x + 3)(2x - 3)$
(c) $(4x^2 - 5)(4x^2 + 5)$ (d) $(x + 4)(x - 3)$

Solution

(a) $x^2 - 25$ (b) $4x^2 - 9$ (c) $16x^4 - 25$

4.2 *Multiplication of Polynomials*

(d) Impossible, but $(x + 4)(x - 3) = x^2 + x - 12$

Often, we will want to find the standard form of the square of a binomial.

Example 6 Find the standard form of $(2x + 3)^2$.

Solution $(2x + 3)^2 = (2x + 3)(2x + 3) = 4x^2 + 12x + 9$. Notice that the standard form of the square of the binomial $2x + 3$ is a trinomial whose second degree term $4x^2$ is the square of the first degree term $2x$, and whose constant term 9 is the square of the constant term 3. But the first degree term of the trinomial, $12x$, is *twice* the product of the first degree term $2x$ and the constant term 3. That is, $2[(2x)(3)] = 12x$. The next theorem shows that this is generally a true result.

Theorem 5

$$(a + b)^2 = a^2 + 2ab + b^2$$

Proof $(a + b)^2 = (a + b)(a + b) = a(a + b) + b(a + b)$
$= a^2 + ab + ba + b^2 = a^2 + 2ab + b^2$

Example 7 Use Theorem 5, if possible, to find the standard form by inspection.
(a) $(x + 2)^2$ (b) $(3x + 5)^2$ (c) $(2x - 3)^2$

Solution

(a) $x^2 + 4x + 4$ (b) $9x^2 + 30x + 25$
(c) $4x^2 - 12x + 9$. Note that twice the product of the two terms of the binomial is $2[(2x)(-3)] = -12x$.

The theorems and techniques of this section may be used to find the standard form for a variety of polynomials.

Example 8 Find the standard form.
(a) $(2x)(3x - 1)(x + 5)$ (b) $(3x + 2)(x + 5)(3x - 2)$
(c) $(x - 4)(2x - 1)^2$ (d) $[2x(x - 3)]^2$

Solution

(a) $(2x)(3x - 1)(x + 5) = 2x[(3x - 1)(x + 5)]$
$= 2x(3x^2 + 14x - 5) = 6x^3 + 28x^2 - 10x$
(b) $(3x + 2)(x + 5)(3x - 2) = (x + 5)[(3x + 2)(3x - 2)]$

130   Polynomials

$$= (x+5)(9x^2-4) = x(9x^2-4)+5(9x^2-4)$$
$$= 9x^3-4x+45x^2-20 = 9x^3+45x^2-4x-20$$

(c) $(x-4)(2x-1)^2 = (x-4)(4x^2-4x+1)$
$$= x(4x^2-4x+1)+(-4)(4x^2-4x+1)$$
$$= 4x^3-4x^2+x-16x^2+16x-4 = 4x^3-20x^2+17x-4$$

(d) $[2x(x-3)]^2 = (2x)^2(x-3)^2 = 4x^2(x^2-6x+9)$
$$= 4x^4-24x^3+36x^2$$

Exercises 4.2

For Exercises 1–20 find the standard form by inspection, using the schematic method described in this section.

1. $(x+2)(x+3)$
2. $(x+4)(x+1)$
3. $(x+5)(x+2)$
4. $(x+7)(x+3)$
5. $(x-2)(x-4)$
6. $(x-3)(x-1)$
7. $(x-5)(x-3)$
8. $(x-3)(x-4)$
9. $(x+4)(x-3)$
10. $(x-5)(x+3)$
11. $(x-7)(x+5)$
12. $(x+6)(x-4)$
13. $(2x+1)(3x+2)$
14. $(3x+5)(2x+1)$
15. $(4x-3)(2x-1)$
16. $(5x-2)(3x-2)$
17. $(4x-1)(3x-2)$
18. $(5x-3)(2x+1)$
19. $(6x-1)(2x-5)$
20. $(4x+3)(3x-2)$

For Exercises 21–28 use Theorem 4 to find the standard form.

$(a+b)(a-b) = a^2-b^2$

21. $(x+5)(x-5)$
22. $(x+2)(x-2)$
23. $(x-7)(x+7)$
24. $(x-1)(x+1)$
25. $(2x+1)(2x-1)$
26. $(3x-1)(3x+1)$
27. $(4x+3)(4x-3)$
28. $(5x-2)(5x+2)$

For Exercises 29–44 use Theorem 5 to find the standard form.

$(a+b)^2 = a^2+2ab+b^2$

29. $(x+3)^2$
30. $(x+7)^2$
31. $(2x+1)^2$
32. $(3x+1)^2$
33. $(x-5)^2$
34. $(x-1)^2$
35. $(3x-1)^2$
36. $(4x-1)^2$
37. $(2x+3)^2$
38. $(3x+2)^2$

39. $(5x + 3)^2$
40. $(4x + 5)^2$
41. $(2x - 3)^2$
42. $(5x - 2)^2$
43. $(7x - 2)^2$
44. $(4x - 3)^2$

For Exercises 45–72 use any method to find the standard form.

45. $(2x + 3)(x^2 + x + 2)$
46. $(x + 5)(2x^2 + 3x + 4)$
47. $(4x + 2)(x^2 - 2x - 1)$
48. $(x^2 - 3x - 2)(2x + 3)$
49. $(3x - 1)(2x^2 + x + 3)$
50. $(x - 6)(x^2 - 3x - 2)$
51. $(2x^2 - 5x - 1)(2x - 3)$
52. $(x^2 + 2x + 1)(3x - 2)$
53. $(2x + 1)(x - 3) + (x + 4)(3x - 1)$
54. $(2x + 1)^2 + (x + 5)(2x - 1)$
55. $(3x - 2)^2 + (x + 2)(x - 2)$
56. $(3x + 1)(4x - 1) + (2x + 1)(2x - 1)$
57. $(3x + 4)(3x - 4) - (x + 4)^2$
58. $(2x + 3)^2 - (4x + 1)(x + 1)$
59. $(3x + 2)^2 - (2x + 3)^2$
60. $(3x + 2)(x - 4) - (2x + 5)^2$
61. $(2x + 1)(x + 3)(x + 2)$
62. $(x + 5)(x - 5)(4x + 3)$
63. $(2x + 3)(x + 7)(2x - 3)$
64. $(x + 4)(2x - 1)^2$
65. $(x^2 + 2x + 1)(x^2 + x + 3)$
66. $(2x^2 - 3x + 1)(x^2 + 4x + 2)$
67. $(x + 5)^2(x + 3)(x - 3)$
68. $(2x - 1)^2(x + 4)^2$
69. $[x(x + 1)]^2$
70. $[-2x(x - 1)]^2$
71. $(x^2 - 2x + 3)^2$
72. $[(x + 4)(x - 2)]^2$

4.3 Factoring Polynomials

When an expression is written as a product, we shall say that it is in **factored** form. Then $(x + 2)(x - 3)$, $-14x^3$, $3x(x + 2)$ and $(2x - 1)^2$ are all in factored form, but neither $3x + 4$ nor $2x(x + 5) + 3$ is in factored form.

132   Polynomials

Example 1   Which of these expressions are in factored form?
(a) $(4x + 5)(4x - 5)$ (b) $2x(3x - 1)^2$ (c) $2x^2$
(d) $3x(x + 2) + 5(x + 2)$ (e) $x(x + 5) - 3$

Solution   Only (a), (b), and (c) are in factored form.

Often we will want to replace an expression which is *not* in factored form by an equivalent expression which *is* in factored form, and we shall call this process "factoring the expression." The distributive axiom may often be used to factor an expression.

Example 2   Factor these expressions.
(a) $3x + 6$ (b) $2x - 8$ (c) $x^2 + 3x$

Solution   (a) $3(x + 2)$ (b) $2(x - 4)$ (c) $x(x + 3)$

Notice that each of the expressions in Example 2 was factored by finding a factor which was repeated in each term. The repeated factor in $3x + 6$ is 3, since 3 is a factor of both $3x$ and 6; the repeated factor in $2x - 8$ is 2; and in $x^2 + 3x$ it is x. Finding a factor which is repeated in each term is a commonly used technique for factoring an expression.

Example 3   Factor each expression by finding a factor which is repeated in each term.
(a) $5x^2 + 10x - 20$ (b) $2x^3 - 5x^2 + x$
(c) $x(x + 1) + 2(x + 1)$

Solution

(a) $5(x^2 + 2x - 4)$ (b) $x(2x^2 - 5x + 1)$
(c) The repeated factor is the binomial $x + 1$. Then
$x(x + 1) + 2(x + 1) = (x + 2)(x + 1)$

Sometimes Theorem 4, $(a + b)(a - b) = a^2 - b^2$, may be used to factor an expression even when there is no repeated factor. If a *binomial* is the difference of squares, we may always factor it.

Example 4   Factor, if possible, using Theorem 4.
(a) $x^2 - 9$ (b) $4x^2 - 1$ (c) $16x^2 - 25$
(d) $x^4 - 4$ (e) $x^2 + 49$

Solution

(a) $(x + 3)(x - 3)$ (b) $(2x + 1)(2x - 1)$
(c) $(4x + 5)(4x - 5)$ (d) $(x^2 + 2)(x^2 - 2)$

(e) Impossible, since $x^2 + 49$ is not the *difference* of squares. Indeed, the only ways we shall be able to factor a binomial are the removal of a repeated factor and by means of Theorem 4.

However, Theorem 5, $(a + b)^2 = a^2 + 2ab + b^2$, can be used to factor a trinomial which is the square of a binomial. For example, the trinomial $x^2 + 6x + 9$ is the square of $x + 3$. Notice that its first term x^2 is the square of x and its constant term 9 is the square of 3. This suggests that it *might* be the square of $x + 3$. And in fact $(x + 3)^2 = x^2 + 6x + 9$. We can also factor the trinomial $x^2 - 10x + 25$ by noting that the first and third terms, x^2 and 25, are the squares of x and 5, respectively. And while $(x + 5)^2 = x^2 + 10x + 25$ is *not* the result we seek, $(x - 5)^2 = x^2 - 10x + 25$, and so the factored form is $(x - 5)^2$.

Example 5 Use Theorem 5, if possible, to factor each of the following:
(a) $4x^2 + 12x + 9$
(b) $9x^2 - 30x + 25$
(c) $x^2 + 49$
(d) $x^2 + 6x - 9$
(e) $x^2 + 4x + 9$
(f) $x^2 + 4x + 3$

Solution

(a) $4x^2$ is the square of $2x$ and 9 is the square of 3
$(2x + 3)^2 = 4x^2 + 12x + 9$
(b) $9x^2$ is the square of $3x$ and 25 is the square of -5
$(3x - 5)^2 = 9x^2 - 30x + 25$
(c) Theorem 5 *cannot* be used to factor $x^2 + 49$, since it is not a trinomial. $(x + 7)^2 \neq x^2 + 49$. In fact $(x + 7)^2 = x^2 + 14x + 49$
(d) Theorem 5 cannot be used to factor $x^2 + 6x - 9$. The constant term -9 is *not* the square of any real number.
(e) Theorem 5 cannot be used to factor $x^2 + 4x + 9$. While x^2 is the square of x and 9 is the square of 3, $(x + 3)^2 = x^2 + 6x + 9$, which is *not* the given trinomial.
(f) Theorem 5 cannot be used to factor $x^2 + 4x + 3$, since 3 is not the square of an integer.

While $x^2 + 4x + 3$ cannot be factored by any of the techniques so far discussed, it is easily seen that $x^2 + 4x + 3 = (x + 1)(x + 3)$. Let us investigate a method which can be used to find factors such as these when it is possible to do so.

We shall first try to factor $x^2 + 8x + 12$. This is a second degree trinomial which has no repeated factors and is not the square of a binomial. We shall try to find a pair of first degree binomial factors. Recall that the second degree term x^2 must be the product of the two first degree terms of the factors. Hence, these terms must both be x. And the constant term 12 must be the product of the two constant terms of the factors. However, there are several choices for the constant terms: 3 and 4, 2 and 6, 1 and 12. We shall try 3 and 4 as a first guess. But $(x + 3)(x + 4) = x^2 + 7x + 12$, and so we have *not* found the factored form. Next we shall try 2 and 6. Since $(x + 2)(x + 6) = x^2 + 8x + 12$, we *have* found the factored form.

Example 6 Factor $3x^2 + 7x + 4$.

Solution The first degree terms must be $3x$ and x since $(3x)(x) = 3x^2$. The constant terms may be 2 and 2, or 1 and 4. We begin to make trials.

$$(3x + 2)(x + 2) = 3x^2 + 8x + 4$$
$$(3x + 1)(x + 4) = 3x^2 + 13x + 4$$
$$(3x + 4)(x + 1) = 3x^2 + 7x + 4$$

Hence the factored form is $(3x + 4)(x + 1)$.

When the first degree term of the trinomial has a negative coefficient, further possibilities must be considered. The constant term may be positive as the following example illustrates.

Example 7 Factor $4x^2 - 8x + 3$.

Solution The first degree terms may be $4x$ and x, or $2x$ and $2x$. Since the first degree term of the trinomial is $-8x$, the constant terms of the factors must be *negative*, and we will try -1 and -3.

$$(4x - 1)(x - 3) = 4x^2 - 13x + 3$$
$$(4x - 3)(x - 1) = 4x^2 - 7x + 3$$
$$(2x - 3)(2x - 1) = 4x^2 - 8x + 3$$

Hence the factored form is $(2x - 3)(2x - 1)$.

Or the constant term of the trinomial may be negative, requiring that even more possibilities be considered.

Example 8 Factor $x^2 + x - 12$.

Solution The first degree terms must be both x. There are several possibilities for the constant terms: 3 and -4, -3 and 4, 2 and -6, -2 and 6, 1 and -12, -1 and 12. By trial and error we hope to find the required factors.

$$(x + 3)(x - 4) = x^2 - x - 12$$
$$(x - 3)(x + 4) = x^2 + x - 12$$

4.3 Factoring Polynomials

Our second trial has proved successful and the factored form is $(x - 3)(x + 4)$. But note that we might have tried many other possibilities before finding the correct one.

As the examples illustrate, factoring a second degree trinomial is a process of trial and error, but we may often limit the trials to only a few. It will be helpful to summarize these limiting conditions on the terms of the binomial factors.

When factoring a nonsquare second degree trinomial into the product of two first degree binomials:

1. The product of the first degree terms of the factors must equal the second degree term of the trinomial.
2. The product of the constant terms of the factors must equal the constant term of the trinomial.
3. When the constant term of the trinomial is positive, the constant terms of the factors will be both positive or both negative, depending on whether the coefficient of the first degree term of the trinomial is positive or negative.
4. When the constant term of the trinomial is negative, one constant term of the factors will be positive and the other will be negative.

Example 9   Factor each of the following trinomials:
(a) $5x^2 - 31x + 6$ (b) $4x^2 + 15x - 4$

Solution

(a) $5x^2 - 31x + 6$. The factors of the first term are $5x$ and x. The factors of the third term might be -3 and -2, or -6 and -1. (We use negative constants since the coefficient of the second term in the trinomial is negative.) There are four cases to consider:

$$(5x - 3)(x - 2) = 5x^2 - 13x + 6$$
$$(5x - 2)(x - 3) = 5x^2 - 17x + 6$$
$$(5x - 6)(x - 1) = 5x^2 - 11x + 6$$
$$(5x - 1)(x - 6) = 5x^2 - 31x + 6$$

The last pair of factors is the correct one.

(b) $4x^2 + 15x - 4$. The factors of the first term might be $2x$ and $2x$, or $4x$ and x. The factors of the constant term might be 2 and -2, 4 and -1, or -4 and 1. There are actually nine possibilities to consider.

Polynomials

Trial and error eventually yield the correct factors.

$$4x^2 + 15x - 4 = (4x - 1)(x + 4)$$

We end this section with a summary of the methods of factoring we have considered.

---

1. When each term of a binomial or a trinomial contains a repeated factor, we can remove this repeated factor. $4x^2 - 6x + 18 = 2(2x^2 - 3x + 9)$.
2. When a binomial is the difference of two squares, we can use Theorem 4 to obtain its two factors. $4x^2 - 49 = (2x + 7)(2x - 7)$.
3. When a trinomial is the square of a binomial, Theorem 5 gives its factors. $9x^2 - 24x + 16 = (3x - 4)^2$.
4. When a trinomial of degree 2 is not a square of a binomial, it may be possible to find two first degree binomials which are its factors. $6x^2 - 7x - 3 = (2x - 3)(3x + 1)$.

In the next section we shall look more closely at the subject of factoring. We shall observe that some polynomials can be factored in more than one way, some polynomials cannot be factored at all in ways which are useful to us, and there is a standard form of factored expressions.

Exercises 4.3

For Exercises 1–12 factor by finding a repeated monomial factor.

1. $3x + 9$
2. $2x - 4$
3. $6x + 8$
4. $12x - 9$
5. $x^2 + 3x$
6. $2x^2 - 5x$
7. $5x^3 - 3x$
8. $3x^3 + 2x$
9. $3x^2 - 3x + 6$
10. $6x^2 + 2x - 6$
11. $5x^3 + 2x^2 - 2x$
12. $2x^4 - 3x^2 + x$

For Exercises 13–64 factor by any method.

13. $x^2 - 4$
14. $9x^2 - 1$
15. $9x^2 - 4$
16. $25x^2 - 9$
17. $x^2 + 4x + 4$
18. $x^2 - 8x + 16$
19. $4x^2 + 12x + 9$
20. $9x^2 - 24x + 16$
21. $x^2 + 3x + 2$
22. $x^2 + 6x + 5$
23. $x^2 + 8x + 7$
24. $x^2 + 12x + 11$

4.4 Complete Factoring

25. $x^2 - 3x + 2$
26. $x^2 - 4x + 3$
27. $x^2 - 6x + 5$
28. $x^2 - 8x + 7$
29. $x^2 + 5x + 6$
30. $x^2 + 7x + 12$
31. $x^2 + 13x + 12$
32. $x^2 + 7x + 10$
33. $x^2 - 7x + 12$
34. $x^2 - 6x + 8$
35. $x^2 - 9x + 18$
36. $x^2 - 9x + 20$
37. $x^2 + 2x - 3$
38. $x^2 - x - 2$
39. $x^2 + 4x - 5$
40. $x^2 + 6x - 7$
41. $x^2 + x - 6$
42. $x^2 - 2x - 8$
43. $x^2 + 3x - 18$
44. $x^2 - 4x - 12$
45. $2x^2 + x - 1$
46. $2x^2 + 5x + 2$
47. $3x^2 - 7x + 2$
48. $5x^2 - 2x - 3$
49. $4x^2 - 8x + 3$
50. $6x^2 - 5x - 6$
51. $5x^2 - 26x + 5$
52. $6x^2 + 25x - 9$
53. $6x^2 + x - 5$
54. $5x^2 + 5x - 15$
55. $4x^2 + 8$
56. $4x^2 - 25$
57. $x(x - 1) + 3(x - 1)$
58. $2x(x + 3) - 3(x + 3)$
59. $5x(2x + 1) + (2x + 1)$
60. $3x(3x - 5) - (3x - 5)$
61. $10x^2 - 19x - 15$
62. $x^2 + 3 - 12$
63. $16x^2 - 66x - 27$
64. $120x^2 - 118x + 9$

4.4 Complete Factoring

Sometimes it is possible to factor an expression in more than one way. For example, the binomial $4x^2 + 6x$ may be factored in at least three ways: $2(2x^2 + 3x)$, $x(4x + 6)$, and $2x(2x + 3)$. In this section we will look more closely at factoring and agree on a best or most complete factorization.

We have learned two methods for factoring a binomial. First, we may remove a common factor: $2x + 4 = 2(x + 2)$. Second, we may factor the difference of squares: $x^2 - 1 = (x + 1)(x - 1)$. There are no other techniques for factoring a binomial which we will consider. Then is it possible to factor $2x + 1$? Surely this is not the difference of squares, and there would seem to be no repeated factor. But notice that $2x + 1 = 2(x + \frac{1}{2})$. Have we factored the expression $2x + 1$? Until now we have considered only polynomials in which all of the coefficients (and the constant term) are *integers*. Notice that $2x + 1$ is such a binomial with integral coefficients, but that $x + \frac{1}{2}$ is not, since $\frac{1}{2}$ is not an integer. Let us agree that the factored form of a polynomial with integral coefficients shall be a product of polynomials all of whose coefficients are integers. Then we will *not* consider $2(x + \frac{1}{2})$ to be the factored form of $2x + 1$. But what *is* the factored form of $2x + 1$? It happens that with the restrictions we have imposed about integral coefficients, the only factored forms for $2x + 1$ are $(1)(2x + 1)$ and the equally trivial $(-1)(-2x - 1)$. We will not usually be interested in either of these trivial factorizations, but will say that $2x + 1$ is a **prime** polynomial, and shall not attempt to factor it. More generally we will say that a polynomial is prime whenever its only allowable factorizations are the two trivial ones.

Let us return to the factorization of $4x^2 + 6x$. We noted that $2(2x^2 + 3x)$, $x(4x + 6)$ and $2x(2x + 3)$ are all factorizations of $4x^2 + 6$. Notice that only in the factorization $2x(2x + 3)$ is the *binomial* factor $2x + 3$ a prime. For both $2x^2 + 3x$ and $4x + 6$ may be factored and so are not prime. Let us agree that $2x(2x + 3)$ is the complete factorization of $4x^2 + 6x$, since the binomial factor $2x + 3$ is prime. More generally, an expression will be considered completely factored only if each of its factors (which are not monomials) are prime polynomials.

Example 1   Find the complete factorization.
(a) $12x^2 - 16x$ (b) $5x^3 + 10x^2$

Solution   (a) $4x(3x - 4)$ (b) $5x^2(x + 2)$

Sometimes the complete factorization of a binomial may be found by first removing a common factor and then factoring the difference of squares. The next example illustrates this.

Example 2   Find the complete factorization
(a) $x^3 - x$ (b) $2x^3 - 50x$ (c) $x^4 - 16$

Solution

(a) $x^3 - x = x(x^2 - 1) = x(x + 1)(x - 1)$. Notice that $x^2 - 1$ was not prime, so we replaced it with $(x + 1)(x - 1)$.
(b) $2x^3 - 50x = 2x(x^2 - 25) = 2x(x + 5)(x - 5)$
(c) $x^4 - 16 = (x^2 + 4)(x^2 - 4) = (x^2 + 4)(x + 2)(x - 2)$. Notice that while $x^2 - 4$ is the *difference* of squares and may be factored, $x^2 + 4$ is not and is a prime polynomial.

We have learned three techniques for factoring a trinomial. First, by removing a common factor:

$$x^3 + 2x^2 + 3x = x(x^2 + 2x + 3)$$

Second, as the square of a binomial:

$$x^2 + 4x + 4 = (x + 2)^2$$

4.4 Complete Factoring

Third, as the product of two different binomials:
$$x^2 + 3x - 4 = (x + 4)(x - 1)$$
These are the only techniques we will consider, and as is the case for binomials, we may have to use two or more of these methods to find the complete factorization of a trinomial.

Example 3 Factor completely.
(a) $2x^3 - 12x^2 + 18x$ (b) $6x^2 + 3x - 9$

Solution

(a) $2x^3 - 12x^2 + 18x = 2x(x^2 - 6x + 9) = 2x(x - 3)^2$
(b) $6x^2 + 3x - 9 = 3(2x^2 + x - 3) = 3(2x + 3)(x - 1)$

Sometimes both binomials and trinomials may be more easily factored if we first remove the trivial factor -1. We will want to do this whenever the coefficient of the highest degree term is negative.

Example 4 Factor completely.
(a) $-x^2 + 16$ (b) $-x^2 + 4x - 4$ (c) $-3x^2 - 3x + 18$

Solution

(a) $-x^2 + 16 = (-1)(x^2 - 16) = -(x + 4)(x - 4)$
(b) $-x^2 + 4x - 4 = (-1)(x^2 - 4x + 4) = -(x - 2)^2$
(c) $-3x^2 - 3x + 18 = (-3)(x^2 + x - 6) = -3(x + 3)(x - 2)$

When the polynomial is not in standard form we will often (but not always) want to first write it in standard form before factoring.

Example 5 Factor completely.
(a) $x(x - 2) - 3$ (b) $3x(x + 1) + 6(x + 1)$

Solution

(a) $x(x - 2) - 3 = x^2 - 2x - 3 = (x + 1)(x - 3)$
(b) Since there is the repeated factor $x + 1$, we will *not* write the polynomial in standard form.
$$3x(x + 1) + 6(x + 1) = (3x + 6)(x + 1) = 3(x + 2)(x + 1)$$

Notice that $3x + 6$ was not prime, so we replaced it with $3(x + 2)$.

Usually we will be unable to factor an expression with more than three terms, but there are two techniques which may sometimes be used. Of

course, we may remove a common monomial factor:
$$2x^4 + 4x^3 - 2x^2 - 6x = 2x(x^3 + 2x^2 - x - 3)$$

And sometimes we may group the terms and thus find a repeated factor. Example 6 illustrates this method.

Example 6 Factor $x^3 + x^2 + 2x + 2$.

Solution We shall group the first two terms and the last two terms. $x^3 + x^2 + 2x + 2 = (x^3 + x^2) + (2x + 2) = x^2(x + 1) + 2(x + 1) = (x^2 + 2)(x + 1)$

Sometimes we will want to factor a polynomial with rational coefficients, some of which are not integers, such as $x + \frac{1}{2}$. Notice that $(\frac{1}{2})(2x + 1) = x + \frac{1}{2}$, and so $(\frac{1}{2})(2x + 1)$ is a factorization. And notice that the binomial factor $2x + 1$ has *integral* coefficients. Let us agree that the complete factorization of a polynomial with rational coefficients will be a product in which every factor except the monomial factor is a prime polynomial with integral coefficients.

Example 7 Factor completely.
(a) $\frac{1}{3}x + 2$
(b) $\frac{1}{2}x^2 - 2$

Solution

(a) $\frac{1}{3}x + 2 = (\frac{1}{3})(x + 6)$
(b) $\frac{1}{2}x^2 - 2 = (\frac{1}{2})(x^2 - 4) = (\frac{1}{2})(x + 2)(x - 2)$

Let us now review all of the techniques we have considered for factoring polynomials.

1. For *any* polynomial we will *first* remove any factor which is repeated in each term. When the coefficient of the highest degree term of the polynomial is negative, we will remove the common factor -1.
2. If a *binomial* is the difference of squares, we will factor it by Theorem 4: $a^2 - b^2 = (a + b)(a - b)$.
3. If a *trinomial* is the square of a binomial, we will factor it by Theorem 5: $a^2 + 2ab + b^2 = (a + b)^2$.
4. If a *trinomial* is the product of two different binomials, we will factor it by trial and error.
5. If a polynomial has more than three terms, we will group the terms and try to find a repeated factor.
6. When a polynomial has rational coefficients which are *not* integers, we will first factor it into the product of a rational number and a polynomial with integral coefficients and then proceed to factor this latter polynomial with the rules stated above.

4.4 Complete Factoring

Let us agree that to "factor" shall mean to "factor completely," and the complete factorization of a polynomial shall be a factorization in which every factor except the monomial factor is a prime polynomial with integral coefficients. Some examples will illustrate these rules for factoring.

Example 8 Factor each of the following.
(a) $-4x^2 + 100$
(b) $3x^3 - 24x^2 + 48x$
(c) $x^2 - \frac{5}{2}x + \frac{3}{2}$
(d) $2x^3 - x^2 - 8x + 4$
(e) $2x^2 + 3x + 2$

Solution

(a) $-4x^2 + 100 = -4(x^2 - 25)$ [Rule 1]
$\qquad\qquad\quad = -4(x+5)(x-5)$ [Rule 2]

(b) $3x^3 - 24x^2 + 48x = 3x(x^2 - 8x + 16)$ [Rule 1]
$\qquad\qquad\qquad\quad = 3x(x-4)^2$ [Rule 3]

(c) $x^2 - \frac{5}{2}x + \frac{3}{2} = (\frac{1}{2})(2x^2 - 5x + 3)$ [Rule 6]
$\qquad\qquad\quad = (\frac{1}{2})(2x - 3)(x - 1)$ [Rule 4]

(d) $2x^3 - x^2 - 8x + 4 = (2x^3 - x^2) + (-8x + 4)$
$\qquad\qquad\qquad\quad = x^2(2x - 1) + (-4)(2x - 1)$
$\qquad\qquad\qquad\quad = (x^2 - 4)(2x - 1)$ [Rule 5]
$\qquad\qquad\qquad\quad = (x + 2)(x - 2)(2x - 1)$ [Rule 2]

(e) $2x^2 + 3x + 2$ cannot be factored by any of the given rules. It is a prime polynomial.

Exercises 4.4

Each of the polynomials in Exercises 1–56 can be factored. Factor each completely.

1. $2x^2 + 6x$
2. $3x^3 + 3x^2$
3. $12x^2 - 8x$
4. $8x^4 - 10x^2$
5. $2x^2 - 2$
6. $3x^2 - 12$
7. $8x^2 - 2$
8. $45x^2 - 125$
9. $x^3 - 4x$
10. $16x^3 - 9x$
11. $18x^3 - 8x$
12. $20x^4 - 45x^2$
13. $3x^2 + 6x + 3$
14. $2x^2 - 8x + 8$
15. $8x^2 + 40x + 50$
16. $27x^2 - 36x + 12$
17. $x^3 - 2x^2 + x$
18. $9x^3 + 6x^2 + x$
19. $3x^3 - 18x^2 + 27x$
20. $8x^3 - 8x^2 + 2x$
21. $3x^2 + 9x + 6$
22. $4x^2 + 10x - 6$

23. $5x^3 - 11x^2 + 2x$
24. $6x^2 + 21x^2 - 12x$
25. $-x^2 + 9$
26. $-4x^2 + 1$
27. $-x^2 + 6x - 9$
28. $-x^2 + 2x + 15$
29. $-4x^2 - 18x + 10$
30. $-12x^2 - 12x - 3$
31. $-6x^3 - 4x^2 + 2x$
32. $-8x^3 + 36x^2 + 20x$
33. $x(x - 2) - 15$
34. $2x(x + 2) + x - 3$
35. $3(x^2 + 1) + 6x(x + 1) - 2$
36. $(6x + 5)(x - 1) + 4$
37. $(3x + 1)^2 - 12x$
38. $(2x + 3)(2x - 3) - 16$
39. $3x^2(2x + 3) + (2x + 3)$
40. $2x^2(x - 5) - (x - 5)$
41. $x^3 + 2x^2 + 3x + 6$
42. $2x^3 - 2x^2 + 5x - 5$
43. $3x^3 + x^2 + 3x + 1$
44. $2x^3 - 2x^2 - 3x + 3$
45. $x^3 + x^2 - 9x - 9$
46. $2x^3 - x^2 - 8x + 4$
47. $2x^3 + 6x^2 + 2x + 6$
48. $x^4 + x^3 - x^2 - x$
49. $2x + \frac{1}{2}$
50. $x - \frac{2}{3}$
51. $\frac{1}{2}x + 3$
52. $\frac{1}{2}x + \frac{3}{2}$
53. $\frac{1}{2}x^2 + x$
54. $\frac{1}{3}x^2 - 3$
55. $2x^2 + 2x + \frac{1}{2}$
56. $\frac{1}{2}x + \frac{1}{5}$

For Exercises 57–64 factor, if possible.

57. $3x^2 + 24x + 6$
58. $x^4 - 1$
59. $x^3 + 4x$
60. $50x^2 - 5x - 1$
61. $3x^2 + x + 1$
62. $2x^2 - x + 3$
63. $x^5 - 16x$
64. $x^4 + 2x^2 - 3$

4.5 Second Degree Equations

In Chapter 2 we observed that the product of zero and any real number is zero. We will now show that whenever the product of two numbers is zero, then at least one of the factors must have been zero.

Theorem 6

> If $ab = 0$, then either $a = 0$ or $b = 0$

Proof We begin with the assumption that $ab = 0$. If $a = 0$, then the theorem is certainly true. But if $a \neq 0$, then a has a multiplicative inverse, $1/a$. Then since $ab = 0$, it follows that

4.5 Second Degree Equations

$$\frac{1}{a} \cdot (ab) = \frac{1}{a} \cdot 0 \qquad \left[\text{Multiplying each expression by } \frac{1}{a}\right]$$

$$\frac{1}{a} \cdot (ab) = 0 \qquad \text{[Since the product of 0 and any real number is 0]}$$

$$\left(\frac{1}{a} \cdot a\right) \cdot b = 0 \qquad \text{[Associative axiom]}$$

$$1 \cdot b = 0 \qquad \text{[Multiplicative inverse axiom]}$$
$$b = 0 \qquad \text{[Multiplicative identity axiom]}$$

Hence if $a \neq 0$, then $b = 0$. This is the result we sought.

In Chapter 3 we learned how to find the solution set for any first degree polynomial equation. Let us briefly review this technique.

Example 1 Find the solution set.
(a) $x - 4 = 0$
(b) $3x + 2 = 0$

Solution

(a) $\quad x - 4 = 0$
$(x - 4) + 4 = 0 + 4$
$x = 4$
$\{4\}$

(b) $\quad 3x + 2 = 0$
$(3x + 2) + (-2) = 0 + (-2)$
$3x = -2$
$\left(\frac{1}{3}\right) 3x = \left(\frac{1}{3}\right)(-2)$
$x = -\frac{2}{3}$
$\left\{-\frac{2}{3}\right\}$

The equation $x^2 - 5x + 6 = 0$ is a second degree polynomial equation since one of its expressions is a polynomial of degree two. We shall often refer to such a second degree equation as a **quadratic** equation. Let us see how Theorem 6 can be used to find the solution sets for some quadratic equations. Notice that $x^2 - 5x + 6$ can be factored: $x^2 - 5x + 6 = (x - 2)(x - 3)$. Hence our equation is equivalent to $(x - 2)(x - 3) = 0$. But by Theorem 6 if $(x - 2)(x - 3) = 0$, then either $x - 2 = 0$ or $x - 3 = 0$. Using the set notation of Section 3.5, we see that the solution set of $x^2 - 5x + 6 = 0$ is $\{x: (x - 2)(x - 3) = 0\} = \{x: x - 2 = 0 \text{ or } x - 3 = 0\} = \{x: x - 2 = 0\} \cup \{x: x - 3 = 0\} = \{2\} \cup \{3\} = \{2, 3\}$. The solution set for this quadratic equation is the *union* of the solution sets of a pair of first degree equations.

Example 2 Use the factoring technique and Theorem 6 to find the solution set for each of these.
(a) $x^2 + 2x - 8 = 0$
(b) $0 = 2x^2 + 3x + 1$
(c) $x^2 + 5x = 0$

Solution

(a) $x^2 + 2x - 8 = 0$
$(x + 4)(x - 2) = 0$
$\{x: x + 4 = 0\} = \{-4\}$
$\{x: x - 2 = 0\} = \{2\}$
$\{-4\} \cup \{2\} = \{-4, 2\}$

(b) $0 = 2x^2 + 3x + 1$
$0 = (2x + 1)(x + 1)$
$\{x: 2x + 1 = 0\} = \{-\frac{1}{2}\}$
$\{x: x + 1 = 0\} = \{-1\}$
$\{-\frac{1}{2}\} \cup \{-1\} = \{-\frac{1}{2}, -1\}$

(c) $x^2 + 5x = 0$
$x(x + 5) = 0$
$\{x: x = 0\} = \{0\}$
$\{x: x + 5 = 0\} = \{-5\}$
$\{0\} \cup \{-5\} = \{0, -5\}$

When neither the left nor the right expression of a quadratic equation is zero, we must first find an equivalent equation which is in zero form–an equation in which one of the expressions is zero.

Example 3 Find the solution set of $(x - 4)(x - 5) = 12$.

Solution

$$(x - 4)(x - 5) = 12$$
$$x^2 - 9x + 20 = 12$$
$$x^2 - 9x + 8 = 0$$
$$(x - 1)(x - 8) = 0$$
$$\{x: x - 1 = 0\} = \{1\}$$
$$\{x: x - 8 = 0\} = \{8\}$$
$$\{1\} \cup \{8\} = \{1, 8\}$$

We should point out two rather tempting errors that we shall want to *avoid* when finding the solution set for a quadratic equation such as the one in Example 3. Theorem 6 requires that the product of two numbers be *zero* before we may make any conclusion about the factors. Then it is *not* correct to assume that for the equation $(x - 4)(x - 5) = 12$ that $x - 4 = 0$ or $x - 5 = 0$. Notice that this would lead to the erroneous solution set $\{4, 5\}$. Neither does Theorem 6 suggest that $x - 4 = 12$ nor $x - 5 = 12$. This false conclusion leads to the equally false solution set $\{16, 17\}$.

Example 4 Find the solution set for $2x(x - 1) = 5(x + 3)$.

Solution

$$2x(x - 1) = 5(x + 3)$$

4.5 Second Degree Equations

$$2x^2 - 2x = 5x + 15$$
$$2x^2 - 2x + (-5x - 15) = 5x + 15 + (-5x - 15)$$
$$2x^2 - 7x - 15 = 0$$
$$(2x + 3)(x - 5) = 0$$
$$\{x: 2x + 3 = 0\} = \{-\tfrac{3}{2}\}$$
$$\{x: x - 5 = 0\} = \{5\}$$
$$\{-\tfrac{3}{2}, 5\}$$

It is, of course, possible to find the graph of the solution set of a quadratic, as well as a first degree equation.

Example 5 Find and graph the solution set of each of the following.
(a) $x^2 = 9$ (b) $x(x + 5) = x - 4$
(c) $4x^2 - 20x + 16 = 0$

Solution

(a)
$$x^2 = 9$$
$$x^2 - 9 = 0$$
$$(x + 3)(x - 3) = 0$$
$$\{-3, 3\}$$

(b)
$$x(x + 5) = x - 4$$
$$x^2 + 5x = x - 4$$
$$x^2 + 4x + 4 = 0$$
$$(x + 2)^2 = 0$$
$$\{-2\}$$

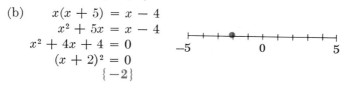

Note that since *both* factors are $x + 2$, there is only *one* member in the solution set.

(c) $4x^2 - 20x + 16 = 0$
$4(x^2 - 5x + 4) = 0$
$4(x - 1)(x - 4) = 0$

Since $4 = 0$ is always false, we need only to consider $x - 1 = 0$ or $x - 4 = 0$. Then the solution set is $\{1, 4\}$.

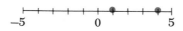

Let us now summarize the factoring technique for finding the solution set for a quadratic equation.

1. First write the equation in zero form.
2. Factor the expression which is a second degree polynomial.
3. Use Theorem 6 to conclude that at least one of the factors must be zero.
4. Solve the resulting pair of first degree equations.
5. The solution set of the quadratic equation is the union of the solution sets of this pair of first degree equations.

Some problems lead to open sentences which are quadratic equations. When solving such problems it is essential to carefully consider the replacement set for the variable as the following examples illustrate.

Example 6 Find three consecutive integers such that the product of the first and third is 23 more than 10 times the second.

Solution Let x represent the first of the three integers, where $x \in I$. Then $x + 1$ and $x + 2$ represent the second and third integers. An equation is
$$x(x + 2) = 23 + 10(x + 1)$$
The product of the first and third is 23 more than 10 times the second.
$$x^2 + 2x = 23 + 10x + 10$$
$$x^2 + 2x = 10x + 33$$
$$x^2 - 8x - 33 = 0$$
$$(x - 11)(x + 3) = 0$$
$$\{11, -3\}$$

Then the integers we seek are either 11, 12, and 13 or -3, -2, and -1.

Example 7 The length of a rectangle is twice its width. If 3 inches are subtracted from its length and added to its width, the new rectangle has an area of 110 square inches. What are the dimensions of the original rectangle?

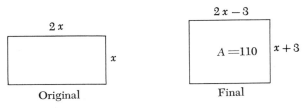

Figure 4.2

4.5 Second Degree Equations

Solution Let x represent the number of inches in the original width, where $x \in (0, \infty)$. Then $2x$ represents the number of inches in the original length. But $2x - 3$ and $x + 3$ represent the number of inches in the final length and width respectively. See Figure 4.2. Since the area of a rectangle is the product of its length and width, the sentence suggested here is

$$(2x - 3)(x + 3) = 110$$
$$2x^2 + 3x - 9 = 110$$
$$2x^2 + 3x - 119 = 0$$
$$(2x + 17)(x - 7) = 0$$

If $x \in R$ then $\{-17/2, 7\}$ would be the solution set. But we have said that $x \in (0, \infty)$, and so we see that $\{7\}$ is the solution set. Then the dimensions are 7 inches by 14 inches.

The technique given in this section may be used to find the solution set for any quadratic equation whose zero form has a second degree expression which may be factored into the product of two first degree polynomials. In Chapter 9 we will give a method for finding the solution set of a quadratic equation which cannot be factored.

Exercises 4.5

For Exercises 1–32 find the solution set.

1. $(x - 2)(x - 1) = 0$
2. $(x + 3)(x + 4) = 0$
3. $(2x - 1)(x + 1) = 0$
4. $(x - 4)(3x + 5) = 0$
5. $(4x - 1)(3x + 2) = 0$
6. $x(x + 5) = 0$
7. $(3x - 5)x = 0$
8. $x(2x + 5) = 0$
9. $x^2 - 3x + 2 = 0$
10. $x^2 + 5x + 6 = 0$
11. $x^2 - 3x - 4 = 0$
12. $x^2 + x - 6 = 0$
13. $2x^2 - 5x - 3 = 0$
14. $5x^2 + 9x - 2 = 0$
15. $2x^2 + 16x + 32 = 0$
16. $6x^2 - 7x - 3 = 0$
17. $x^2 - 25 = 0$
18. $x^2 - 6x + 9 = 0$
19. $4x^2 - 4x + 1 = 0$
20. $25x^2 - 16 = 0$
21. $3x^2 - 6x = 0$
22. $9x^2 + 12x + 4 = 0$
23. $2x^2 - 5x = 0$
24. $4x^2 + 3x = 0$
25. $2x^2 = x + 1$
26. $3x^2 + 2 = 7x$
27. $x(x - 1) = 6$
28. $x(x + 4) = 5$

148 Polynomials

29. $(x - 2)(x + 1) = 4$
30. $(x + 6)(x + 1) = -4$
31. $(x - 5)^2 = 10 - 2x$
32. $(x - 5)(x + 3) = 9$

For Exercises 33–40 find the solution set and graph it.

33. $x(x + 1) = 6$
34. $x^2 + 7x = 3(x - 1)$
35. $(x + 2)^2 = 8x$
36. $x^2 = x$
37. $2x(x + 3) = x^2 - 8$
38. $3x(x + 1) = (2x + 3)(x + 1)$
39. $(2x - 3)(x + 1) = (x - 1)(x + 3)$
40. $(3x + 1)(x + 1) = 2x(x + 3)$

For Exercises 41–48 find the zeros of the function.

41. $f(x) = (x + 3)(x - 2)$
42. $f(x) = 2x(x + 4)$
43. $f(x) = (2x + 1)^2$
44. $f(x) = x^2 - 1$
45. $f(x) = x^2 - x - 2$
46. $f(x) = x^2 + x - 6$
47. $f(x) = 2x^2 - x - 1$
48. $f(x) = (x + 1)^2 - 1$

For Exercises 49–56 find an open sentence and use it to solve the given problem.

49. The product of two consecutive natural numbers is 3 more than 3 times their sum. What are they?
50. The product of two consecutive even integers is 30 more than the smaller integer. What are they?
51. One number is 3 more than another. If the larger number is twice their product, what are they?
52. There are three consecutive even integers such that twice their sum equals the product of the first and second. What are they?
53. The length of a certain rectangle is 6 inches more than its width. If its area is 27 square inches, what are its dimensions?
54. The length of a rectangle is twice its width. If the rectangle were only 2 inches wider, its area would be 240 square inches. Find the dimensions of the given rectangle.
55. The length of a rectangle is one inch more than twice its width. If the width is decreased by 3 inches, the area of the resulting rectangle will be 60 square inches. What are the dimensions of the original rectangle?
56. One rectangle is three times as long as it is wide. Another is one inch

wider and twice as long as the first. If the area of the second rectangle is 24 square inches more than the area of the first, what are the dimensions of the first rectangle?

4.6 Second Degree Inequalities

In Chapter 3 we learned how to find the solution set for any first degree inequality. Let us briefly review that technique.

Example 1 Find the solution set and graph it.
(a) $x - 2 < 0$ (b) $x + 1 \geq 0$

Solution

(a)
$$x - 2 < 0$$
$$(x - 2) + 2 < 0 + 2$$
$$x < 2 \qquad \langle -\infty, 2 \rangle$$

(b)
$$x + 1 \geq 0$$
$$x + 1 + (-1) \geq 0 + (-1)$$
$$x \geq -1 \qquad [-1, \infty)$$

Often we will want to find the intersection of two sets each of which is the solution set for a first degree inequality.

Example 2 Find this set and graph it:

$$\{x: x - 1 > 0 \text{ and } x + 1 > 0\}$$

Solution
$$\{x: x - 1 > 0 \text{ and } x + 1 > 0\}$$
$$= \{x: x - 1 > 0\} \cap \{x: x + 1 > 0\}$$
$$= \{x: x > 1\} \cap \{x: x > -1\}$$
$$= \langle 1, \infty \rangle \cap \langle -1, \infty \rangle = \langle 1, \infty \rangle$$

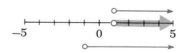

And we will want to find the union of two sets each of which is the solution set for a first degree inequality.

Example 3 Find this set and graph it:

$$\{x: x \leq -1 \text{ or } x \geq 2\}$$

Solution

$$\{x\colon x \leq -1 \text{ or } x \geq 2\}$$
$$= \{x\colon x \leq -1\} \cup \{x\colon x \geq 2\}$$

In Chapter 2 we observed that the product of two positive numbers is positive and the product of two negative numbers is positive, but the product of one positive and one negative number is negative. From this it is easily seen that *whenever* the product of two numbers is positive, then either both factors are positive or both factors are negative. And *whenever* the product of two numbers is negative, then one factor is positive and the other factor is negative. We shall symbolize these observations in the next two theorems.

Theorem 7

If $ab > 0$, then either $a > 0$ and $b > 0$ **or** $a < 0$ and $b < 0$

Theorem 8

If $ab < 0$, then either $a > 0$ and $b < 0$ **or** $a < 0$ and $b > 0$

These theorems form the basis of our technique for finding the solution set of a second degree inequality. Let us consider the inequality $x^2 - x - 2 > 0$. By factoring we see that this is equivalent to $(x - 2)(x + 1) > 0$. But this will become a true statement only when we choose replacements for the variable which will make the product of two factors, $x - 2$ and $x + 1$, positive. Hence, by Theorem 7 either both factors are positive, $x - 2 > 0$ and $x + 1 > 0$, or both factors are negative, $x - 2 < 0$ and $x + 1 < 0$. Using set notation we may write

$$\{x\colon (x - 2)(x + 1) > 0\}$$
$$= \{x\colon x - 2 > 0 \text{ and } x + 1 > 0 \quad \textbf{or} \quad x - 2 < 0 \text{ and } x + 1 < 0\}$$

But now we see that we must find the *union* of two sets:

$$A = \{x\colon x - 2 > 0 \text{ and } x + 1 > 0\}$$
$$B = \{x\colon x - 2 < 0 \text{ and } x + 1 < 0\}$$

Let us first find set A. This set is itself the *intersection* of two sets: $A = \{x\colon x - 2 > 0\} \cap \{x\colon x + 1 > 0\}$. Figure 4.3 shows set $A =$

$\langle 2, \infty \rangle$. Similarly, set B is also an *intersection*: $B = \{x: x - 2 < 0\} \cap \{x: x + 1 < 0\} = \langle -\infty, -1 \rangle$ as is shown in the figure. Finally, the solution set we seek is the *union* of sets A and B, which Figure 4.3 shows is $\langle -\infty, -1 \rangle \cup \langle 2, \infty \rangle$.

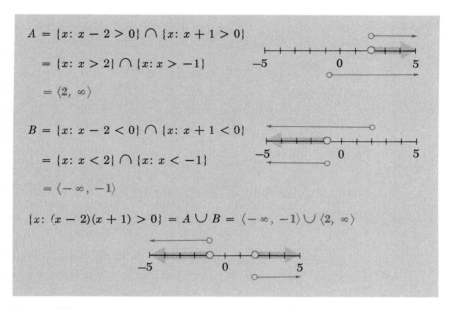

Figure 4.3

Next let us consider the inequality $x^2 - 3x + 2 < 0$ which is equivalent to $(x - 1)(x - 2) < 0$. Since we seek replacements for the variable which will make the product of two factors *negative*, we will use Theorem 8 to conclude that either the first factor is positive and the second is negative, $x - 1 > 0$ and $x - 2 < 0$, or the first factor is negative and the second is positive, $x - 1 < 0$ and $x - 2 > 0$. Then as before we will find the *union* of two sets: $A = \{x: x - 1 > 0 \text{ and } x - 2 < 0\}$, $B = \{x: x - 1 < 0 \text{ and } x - 2 > 0\}$. Again, each of these sets is itself the *intersection* of two sets which are solution sets of first degree inequalities. Figure 4.4 shows the sets A and B and their union. Notice that set B is the empty set and that $A \cup B = A$. Hence, the solution set we seek here is just set $A = \langle 1, 2 \rangle$.

It is generally true that the solution set for a second degree inequality is the *union* of two sets, each of which is the *intersection* of the solution sets for a pair of *first* degree inequalities. Of course, we must first write

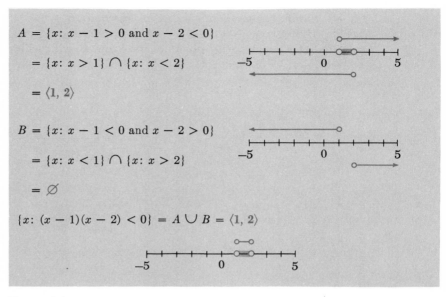

Figure 4.4

the inequality in zero factored form: the left expression a product of first degree polynomials and the right expression zero. Then we may use either Theorem 7 or Theorem 8 to find the solution set. Some examples will further illustrate the technique.

Example 4 Find the solution set and graph it. $x^2 + x > 6$.

Solution

$$x^2 + x > 6$$
$$x^2 + x - 6 > 0$$
$$(x + 3)(x - 2) > 0$$

Since we seek replacements for x which make a product *greater* than zero, we shall use Theorem 7. Either

$x + 3 > 0$ and $x - 2 > 0$ **or** $x + 3 < 0$ and $x - 2 < 0$

$A = \{x: x + 3 > 0\} \cap \{x: x - 2 > 0\}$

$ = \{x: x > -3\} \cap \{x: x > 2\}$

$ = \langle 2, \infty \rangle$

152

4.6 Second Degree Inequalities

$$B = \{x: x + 3 < 0\} \cap \{x: x - 2 < 0\}$$
$$= \{x: x < -3\} \cap \{x: x < 2\}$$
$$= \langle -\infty, -3 \rangle$$
$$\{x: x^2 + x > 6\} = A \cup B = \langle -\infty, -3 \rangle \cup \langle 2, \infty \rangle$$

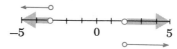

Recall that the complement of a set is the set of all members which are *not* in that set, but which *are* in some universe. If we agree that R is the universe, then the complement of the set $[-3, 2]$ is exactly the union of $\langle -\infty, -3 \rangle$ and $\langle 2, \infty \rangle$. Hence, we may indicate the solution set to the inequality of Example 4 by writing $[-3, 2]$. Notice that $[-3, 2]$ contains the end points -3 and 2, while its complement, $\langle -\infty, -3 \rangle \cup \langle 2, \infty \rangle = [-3, 2]$, does *not* contain these end points.

Example 5 Find the solution set and graph it. $x^2 < 3x$.

Solution
$$x^2 < 3x$$
$$x^2 - 3x < 0$$
$$x(x - 3) < 0$$

Since the product is *less* than zero, we shall use Theorem 8. $x > 0$ and $x - 3 < 0$ **or** $x < 0$ and $x - 3 > 0$.

$A = \{x: x > 0\} \cap \{x: x < 3\}$ $B = \{x: x < 0\} \cap \{x: x > 3\}$
$ = \langle 0, 3 \rangle$ $ = \emptyset$

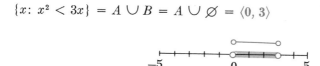

$\{x: x^2 < 3x\} = A \cup B = A \cup \emptyset = \langle 0, 3 \rangle$

154   Polynomials

Example 6   Find the solution set and graph it. $(x+1)(x+3) \geq 0$.

Solution   Here we may use both Theorem 6 and Theorem 7 to conclude that

$x + 1 \geq 0$ and $x + 3 \geq 0$ **or** $x + 1 \leq 0$ and $x + 3 \leq 0$

$A = \{x: x+1 \geq 0\} \cap \{x: x+3 \geq 0\}$
$ = \{x: x \geq -1\} \cap \{x: x \geq -3\}$
$ = [-1, \infty)$

$B = \{x: x+1 \leq 0\} \cap \{x: x+3 \leq 0\}$
$ = \{x: x \leq -1\} \cap \{x: x \leq -3\}$
$ = \langle -\infty, -3]$

$\{x: (x+1)(x+3) \geq 0\} = A \cup B = [-1, \infty) \cup \langle -\infty, -3] = \overline{\langle -3, -1 \rangle}$

Let us now summarize the technique presented in this section for finding the solution set for a second degree inequality.

> 1. Write the inequality in zero factored form.
> 2. If the product is *greater* than zero, use Theorem 7 to conclude that either both factors are positive or both factors are negative.
> 3. If the product is *less* than zero, use Theorem 8 to conclude that either the first factor is positive and the second is negative, or the first factor is negative and the second is positive.
> 4. Use graphs to find the *union* of two sets, each of which is the *intersection* of a pair of solution sets of first degree inequalities.

As we see, the technique for finding the solution set for a second degree inequality is quite involved. It would be helpful if there were some convenient method to check the solution set which we find. Unfortunately, since the set will often be infinite, we cannot make every possible replacement. However, we may make a partial check by selecting several numbers at random and making these replacements for the variable. The next example will illustrate this.

Example 7   Find the solution set and partially check it. $x^2 < 3x$.

4.6 Second Degree Inequalities

Solution In Example 5 we found the solution set. $\{x: x^2 < 3x\} = \langle 0, 3 \rangle$.

Let us first choose a number which is in $\langle 0, 3 \rangle$.
If $x = 2$, then $x^2 < 3x$ becomes $4 < 6$, which is a *true* statement. Next we shall choose several numbers at random which are *not* in $\langle 0, 3 \rangle$.
If $x = 4$, $x^2 < 3x$ becomes $16 < 12$, *false*.
If $x = -2$, $x^2 < 3x$ becomes $4 < -6$, *false*.
If $x = 0$, $x^2 < 3x$ becomes $0 < 0$, *false*.

These random choices for the variable do *not* prove that the solution set is correct, but they suggest that it *might* be correct. If we have made no mistake in the use of Theorem 7, of course, there is no need to check our solution set at all. But the partial check illustrated here can often detect a *misuse* of Theorem 7 or Theorem 8.

Exercises 4.6

For Exercises 1–16 find the indicated set and graph it.

1. $\{x: x + 2 < 0\}$
2. $\{x: x - 3 \leq 0\}$
3. $\{x: x + 1 > 0\}$
4. $\{x: x - 1 > 0\}$
5. $\{x: x + 1 > 0 \text{ and } x - 3 > 0\}$
6. $\{x: x - 2 > 0 \text{ and } x - 1 > 0\}$
7. $\{x: x < 0 \text{ and } x + 2 < 0\}$
8. $\{x: x + 2 \leq 0 \text{ and } x - 4 \leq 0\}$
9. $\{x: x + 1 > 0 \text{ and } x - 2 < 0\}$
10. $\{x: x + 1 < 0 \text{ and } x - 2 > 0\}$
11. $\{x: x + 3 \geq 0 \text{ and } x + 1 \leq 0\}$
12. $\{x: x + 3 < 0 \text{ and } x + 1 > 0\}$
13. $\{x: x + 2 > 0 \text{ and } x - 1 > 0\} \cup \{x: x + 2 < 0 \text{ and } x - 1 < 0\}$
14. $\{x: x - 1 > 0 \text{ and } x - 3 > 0\} \cup \{x: x - 1 < 0 \text{ and } x - 3 < 0\}$
15. $\{x: x + 2 \geq 0 \text{ and } x - 1 \leq 0\} \cup \{x: x + 2 \leq 0 \text{ and } x - 1 \geq 0\}$
16. $\{x: x > 0 \text{ and } x - 2 < 0\} \cup \{x: x < 0 \text{ and } x - 2 > 0\}$

For Exercises 17–32 express, where possible, the solution set as an interval or the complement of an interval and graph it.

17. $x^2 - 1 > 0$
18. $x^2 - x - 2 > 0$
19. $x^2 + x - 6 \geq 0$
20. $x^2 - 4 > 0$
21. $x^2 - 3x - 4 < 0$
22. $x^2 - 2x - 8 < 0$
23. $x^2 - 9 < 0$
24. $x^2 + 3x + 2 \leq 0$
25. $x^2 < 2x$
26. $x^2 > x$
27. $x^2 \leq 4$
28. $x^2 \geq 1$
29. $x^2 + 2x < 8$
30. $x^2 > 0$
31. $x^2 - 4x + 4 \leq 0$
32. $x^2 + 1 \geq 2x$

Rational Algebraic Expressions

Chapter FIVE

5.1 Products and Quotients

In Section 1.2 we agreed that two algebraic expressions are equivalent over a domain if, for each replacement of the variable by a number in that domain, the expressions have the same value. When the domain is finite, such replacements can actually be made to determine whether or not the expressions are equivalent.

Example 1 Make replacements to determine whether or not these expressions are equivalent over the domain $\{0, 1\}$.

(a) $x^2 - 5x + 6$ and $(x-3)(x-2)$
(b) $(x+2)^2$ and $x^2 + 4$
(c) $\dfrac{x}{x+3}$ and $\dfrac{x^2 - 2x}{x^2 + x - 6}$

Solution

(a) If $x = 0$, then $x^2 - 5x + 6 = 6$ and $(x-3)(x-2) = 6$. If $x = 1$, then $x^2 - 5x + 6 = 2$ and $(x-3)(x-2) = 2$. Equivalent over $\{0, 1\}$.

(b) If $x = 0$, then $(x + 2)^2 = 4$ and $x^2 + 4 = 4$. But if $x = 1$, then $(x + 2)^2 = 9$ and $x^2 + 4 = 5$. Not equivalent over $\{0, 1\}$.

(c) If $x = 0$, then

$$\frac{x}{x+3} = \frac{0}{3} = 0 \quad \text{and} \quad \frac{x^2 - 2x}{x^2 + x - 6} = \frac{0}{-6} = 0$$

If $x = 1$, then

$$\frac{x}{x+3} = \frac{1}{4} \quad \text{and} \quad \frac{x^2 - 2x}{x^2 + x - 6} = \frac{-1}{-4} = \frac{1}{4}$$

Equivalent over $\{0, 1\}$.

However, we shall often want to know whether or not two expressions are equivalent over an *infinite* domain, and we will not be able to make all replacements to decide this question. In Chapter 4 we used the distributive and other axioms to show that some polynomial expressions are equivalent over R. In fact, the polynomials of Example 1(a) are equivalent over R since $(x - 3)(x - 2)$ is merely the factored form of $x^2 - 5x + 6$. But, of course, the polynomials of Example 1(b) are not equivalent over R, since they are not equivalent over $\{0, 1\}$. And it is easy to see that the expressions of Example 1(c) (which are not polynomials) are *not* equivalent over R. If $x = 2$, the expression $\frac{x}{x+3}$ becomes $\frac{2}{5}$, but $\frac{x^2 - 2x}{x^2 + x - 6}$ becomes $\frac{0}{0}$, a meaningless symbol (and certainly not a symbol which represents $\frac{2}{5}$), since the *denominator* is zero.

As we shall see in Example 3, there is *some* infinite domain over which the expressions of Example 1(c) are equivalent. Let us first observe that the expression $\frac{x}{x+3}$ will have a real value for every real replacement of the variable except -3, for with that replacement the expression becomes $-\frac{3}{0}$. Then the domain of $\frac{x}{x+3}$ is $\{x \colon x \neq -3\} = \overline{\{-3\}}$. Similarly, the domain of $\frac{x^2 - 2x}{x^2 + x - 6}$ is the set of all real numbers except those that make the *denominator* of the fraction have a value of *zero*. And since the denominator $x^2 + x - 6$ has the factored form $(x + 3)(x - 2)$, we see that we must exclude both -3 and 2 from the domain of this expression. Then its domain is $\overline{\{-3, 2\}}$.

Example 2 Find the domain for each of these expressions.

5.1 *Products and Quotients*

(a) $\dfrac{x}{x-3}$ (b) $\dfrac{x-2}{(x-1)(x+1)}$ (c) $\dfrac{x^2-2x-8}{x^2-5x+4}$

Solution

(a) $\overline{\{x \colon x \neq 3\}} = \overline{\{3\}}$
(b) $\overline{\{x \colon x \neq 1 \text{ and } x \neq -1\}} = \overline{\{-1, 1\}}$. Either -1 or 1 gives the denominator a value of 0.
(c) We shall factor the denominator $x^2 - 5x + 4 = (x-1)(x-4)$. The domain is $\overline{\{1, 4\}}$.

In Chapter 2 we used the canceling theorem $\dfrac{ax}{bx} = \dfrac{a}{b}$ to show that $\dfrac{6}{8} = \dfrac{2 \cdot 3}{2 \cdot 4} = \dfrac{3}{4}$. We may sometimes use this theorem to cancel an *algebraic* factor from the numerator and denominator of a fraction, as the next example illustrates.

Example 3   Use the canceling theorem to show that $\dfrac{x}{x+3}$ and $\dfrac{x^2-2x}{x^2+x-6}$ are equivalent over $\overline{\{-3, 2\}}$.

Solution   We shall first factor:

$$\frac{x^2 - 2x}{x^2 + x - 6} = \frac{x(x-2)}{(x+3)(x-2)}$$

Then we shall cancel the common factor $x - 2$:

$$\frac{x(x-2)}{(x+3)(x-2)} = \frac{x}{x+3}$$

We should observe that each of the expressions above has the same real value for each replacement from $\overline{\{-3, 2\}}$ and that this set is the *intersection* of $\overline{\{-3\}}$ and $\overline{\{-3, 2\}}$, the domains of the given expressions $\dfrac{x}{x+3}$ and $\dfrac{x^2-2x}{x^2+x-6}$.

Let us call expressions such as those in Examples 2 and 3 **rational algebraic** expressions. Notice that each is a fraction whose numerator and denominator is a polynomial. While the domain of every polynomial may be R and we say that two polynomials are equivalent if they are equivalent over R, as we have seen, the domain of a rational expression is often some infinite subset of R. Let us agree to call any two expressions equivalent if they are equivalent over some infinite set which is the *inter-*

section of their domains. And we shall call any expression which is equivalent to a fraction whose numerator and denominator are polynomials a rational algebraic expression.

Example 4 Use the canceling theorem to decide whether or not these rational expressions are equivalent, and if so over what domain.

(a) $\dfrac{(x+3)(x-2)}{(x-3)(x-2)}$ and $\dfrac{x+3}{x-3}$

(b) $\dfrac{x^2-x-12}{x^2+x-20}$ and $\dfrac{x+3}{x+5}$

(c) $\dfrac{x^2+4x-5}{x^2-x-2}$ and $\dfrac{x+5}{x-2}$

Solution

(a) We may cancel $x-2$. Equivalent over $\overline{\{2,3\}}$

(b) $\dfrac{x^2-x-12}{x^2+x-20} = \dfrac{(x-4)(x+3)}{(x-4)(x+5)} = \dfrac{x+3}{x+5}$

We have canceled $x-4$. Equivalent over $\overline{\{4,-5\}}$.

(c) $\dfrac{x^2+4x-5}{x^2-x-2} = \dfrac{(x+5)(x-1)}{(x-2)(x+1)}$. Not equivalent. There is no common factor which we may cancel.

If two polynomials have no factors in common (except the trivial 1 or -1), we shall say that they are **relatively prime**. Then x^2+4x-5 and x^2-x-2 of Example 4(c) are relatively prime since they have no common factors, but x^2-x-12 and x^2+x-20 of Example 4(b) are *not* relatively prime since they have the common factor $x-4$. We shall say that a **standard form** of a rational expression is an equivalent fraction whose numerator and denominator are relatively prime polynomials, where these polynomials may be written in either standard or factored form. Then $\dfrac{x^2+4x-5}{x^2-x-2}$ of Example 4(c) is already in standard form, but a standard form of $\dfrac{x^2-x-12}{x^2+x-20}$ of Example 4(b) is $\dfrac{x+3}{x+5}$.

We may always find a standard form of a rational expression if we first factor its numerator and denominator and then cancel all factors which are common to both the numerator and denominator. Some examples will illustrate the technique.

Example 5 Express each of the following in standard form.

(a) $\dfrac{6x}{8x^2}$

(b) $\dfrac{3x+3}{x^2-2x-3}$

5.1 Products and Quotients

(c) $\dfrac{x^2 + 9x + 18}{x^2 - 2x - 15}$ (d) $\dfrac{x + 5}{x^2 + 4x - 5}$

(e) $\dfrac{4x^2 - 4x - 24}{2x - 6}$ (f) $\dfrac{3(x + 1) + 4(x + 2)}{(x + 1)(x + 2)}$

Solution

(a) $\dfrac{6x}{8x^2} = \dfrac{3(2x)}{(4x)(2x)} = \dfrac{3}{4x}$

(b) $\dfrac{3x + 3}{x^2 - 2x - 3} = \dfrac{3(x + 1)}{(x - 3)(x + 1)} = \dfrac{3}{x - 3}$

(c) $\dfrac{x^2 + 9x + 18}{x^2 - 2x - 15} = \dfrac{(x + 3)(x + 6)}{(x + 3)(x - 5)} = \dfrac{x + 6}{x - 5}$

(d) $\dfrac{x + 5}{x^2 + 4x - 5} = \dfrac{(1)(x + 5)}{(x - 1)(x + 5)} = \dfrac{1}{x - 1}$

Note that we made the trivial factorization of the numerator $x + 5 = (1)(x + 5)$ so that we might use the canceling theorem.

(e) $\dfrac{4x^2 - 4x - 24}{2x - 6} = \dfrac{4(x^2 - x - 6)}{2(x - 3)} = \dfrac{2 \cdot 2(x - 3)(x + 2)}{1 \cdot 2(x - 3)}$

$= \dfrac{2(x + 2)}{1} = 2(x + 2) = 2x + 4$

When the denominator of a fraction is 1, we shall usually not write it, and $\dfrac{4x^2 - 4x - 24}{2x - 6}$ is, in fact, equivalent to the polynomial $2x + 4$. Indeed, any polynomial is equivalent to a rational expression whose denominator is 1 (and whose numerator is merely that polynomial).

(f) $\dfrac{3(x + 1) + 4(x + 2)}{(x + 1)(x + 2)} = \dfrac{3x + 3 + 4x + 8}{(x + 1)(x + 2)} = \dfrac{7x + 11}{(x + 1)(x + 2)}$

Note that *no* canceling was possible in this example. Neither $(x + 1)$ nor $(x + 2)$ is a *factor* of the numerator.

It is important to remember that when finding a standard form of a rational expression, we may do no canceling until *both the numerator and denominator of the fraction are in factored form*. Otherwise the resulting expression will not be equivalent to the original. For example consider the rational expression $\dfrac{x(x + 1) - 2}{(x + 1)(x - 1)}$, whose numerator is *not* in factored form. Canceling $x + 1$ results in $\dfrac{x - 2}{x - 1}$ which is *not* equivalent to the original expression. For if we replace x by 2, then $\dfrac{x(x + 1) - 2}{(x + 1)(x - 1)}$ has

162 Rational Algebraic Expressions

value $\frac{4}{3}$, while $\dfrac{x-2}{x-1}$ has value 0. Proper use of the canceling theorem shows that

$$\frac{x(x+1)-2}{(x+1)(x-1)} = \frac{x^2+x-2}{(x+1)(x-1)} = \frac{(x+2)(x-1)}{(x+1)(x-1)} = \frac{x+2}{x+1}$$

We now note that both $\dfrac{x(x+1)-2}{(x+1)(x-1)}$ and $\dfrac{x+2}{x+1}$ have value $\frac{4}{3}$ when $x = 2$. In fact, these expressions are equivalent over $\{-1, 1\}$.

In Chapter 2 we adopted a rule for multiplying fractions $\dfrac{a}{b} \cdot \dfrac{c}{d} = \dfrac{ac}{bd}$. This same rule will make it possible for us to find a standard form for the product of two rational expressions as the following example illustrates.

Example 6   Find a standard form.

(a) $\dfrac{3x}{13} \cdot \dfrac{2x}{17}$

(b) $\dfrac{x-3}{x+5} \cdot \dfrac{x-5}{x+6}$

(c) $\dfrac{x-5}{x+2} \cdot \dfrac{x+1}{x-5}$

(d) $\dfrac{x^2-4}{x+3} \cdot \dfrac{3}{x+2}$

(e) $\dfrac{x^2-3x+2}{x-1} \cdot \dfrac{x-6}{x^2-5x+6}$

Solution

(a) $\dfrac{3x}{13} \cdot \dfrac{2x}{17} = \dfrac{(3x)(2x)}{13 \cdot 17} = \dfrac{6x^2}{13 \cdot 17}$

(b) $\dfrac{x-3}{x+5} \cdot \dfrac{x-5}{x+6} = \dfrac{(x-3)(x-5)}{(x+5)(x+6)}$

(c) $\dfrac{x-5}{x+2} \cdot \dfrac{x+1}{x-5} = \dfrac{(x-5)(x+1)}{(x+2)(x-5)} = \dfrac{x+1}{x+2}$

Note that we canceled the factor $x - 5$.

(d) $\dfrac{x^2-4}{x+3} \cdot \dfrac{3}{x+2} = \dfrac{(x+2)(x-2)(3)}{(x+3)(x+2)} = \dfrac{3(x-2)}{x+3}$

(e) $\dfrac{x^2-3x+2}{x-1} \cdot \dfrac{x-6}{x^2-5x+6} = \dfrac{(x-2)(x-1)(x-6)}{(x-1)(x-2)(x-3)} = \dfrac{x-6}{x-3}$

Here we canceled both $x - 2$ and $x - 1$.

The definition of division given in Chapter 2, $\dfrac{a}{b} \div \dfrac{c}{d} = \dfrac{a}{b} \cdot \dfrac{d}{c}$, can be used to obtain a standard form for the quotient of two rational expressions. Consider the following examples.

Example 7 Find a standard form.

(a) $\dfrac{8}{3x} \div \dfrac{5x}{9}$

(b) $\dfrac{3}{x+1} \div \dfrac{21}{2x+2}$

(c) $\dfrac{x+4}{x^2+x-12} \div \dfrac{x-1}{x-3}$

(d) $\dfrac{x^2-x-2}{x^2+2x+1} \div \dfrac{x-2}{x+1}$

Solution

(a) $\dfrac{8}{3x} \div \dfrac{5x}{9} = \dfrac{8}{3x} \cdot \dfrac{9}{5x} = \dfrac{8 \cdot 3 \cdot 3}{(3 \cdot 5)x^2} = \dfrac{8 \cdot 3}{5x^2} = \dfrac{24}{5x^2}$

(b) $\dfrac{3}{x+1} \div \dfrac{21}{2x+2} = \dfrac{3}{x+1} \cdot \dfrac{2x+2}{21} = \dfrac{3 \cdot 2(x+1)}{(x+1) \cdot 3 \cdot 7} = \dfrac{2}{7}$

(c) $\dfrac{x+4}{x^2+x-12} \div \dfrac{x-1}{x-3} = \dfrac{x+4}{x^2+x-12} \cdot \dfrac{x-3}{x-1}$

$= \dfrac{(x+4)(x-3)(1)}{(x+4)(x-3)(x-1)} = \dfrac{1}{x-1}$

(d) $\dfrac{x^2-x-2}{x^2+2x+1} \div \dfrac{x-2}{x+1} = \dfrac{x^2-x-2}{x^2+2x+1} \cdot \dfrac{x+1}{x-2}$

$= \dfrac{(x-2)(x+1)(x+1)(1)}{(x+1)(x+1)(x-2)(1)} = 1.$ Every factor except the trivial factor 1 was canceled.

We shall end this section with a summary of the technique for finding a standard form of some rational expressions.

1. If necessary, use the rules for multiplying and dividing fractions to write the expression as a single fraction.
2. Completely factor the numerator and denominator.
3. Cancel any *factors* which are common to both the numerator and denominator.

Exercises 5.1

For Exercises 1–8 determine the domain of the given rational expression.

1. $\dfrac{2}{x-5}$

2. $\dfrac{x+1}{3x+4}$

3. $\dfrac{x+5}{x}$

4. $\dfrac{x-1}{3x(x-2)}$

5. $\dfrac{x}{x^2-4}$

6. $\dfrac{x^2-9}{x^2+x-2}$

7. $\dfrac{2x+5}{3}$

8. x^2+5x+4

For Exercises 9–16 find a standard form for the given rational expression and determine the domain over which the expression and its standard form are equivalent.

9. $\dfrac{3(x-1)}{5(x-1)}$

10. $\dfrac{2(x+1)}{(x+2)(x+1)}$

11. $\dfrac{(x+3)(x-2)}{(x+2)(x-2)}$

12. $\dfrac{6(x+5)}{3(x+5)(x-1)}$

13. $\dfrac{2x+4}{x^2-4}$

14. $\dfrac{x+4}{x^2+x-12}$

15. $\dfrac{x^2-16}{x-4}$

16. $\dfrac{x^2+6x+9}{x^2-9}$

For Exercises 17–28 find an equivalent expression which is in standard form.

17. $\dfrac{5x}{10x^2}$

18. $\dfrac{3x^2}{9x^3}$

19. $\dfrac{8x}{12x^3}$

20. $\dfrac{25x^2}{20x}$

21. $\dfrac{2x+6}{x^2+4x+3}$

22. $\dfrac{x^2+3x+2}{x^2+2x+1}$

23. $\dfrac{x^2+4x-5}{x^2-1}$

24. $\dfrac{x^2+5x+4}{3x+12}$

25. $\dfrac{x^2+x}{x^2-x-2}$

26. $\dfrac{2x^2+6x}{4x^2-36}$

27. $\dfrac{2x^2-x-1}{3x^2-6x+3}$

28. $\dfrac{2x^2+9x-5}{x^2+10x+25}$

For Exercises 29–40 find a standard form for these products and quotients.

29. $\dfrac{3x}{5} \cdot \dfrac{15}{9x^2}$

30. $\dfrac{35x^2}{8} \cdot \dfrac{36}{49x}$

5.2 *Rational Expressions: Sums and Differences*

31. $\dfrac{x+2}{x-1} \cdot \dfrac{x+2}{x+1}$

32. $\dfrac{3x-3}{x+1} \cdot \dfrac{2x+2}{x+2}$

33. $\dfrac{x^2-2x}{x+2} \cdot \dfrac{2x+4}{x-2}$

34. $\dfrac{x-2}{x^2+3x} \cdot \dfrac{x^2-9}{2x-4}$

35. $\dfrac{x^2-x-2}{x^2-2x-3} \cdot \dfrac{x^2-x-6}{x-2}$

36. $\dfrac{x^2-x-2}{x^2-2x-3} \cdot \dfrac{x^2-x-6}{x^2-4}$

37. $\dfrac{3x^2}{2} \div \dfrac{x}{4}$

38. $\dfrac{2x^2+2x}{2x-6} \div \dfrac{x+1}{x-3}$

39. $\dfrac{(x+1)^2}{x-1} \div \dfrac{x+1}{x^2-x}$

40. $\dfrac{x^2+4x+3}{x^2-1} \div \dfrac{x+3}{x^2-3x+2}$

For Exercises 41–48 find a standard form for the given rational expression.

41. $\dfrac{x(x+3)+2}{x(x+2)+1}$

42. $\dfrac{x(x-2)-8}{(x-2)(x-6)+4}$

43. $\dfrac{x(x+2)-3}{(x+2)(x+4)+1}$

44. $\dfrac{x^2-1}{x+4} \cdot \dfrac{x(x+4)+4}{x^2+3x+2}$

45. $\dfrac{(x+4)x-5}{(x-5)(x+5)}$

46. $\dfrac{(x+2)(x+1)-2}{(x-2)(x+1)} \div \dfrac{x+3}{2(x-2)-x+5}$

47. $\dfrac{(x+3)(x-1)}{(x+3)x-4}$

48. $\dfrac{(x+1)^2-4x}{(x+1)(x-1)}$

5.2 Rational Expressions: Sums and Differences

We have agreed that when a rational expression is in standard form its numerator and denominator are relatively prime polynomials which may be in either standard or factored form. And we have observed that every polynomial is also a rational expression, since every polynomial is equivalent to a fraction whose denominator is 1 (and whose numerator is the given polynomial). Let us also agree that, where possible, the polynomials in the numerator and denominator of a standard form rational expression shall have integral coefficients.

Example 1 Find a standard form.

(a) $\frac{2}{3}x$

(b) $(\frac{1}{3})(2x^2-x+1)$

(c) $\dfrac{-2x}{5} \cdot (x^2+4)$

(d) $\frac{1}{2}x + \frac{1}{4}$

166   Rational Algebraic Expressions

(e) $(x^2 - 4) \cdot \dfrac{1}{x - 2}$

Solution

(a) $\dfrac{2}{3} x = \dfrac{2}{3} \cdot \dfrac{x}{1} = \dfrac{2x}{3}$

(b) $\dfrac{1}{3}(2x^2 - x + 1) = \dfrac{1}{3} \cdot \dfrac{2x^2 - x + 1}{1} = \dfrac{2x^2 - x + 1}{3}$

(c) $\dfrac{-2x}{5} \cdot (x^2 + 4) = \dfrac{-2x}{5} \cdot \dfrac{x^2 + 4}{1} = \dfrac{-2x(x^2 + 4)}{5}$

(d) $\dfrac{1}{2} x + \dfrac{1}{4} = \dfrac{1}{4}(2x + 1) = \dfrac{2x + 1}{4}$

(e) $(x^2 - 4) \cdot \dfrac{1}{x - 2} = \dfrac{(x + 2)(x - 2)}{1} \cdot \dfrac{1}{(x - 2)} = x + 2$

When rational expressions contain more than one symbol of operation we shall agree to perform these operations according to the hierarchy established in Chapter 1: first multiplications and divisions from left to right unless there are symbols of grouping. The following example shows how to simplify more complicated rational expressions.

Example 2 Find a standard form.

(a) $\dfrac{2x + 1}{x - 3} \cdot \dfrac{x - 5}{2x - 1} \cdot \dfrac{x - 3}{2x + 1}$

(b) $\dfrac{2x - 2}{3x + 3} \cdot \dfrac{2x^2 + 2x}{x + 3} \div \dfrac{x - 1}{3x + 9}$

(c) $\dfrac{2x + 1}{x^2 - 6x + 9} \div \dfrac{4x^2 - 1}{x^2 - 7x + 12} \cdot \dfrac{2x^2 - 7x + 3}{3x - 12}$

Solution

(a) $\dfrac{2x + 1}{x - 3} \cdot \dfrac{x - 5}{2x - 1} \cdot \dfrac{x - 3}{2x + 1}$

$= \dfrac{(2x + 1)(x - 5)(x - 3)}{(x - 3)(2x - 1)(2x + 1)} = \dfrac{x - 5}{2x - 1}$

We shall sometimes draw lines to indicate which factors have been canceled.

(b) $\dfrac{2x - 2}{3x + 3} \cdot \dfrac{2x^2 + 2x}{x + 3} \div \dfrac{x - 1}{3x + 9}$

$= \dfrac{2x - 2}{3x + 3} \cdot \dfrac{2x^2 + 2x}{x + 3} \cdot \dfrac{3x + 9}{x - 1}$

5.2 Rational Expressions: Sums and Differences

$$= \frac{2(x-1)(2x)(x+1)(3)(x+3)}{3(x+1)(x+3)(x-1)} = 4x$$

(c) $\dfrac{2x+1}{x^2-6x+9} \div \dfrac{4x^2-1}{x^2-7x+12} \cdot \dfrac{2x^2-7x+3}{3x-12}$

$= \dfrac{2x+1}{x^2-6x+9} \cdot \dfrac{x^2-7x+12}{4x^2-1} \cdot \dfrac{2x^2-7x+3}{3x-12}$

$= \dfrac{(2x+1)(x-4)(x-3)(2x-1)(x-3)(1)}{(x-3)(x-3)(2x+1)(2x-1)(3)(x-4)} = \dfrac{1}{3}$

To find a standard form for the sum (or difference) of two rational expressions we use a technique which is analogous to that for finding the simplest form for the sum (or difference) of two rational numbers. If the denominators of the fractions are the same, we add (or subtract) the numerators and copy the common denominator as in $\frac{2}{7} + \frac{3}{7} = \frac{5}{7}$.

Example 3 Find a standard form of the following.

(a) $\dfrac{x}{2} + \dfrac{3}{2}$

(b) $\dfrac{3x}{5} - \dfrac{7}{5}$

(c) $\dfrac{3x}{8} + \dfrac{9x}{8}$

(d) $\dfrac{3x}{10} - \dfrac{x}{10}$

Solution

(a) $\dfrac{x}{2} + \dfrac{3}{2} = \dfrac{x+3}{2}$ (b) $\dfrac{3x}{5} - \dfrac{7}{5} = \dfrac{3x-7}{5}$

(c) $\dfrac{3x}{8} + \dfrac{9x}{8} = \dfrac{3x+9x}{8} = \dfrac{12x}{8} = \dfrac{4(3x)}{4(2)} = \dfrac{3x}{2}$

(d) $\dfrac{3x}{10} - \dfrac{x}{10} = \dfrac{3x-x}{10} = \dfrac{2x}{10} = \dfrac{x}{5}$

But when the numerators of the fractions are not monomials we must be careful to remember that the fraction bar is a symbol of grouping. This can play a crucial role in simplifying such expressions, as Example 4 illustrates.

Example 4 Find a standard form.

(a) $\dfrac{x-4}{5} + \dfrac{4x-1}{5}$

(b) $\dfrac{x+1}{6} - \dfrac{2x+7}{6}$

(c) $\dfrac{2x-3}{2} - \dfrac{2x-5}{2}$

Solution

(a) $\dfrac{x-4}{5} + \dfrac{4x-1}{5} = \dfrac{(x-4)+(4x-1)}{5}$

$= \dfrac{5x-5}{5} = \dfrac{5(x-1)}{5} = x-1$

(b) $\dfrac{x+1}{6} - \dfrac{2x+7}{6} = \dfrac{(x+1)-(2x+7)}{6} = \dfrac{x+1-2x-7}{6}$

$= \dfrac{-x-6}{6}.$ Note that we *must* use parentheses in the numerator $(x+1) - (2x+7)$ to preserve the original grouping. For without parentheses we would have $x+1-2x+7 = -x+8$ which is *not* equivalent to $-x-6$.

(c) $\dfrac{2x-3}{2} - \dfrac{2x-5}{2} = \dfrac{(2x-3)-(2x-5)}{2}$

$= \dfrac{2x-3-2x+5}{2} = \dfrac{2}{2} = 1$

When finding a standard form for the sum (or difference) of two rational expressions which do *not* have the same denominator, we must first find equivalent fractions which have a common denominator. This is accomplished by multiplying both the numerator and denominator of each of the given fractions by some factor.

Example 5 Find expressions with a common denominator of 24 which are equivalent to each of these: $5x/12$ and $3x/8$.

Solution

$$\dfrac{5x}{12} = \dfrac{(5x)(2)}{(12)(2)} = \dfrac{10x}{24}$$

$$\dfrac{3x}{8} = \dfrac{(3x)(3)}{(8)(3)} = \dfrac{9x}{24}$$

The best common denominator to choose is the LCM (least common multiple) of the given denominators. Notice that the LCM of 12 and 8 is 24, since 24 is the smallest natural number which is a multiple of both 12 and 8. That is why we chose 24 as the common denominator for the fractions in Example 5. We may be able to select the LCM of two numbers by inspection if they are small, but when their LCM is not obvious, we will want to first find their prime factorizations and then use these to select the LCM. The next example illustrates this.

5.2 Rational Expressions: Sums and Differences

Example 6 Find the LCM.
(a) 14 and 22   (b) 25 and 35   (c) 16 and 40

Solution

(a) $14 = 2 \cdot 7$; $22 = 2 \cdot 11$. The LCM is the product of all the prime factors. LCM $= 2 \cdot 7 \cdot 11 = 154$.
(b) $25 = 5^2$; $35 = 5 \cdot 7$. Since 5^2 is in the factorization of 25, the LCM must also have the factor 5^2. LCM $= 5^2 \cdot 7 = 175$.
(c) $16 = 2^4$; $40 = 2^3 \cdot 5$. The LCM contains 2^4, the larger of the two powers of 2. LCM $= 2^4 \cdot 5 = 80$.

After selecting the LCM of the denominators, we will next want to find equivalent fractions which have this LCM as their common denominator.

Example 7 For each pair of fractions first find the LCM of their denominators and then replace each fraction with an equivalent one which has the LCM as its denominator.

(a) $\dfrac{x}{2}$ and $\dfrac{3x}{4}$ (b) $\dfrac{5x}{6}$ and $\dfrac{3}{20}$

(c) $\dfrac{x}{25}$ and $\dfrac{2x}{35}$ (d) $\dfrac{x}{7}$ and $3x$

Solution

(a) LCM of 2 and 4 is 4
$$\frac{x}{2} = \frac{x \cdot 2}{2 \cdot 2} = \frac{2x}{4}$$
$\dfrac{3x}{4}$ already has the denominator 4.

(b) LCM of 6 and 20 is 60.
$$\frac{5x}{6} = \frac{(5x)(10)}{(6)(10)} = \frac{50x}{60} \qquad \frac{3}{20} = \frac{3 \cdot 3}{20 \cdot 3} = \frac{9}{60}$$

(c) LCM of 25 and 35 is $5^2 \cdot 7$
$$\frac{x}{25} = \frac{x}{5^2} = \frac{(x)(7)}{(5^2)(7)} = \frac{7x}{5^2 \cdot 7} \qquad \frac{2x}{35} = \frac{2x}{5 \cdot 7} = \frac{2x(5)}{5 \cdot 7(5)} = \frac{10x}{5^2 \cdot 7}$$

(d) $3x = 3x/1$. LCM of 7 and 1 is 7; $x/7$ already has the denominator 7.
$$\frac{3x}{1} = \frac{(3x)(7)}{(1)(7)} = \frac{21x}{7}$$

170 Rational Algebraic Expressions

After each fraction has been replaced by an equivalent one with a common denominator we may then proceed as in arithmetic: Replace the sum (or difference) by a single fraction whose numerator is the sum (or difference) of the numerators and whose denominator is the common denominator. A standard form for this single fraction may be found by the method of Section 5.1.

Example 8 Find a standard form. Refer to Example 7 for the equivalent fractions with a common denominator.

(a) $\dfrac{x}{2} + \dfrac{3x}{4}$ (b) $\dfrac{5x}{6} - \dfrac{3}{20}$

(c) $\dfrac{x}{25} + \dfrac{2x}{35}$ (d) $\dfrac{x}{7} + 3x$

Solution

(a) $\dfrac{x}{2} + \dfrac{3x}{4} = \dfrac{2x}{4} + \dfrac{3x}{4} = \dfrac{5x}{4}$

(b) $\dfrac{5x}{6} - \dfrac{3}{20} = \dfrac{50x}{60} - \dfrac{9}{60} = \dfrac{50x - 9}{60}$

(c) $\dfrac{x}{25} + \dfrac{2x}{35} = \dfrac{7x}{5^2 \cdot 7} + \dfrac{10x}{5^2 \cdot 7} = \dfrac{17x}{5^2 \cdot 7}$

(d) $\dfrac{x}{7} + 3x = \dfrac{x}{7} + \dfrac{21x}{7} = \dfrac{22x}{7}$

When the denominators are *algebraic* expressions we will *always* want to examine the factors in order to select the LCM. For monomials this is easily done. The LCM of $4x$ and $6x$ is $12x$. The LCM of $2x^3$ and $3x^2$ is $6x^3$. Notice that the LCM contains the larger of the two powers of x. After selecting the LCM we may proceed to find a standard form as before.

Example 9 Find a standard form of the following.

(a) $\dfrac{3}{2x} + \dfrac{1}{4x}$ (b) $\dfrac{5}{x} - \dfrac{3}{2x^2}$

(c) $\dfrac{2x - 1}{12x} + \dfrac{x - 1}{8}$ (d) $\dfrac{x - 3}{2x} - \dfrac{x + 1}{x^2}$

Solution

(a) LCM of $2x$ and $4x$ is $4x$

$$\dfrac{3}{2x} + \dfrac{1}{4x} = \dfrac{6}{4x} + \dfrac{1}{4x} = \dfrac{7}{4x}$$

5.2 Rational Expressions: Sums and Differences

(b) LCM of x and $2x^2$ is $2x^2$
$$\frac{5}{x} - \frac{3}{2x^2} = \frac{5(2x)}{x(2x)} - \frac{3}{2x^2} = \frac{10x - 3}{2x^2}$$

(c) LCM is $24x$
$$\frac{2x - 1}{12x} + \frac{x - 1}{8} = \frac{(2x - 1)(2)}{(12x)(2)} + \frac{(x - 1)(3x)}{(8)(3x)}$$
$$= \frac{(4x - 2) + (3x^2 - 3x)}{24x} = \frac{3x^2 + x - 2}{24x}$$

(d) LCM is $2x^2$
$$\frac{x - 3}{2x} - \frac{x + 1}{x^2} = \frac{(x - 3)(x)}{(2x)(x)} - \frac{(x + 1)(2)}{(x^2)(2)}$$
$$= \frac{(x^2 - 3x) - (2x + 2)}{2x^2} = \frac{x^2 - 5x - 2}{2x^2}$$

Exercises 5.2

For Exercises 1–12 find a standard form in which the numerator and denominator are polynomials with integral coefficients.

1. $\frac{1}{3}x$
2. $\frac{3}{4}x$
3. $(\frac{2}{5})(3x^2)$
4. $(\frac{1}{2})(x + 1)$
5. $\left(\frac{-2}{3}\right)(3x + 5)$
6. $\frac{2x}{5} \cdot (x + 6)$
7. $\frac{x}{3} \div (x + 1)$
8. $\frac{x^2 - 4}{2} \div (x - 2)$
9. $x + (\frac{1}{3})$
10. $(\frac{1}{2})x + 1$
11. $\frac{3}{4}x + \frac{1}{2}$
12. $\frac{2}{3}x - \frac{1}{2}$

For Exercises 13–24 find a standard form.

13. $(x + 3) \cdot \frac{x - 1}{x + 3}$
14. $\frac{x + 2}{x - 1} \cdot (x - 1)(x + 3)$
15. $\frac{x}{x + 2} \cdot (x^2 - 4)$
16. $(x^2 - x - 2) \cdot \frac{2}{x^2 + 2x + 1}$
17. $\frac{x^2 + x}{x - 1} \cdot \frac{x - 2}{x^2 + 2x + 1} \cdot \frac{x^2 - 1}{x^2 + 2x}$
18. $\frac{x^2 - x}{x^2 - x - 2} \cdot \frac{x^2 - 4}{x^3 - x^2} \cdot \frac{x^2 + x}{x^2 - x - 6}$

19. $\dfrac{x^2 - x - 2}{x^2 + 2x - 3} \cdot \dfrac{x^2 + 5x + 6}{x^2 - 4} \cdot \dfrac{x^2 + 3x - 4}{x^2 + 5x + 4}$

20. $\dfrac{2x^2 - 3x - 2}{2x^2 + 9x + 4} \cdot \dfrac{3x^2 + 11x - 4}{2x^2 - 5x + 2} \cdot \dfrac{2x - 1}{3x^2 - x}$

21. $\dfrac{x^2 + 3x - 10}{x^2 - 2x - 3} \div \dfrac{x^2 - x - 2}{x^2 + 2x + 1}$

22. $\dfrac{x^2 + 3x}{x^2 - 1} \cdot \dfrac{x^2 + 3x + 2}{x^2 + 2x - 3} \div \dfrac{x}{x^2 - 2x + 1}$

23. $(2x^2 - 5x - 3) \div \dfrac{2x^2 + 3x + 1}{x + 3} \cdot \dfrac{x - 1}{x^2 - 9}$

24. $\dfrac{4x^2 - 4x + 1}{x^2 + 2x + 1} \div \dfrac{2x^2 + 3x - 2}{x^2 - 16} \div \dfrac{2x^2 - 9x + 4}{x^2 + 3x + 2}$

For Exercises 25–36 find a standard form for the given sum or difference.

25. $\dfrac{2x}{5} + \dfrac{3}{5}$

26. $\dfrac{x}{6} - \dfrac{5}{6}$

27. $\dfrac{4x}{7} - \dfrac{2x}{7}$

28. $\dfrac{3x}{5} + \dfrac{2x}{5}$

29. $\dfrac{x}{3} + \dfrac{2x + 1}{3}$

30. $\dfrac{x - 1}{7} + \dfrac{6x + 8}{7}$

31. $\dfrac{3x + 1}{2} - \dfrac{x + 1}{2}$

32. $\dfrac{x + 4}{6} + \dfrac{2x - 1}{6}$

33. $\dfrac{2x - 6}{5} - \dfrac{x - 1}{5}$

34. $\dfrac{6x - 1}{7} + \dfrac{x + 1}{7}$

35. $\dfrac{4x - 5}{3} - \dfrac{x - 2}{3}$

36. $\dfrac{x + 5}{4} - \dfrac{1 - 3x}{4}$

For Exercises 37–64 first find equivalent fractions with a common denominator and then find a standard form for the indicated sum or difference.

37. $\dfrac{x}{3} + \dfrac{1}{9}$

38. $\dfrac{2x}{15} + \dfrac{3}{5}$

39. $\dfrac{4x}{3} + \dfrac{x}{2}$

40. $\dfrac{3x}{5} - \dfrac{x}{2}$

41. $\dfrac{x + 1}{6} + \dfrac{3x}{2}$

42. $\dfrac{2x - 1}{12} - \dfrac{3x}{8}$

43. $\dfrac{x+2}{5} - \dfrac{x+1}{10}$ 44. $\dfrac{3x+1}{8} + \dfrac{x-1}{6}$

45. $\dfrac{x+6}{8} - \dfrac{2x-1}{20}$ 46. $\dfrac{x}{14} + \dfrac{x+3}{35}$

47. $\dfrac{x-3}{20} - \dfrac{x+1}{28}$ 48. $\dfrac{x+1}{49} + \dfrac{x-1}{35}$

49. $\dfrac{3}{x} + \dfrac{2}{3x}$ 50. $\dfrac{5}{2x} + \dfrac{1}{3x}$

51. $\dfrac{2}{5x} - \dfrac{3}{2x}$ 52. $\dfrac{4}{9x} - \dfrac{1}{12x}$

53. $\dfrac{1}{x^2} + \dfrac{2}{x}$ 54. $\dfrac{5}{2x} + \dfrac{3}{x^2}$

55. $\dfrac{2}{3x^2} - \dfrac{1}{2x}$ 56. $\dfrac{4}{5x^3} + \dfrac{3}{2x^2}$

57. $\dfrac{x+1}{x} + \dfrac{x-1}{2x}$ 58. $\dfrac{x+4}{x} - \dfrac{3x+2}{x^2}$

59. $\dfrac{x+1}{2x^2} + \dfrac{2}{3x}$ 60. $\dfrac{4x+1}{5x} - \dfrac{x-1}{2x^2}$

61. $2 + \dfrac{3}{x}$ 62. $\dfrac{x+1}{x} + x$

63. $x - \dfrac{2x+3}{x}$ 64. $4x - \dfrac{3x-1}{2x}$

5.3 Sums and Differences, Continued

In this section we shall consider the problem of finding a standard form for the sum or difference of two rational expressions when their denominators are binomials or trinomials. When the fractions have a common denominator we may proceed exactly as we did with common monomial denominators: copy the denominator and add or subtract the numerators.

Example 1   Find a standard form

(a) $\dfrac{2x}{x-3} + \dfrac{5}{x-3}$ (b) $\dfrac{3x}{2x-3} - \dfrac{x+3}{2x-3}$

(c) $\dfrac{x^2-2}{3x-2} + \dfrac{x(2x+1)}{3x-2}$

Solution

(a) $\dfrac{2x}{x-3} + \dfrac{5}{x-3} = \dfrac{2x+5}{x-3}$

(b) $\dfrac{3x}{2x-3} - \dfrac{x+3}{2x-3} = \dfrac{3x-(x+3)}{2x-3} = \dfrac{3x-x-3}{2x-3} = \dfrac{2x-3}{2x-3} = 1$

(c) $\dfrac{x^2-2}{3x-2} + \dfrac{x(2x+1)}{3x-2} = \dfrac{(x^2-2)+(2x^2+x)}{3x-2}$

$= \dfrac{3x^2+x-2}{3x-2} = \dfrac{(3x-2)(x+1)}{3x-2} = x+1$

But when the denominators are not common we will want to replace each fraction by an equivalent one with a common denominator which is the LCM of the original denominators. This can be a rather difficult problem with binomial or trinomial denominators. To find the LCM we must always factor (where possible) and then select the higher power of each prime factor for the factors of the LCM.

Example 2 Find a factored form of the LCM for each pair of polynomials.
(a) $2x+2$ and $3x+3$ (b) $3x+4$ and $3x^2+x-4$
(c) x^2+x-6 and x^2-4x+4

Solution

(a) $2x+2 = 2(x+1)$
  $3x+3 = 3(x+1)$; LCM $= 2 \cdot 3(x+1)$
(b) $3x+4$
  $3x^2+x-4 = (3x+4)(x-1)$; LCM $= (3x+4)(x-1)$
(c) $x^2+x-6 = (x+3)(x-2)$
  $x^2-4x+4 = (x-2)^2$; LCM $= (x+3)(x-2)^2$

Notice that we chose the higher power of the factor $x-2$.

Even after the LCM of the denominators is found, replacing each fraction by an equivalent one with the common denominator requires careful attention to technique. As before, this is accomplished by multiplying the numerator and denominator by whatever factor is needed to give the common denominator.

Example 3 Replace each fraction in each pair by an equivalent one with a common denominator. In each case the common denominator is the LCM of the original denominators.

(a) $\dfrac{5}{2x}$ and $\dfrac{1}{2(x-1)}$ (b) $\dfrac{x}{(x+2)(x-2)}$ and $\dfrac{x+1}{(x-2)^2}$

5.3 Sums and Differences, Continued

Solution

(a) LCM is $2x(x-1)$

$$\frac{5}{2x} = \frac{5(x-1)}{2x(x-1)} = \frac{5x-5}{2x(x-1)}$$

Note that we multiplied the numerator and denominator by $x-1$ so that the denominator would be the LCM $2x(x-1)$.

$$\frac{1}{2(x-1)} = \frac{1 \cdot x}{2(x-1)x} = \frac{x}{2x(x-1)}$$

Here we multiplied by x so that the denominator becomes the common denominator.

(b) LCM $= (x+2)(x-2)^2$

$$\frac{x}{(x+2)(x-2)} = \frac{x(x-2)}{(x+2)(x-2)(x-2)} = \frac{x^2-2x}{(x+2)(x-2)^2}$$

We multiplied by $x-2$ to give the common denominator.

$$\frac{x+1}{(x-2)^2} = \frac{(x+1)(x+2)}{(x-2)^2(x+2)} = \frac{x^2+3x+2}{(x+2)(x-2)^2}$$

Here we multiplied by $x+2$.

After each fraction has been replaced by an equivalent one with the common denominator we may easily find a standard form for their sum (or difference) by adding (or subtracting) their numerators as in Example 1. Let us now summarize the technique discussed in this section and illustrate it with several examples.

To find a standard form for the *sum* (or difference) of two rational expressions.
1. First factor the denominator of each fraction.
2. Select the factored form of the LCM of these denominators as a common denominator.
3. Replace each fraction by an equivalent one which has this common denominator. (Leave the denominator in factored form.)
4. Write a single fraction whose denominator is the common denominator (in factored form) and whose numerator is the sum (or difference) of the numerators.
5. Use the technique of Section 5.1 to find a standard form for this single fraction.

Example 4 Find a standard form for each of these.

(a) $\dfrac{x}{2x-4} + \dfrac{4}{5x-10}$ (b) $\dfrac{3}{x^2+x} - \dfrac{x+2}{2x}$

(c) $\dfrac{x}{x^2-2x+1} - \dfrac{3}{x^2+x-2}$

Solution

(a)

$$\dfrac{x}{2x-4} + \dfrac{4}{5x-10} = \dfrac{x}{2(x-2)} + \dfrac{4}{5(x-2)} \quad \text{[Step 1]}$$

$$\text{LCM} = 2 \cdot 5(x-2) \quad \text{[Step 2]}$$

$$\dfrac{x}{2(x-2)} + \dfrac{4}{5(x-2)} = \dfrac{x \cdot 5}{2(x-2)\,5} + \dfrac{4 \cdot 2}{5(x-2)\,2} \quad \text{[Step 3]}$$

$$= \dfrac{x \cdot 5 + 4 \cdot 2}{2 \cdot 5(x-2)} \quad \text{[Step 4]}$$

$$= \dfrac{5x+8}{10(x-2)} \quad \text{[Step 5]}$$

(b)

$$\dfrac{3}{x^2+x} - \dfrac{x+2}{2x} = \dfrac{3}{x(x+1)} - \dfrac{x+2}{2x} \quad \text{[Step 1]}$$

$$\text{LCM} = 2x(x+1). \quad \text{[Step 2]}$$

$$\dfrac{3}{x(x+1)} - \dfrac{x+2}{2x} = \dfrac{3 \cdot 2}{x(x+1)\,2} - \dfrac{(x+2)(x+1)}{2x(x+1)} \quad \text{[Step 3]}$$

$$= \dfrac{6 - (x+2)(x+1)}{2x(x+1)} \quad \text{[Step 4]}$$

$$= \dfrac{6 - (x^2+3x+2)}{2x(x+1)} = \dfrac{-x^2-3x+4}{2x(x+1)} \quad \text{[Step 5]}$$

(c)

$$\dfrac{x}{x^2-2x+1} - \dfrac{3}{x^2+x-2} = \dfrac{x}{(x-1)^2} - \dfrac{3}{(x-1)(x+2)} \quad \text{[Step 1]}$$

$$\text{LCM} = (x-1)^2(x+2). \quad \text{[Step 2]}$$

$$\dfrac{x}{(x-1)^2} - \dfrac{3}{(x-1)(x+2)}$$

$$= \dfrac{x(x+2)}{(x-1)^2(x+2)} - \dfrac{3(x-1)}{(x-1)(x+2)(x-1)} \quad \text{[Step 3]}$$

5.3 *Sums and Differences, Continued*

$$= \frac{(x^2 + 2x) - (3x - 3)}{(x - 1)^2(x + 2)} \qquad \text{[Step 4]}$$

$$= \frac{x^2 - x + 3}{(x - 1)^2(x + 2)} \qquad \text{[Step 5]}$$

If we recall the fact that any polynomial can be considered a rational expression and written as a fraction with denominator 1, this same general technique can be used to simplify expressions such as those in the next example.

Example 5 Find a standard form of $x + 2 - \dfrac{1}{x + 1}$.

Solution $x + 2 - \dfrac{1}{x + 1} = \dfrac{x + 2}{1} - \dfrac{1}{x + 1}$

LCM of 1 and $x + 1$ is $x + 1$

$$\frac{x + 2}{1} - \frac{1}{x + 1} = \frac{(x + 2)(x + 1)}{1 \cdot (x + 1)} - \frac{1}{x + 1}$$

$$= \frac{(x^2 + 3x + 2) - 1}{x + 1} = \frac{x^2 + 3x + 1}{x + 1}$$

Sometimes a common denominator may be more easily found by using the techniques illustrated in the following example.

Example 6 Find a standard form.

(a) $\dfrac{x}{x - 3} + \dfrac{2}{3 - x}$ \qquad (b) $\dfrac{2x}{3x + 4} - \dfrac{x - 1}{3x^2 + x - 4}$

Solution

(a) We notice that $x - 3$ is the additive inverse of $3 - x$, and hence $(3 - x)(-1) = x - 3$.

$$\frac{x}{x - 3} + \frac{2}{3 - x} = \frac{x}{x - 3} + \frac{(2)(-1)}{(3 - x)(-1)} = \frac{x}{x - 3} + \frac{-2}{x - 3}$$

$$= \frac{x - 2}{x - 3}$$

(b) $\dfrac{2x}{3x + 4} - \dfrac{x - 1}{3x^2 + x - 4} = \dfrac{2x}{3x + 4} - \dfrac{(x - 1)(1)}{(3x + 4)(x - 1)}$

$$= \frac{2x}{3x + 4} - \frac{1}{3x + 4} = \frac{2x - 1}{3x + 4}$$

178   Rational Algebraic Expressions

Notice that we reduced the second fraction $\dfrac{x-1}{3x^2+x-4}$ to a standard form before attempting to add.

The next example shows how to find a standard form when there are three or more terms.

Example 7 Find a standard form.
$$\frac{3}{x-1} - \frac{2}{x+1} - \frac{6}{x^2-1}$$

Solution $\dfrac{3}{x-1} - \dfrac{2}{x+1} - \dfrac{6}{x^2-1}$

$= \dfrac{3}{x-1} - \dfrac{2}{x+1} - \dfrac{6}{(x+1)(x-1)}$

LCM $= (x+1)(x-1)$.

$\dfrac{3}{x-1} - \dfrac{2}{x+1} - \dfrac{6}{(x+1)(x-1)}$

$= \dfrac{3(x+1)}{(x-1)(x+1)} - \dfrac{2(x-1)}{(x+1)(x-1)} - \dfrac{6}{(x+1)(x-1)}$

$= \dfrac{(3x+3)-(2x-2)-6}{(x+1)(x-1)} = \dfrac{x-1}{(x+1)(x-1)} = \dfrac{1}{x+1}$

Sometimes we will want to use the distributive axiom to find a standard form for rational expressions. The next example illustrates this.

Example 8 Find a standard form for each of the following.

(a) $4\left(\dfrac{3x}{4} - 5\right)$. (b) $(x+1)\left(\dfrac{2}{x+1} + x\right)$.

(c) $(x^2 - 4)\left(\dfrac{1}{x+2} - \dfrac{2}{x-2}\right)$.

Solution

(a) $4\left(\dfrac{3x}{4} - 5\right) = \cancel{4} \cdot \dfrac{3x}{\cancel{4}} - 4 \cdot 5 = 3x - 20$.

(b) $(x+1)\left(\dfrac{2}{x+1} + x\right) = \dfrac{\cancel{x+1}}{1} \cdot \dfrac{2}{\cancel{x+1}} + (x+1)(x)$

5.3 Sums and Differences, Continued

$$= 2 + (x+1)(x) = x^2 + x + 2$$

(c) $(x^2 - 4)\left(\dfrac{1}{x+2} - \dfrac{2}{x-2}\right) = (x^2 - 4) \cdot \dfrac{1}{x+2} - (x^2 - 4) \cdot \dfrac{2}{x-2}$

$$= \dfrac{\cancel{(x+2)}(x-2)}{1} \cdot \dfrac{1}{\cancel{x+2}} - \dfrac{(x+2)\cancel{(x-2)}}{1} \cdot \dfrac{2}{\cancel{x-2}}$$

$$= (x-2) - 2(x+2) = -x - 6$$

Exercises 5.3

For Exercises 1–8 find the factored form of the LCM of the given pair of polynomials.

1. $x(x+2)$ and $2x$
2. $(x+2)(x-3)$ and $(x+2)(x-2)$
3. $(x+1)^2$ and $2(x+1)$
4. $(x-2)^2(x+1)$ and $(x-2)(x+2)$
5. $x^2 + 3x$ and $2x + 6$
6. $x^3 - x^2$ and $x^2 + x$
7. $x^2 - 1$ and $x^2 + x - 2$
8. $x^2 - 6x + 9$ and $x^2 - 2x - 3$

For Exercises 9–16 replace each pair of fractions by an equivalent pair with a common denominator in factored form.

9. $\dfrac{2}{x+1}$ and $\dfrac{1}{3x+3}$
10. $\dfrac{x}{2x-2}$ and $\dfrac{x-2}{3x-3}$
11. $\dfrac{3}{2x}$ and $\dfrac{5}{x^2-x}$
12. $\dfrac{x+2}{2x+2}$ and $\dfrac{3}{x^2+x}$
13. $\dfrac{3}{x+1}$ and $\dfrac{2}{x^2-1}$
14. $\dfrac{1}{(x+2)(x-1)}$ and $\dfrac{2}{(x-1)(x+3)}$
15. $\dfrac{x+1}{x^2-3x+2}$ and $\dfrac{x}{x^2-4}$
16. $\dfrac{x}{x^2+6x+9}$ and $\dfrac{x-2}{x^2+3x}$

For Exercises 17–44 find a standard form for the indicated sum or difference.

17. $\dfrac{x}{x+3} + \dfrac{3}{x+3}$
18. $\dfrac{2x^2}{2x-1} + \dfrac{x-1}{2x-1}$

19. $\dfrac{x^2+2x}{x^2-2x} - \dfrac{x+6}{x(x-2)}$
20. $\dfrac{(x+1)^2}{(x+2)(x-1)} - \dfrac{4}{x^2+x-2}$
21. $\dfrac{3}{x+2} + \dfrac{2}{x(x+2)}$
22. $\dfrac{x}{2(x-1)} + \dfrac{x+1}{x-1}$
23. $\dfrac{1}{x^2+x} - \dfrac{x+2}{x}$
24. $\dfrac{x}{x+2} - \dfrac{3}{x^2+x-2}$
25. $\dfrac{2}{x^2-x} + \dfrac{1}{3x}$
26. $\dfrac{5}{2x^2-x} + \dfrac{3}{4x-2}$
27. $\dfrac{x+1}{3x-3} - \dfrac{5}{2x-2}$
28. $\dfrac{2x+1}{2x+4} - \dfrac{x-1}{x^2+2x}$
29. $\dfrac{1}{x^2+5x} + \dfrac{2}{x^2-x}$
30. $\dfrac{4}{2x^2-x} - \dfrac{1}{3x^2+x}$
31. $\dfrac{1}{x^2-1} + \dfrac{1}{x^2+x-2}$
32. $\dfrac{1}{x^2-x-6} - \dfrac{1}{x^2-4}$
33. $\dfrac{x}{x^2+3x+2} + \dfrac{3}{x^2+5x+6}$
34. $\dfrac{x-3}{x^2+2x+1} + \dfrac{2x-5}{x+1}$
35. $\dfrac{1}{x^2-4x+4} - \dfrac{1}{x^2-x-2}$
36. $\dfrac{x-1}{x^2+6x+9} - \dfrac{2x-1}{x^2-9}$
37. $\dfrac{x+4}{x-5} + \dfrac{3}{5-x}$
38. $\dfrac{3}{x+2} - \dfrac{x}{4-x^2}$
39. $\dfrac{x^2-4x+4}{x^2+x-6} + \dfrac{x+8}{x+3}$
40. $\dfrac{x^2+x-6}{x^2+2x-3} - \dfrac{x+1}{1-x^2}$
41. $\dfrac{x-2}{x} + \dfrac{2x+6}{3x} - \dfrac{4}{3}$
42. $\dfrac{4}{5x} - \dfrac{2x}{3} + \dfrac{7x^2-12}{15x}$
43. $\dfrac{2}{x-2} - \dfrac{1}{x+3} - \dfrac{10}{x^2+x-6}$
44. $\dfrac{x-1}{x^2} - \dfrac{x-1}{x+1} + \dfrac{x^3+2x+1}{x^2(x+1)}$

For Exercises 45–52 use the distributive axiom and express each of the following as a polynomial in standard form.

45. $3\left(\dfrac{x+1}{3} + 2\right)$
46. $8\left(\dfrac{x^2}{8} + x\right)$
47. $2x\left(\dfrac{x+1}{2x} + 3x\right)$
48. $8x\left(\dfrac{1}{2x} + \dfrac{x-1}{8x}\right)$
49. $(x+1)\left(\dfrac{1}{x+1} + 3\right)$
50. $(x-1)\left(\dfrac{x+2}{x-1} + 1\right)$

51. $(x+2)(x-1)\left(\dfrac{1}{x-1} - \dfrac{1}{x+2}\right)$

52. $(x^2-1)\left(\dfrac{1}{x+1} + \dfrac{1}{x-1}\right)$

5.4 Equations and Inequalities

In Chapter 3 we considered techniques which can be used to find solution sets of first degree equations and inequalities. And in Chapter 4 we saw how factoring could be used to find solution sets of some second degree open sentences. The general principles presented in those chapters can be used to find solution sets of open sentences in which rational expressions occur. Indeed, it will often be the case that one application of the multiplication axiom (for equations or inequalities) results in a first or second degree open sentence of the sort we have previously considered.

Example 1 Find the solution set of each of the following.

(a) $3x - 2 = \dfrac{4x}{3} + 1$

(b) $x^2 - 1 = \dfrac{3x}{2}$

(c) $\dfrac{2x}{5} - 1 \leq 4 - 3x$

(d) $\dfrac{7x}{4} - 1 = x - \dfrac{5}{4}$

Solution

(a) $3x - 2 = \dfrac{4x}{3} + 1$

We multiply both expressions by 3.

$$3(3x - 2) = 3\left(\dfrac{4x}{3} + 1\right)$$

Using the distributive axiom we obtain

$$3 \cdot 3x - 3 \cdot 2 = 3 \cdot \dfrac{4x}{3} + 3 \cdot 1$$

$$9x - 6 = 4x + 3$$

Note that we obtained a first degree polynomial equation with integral coefficients by multiplying by the denominator, 3.

$$(9x - 6) + (-4x) = (4x + 3) + (-4x)$$

$$5x - 6 = 3$$

Rational Algebraic Expressions

$$(5x - 6) + 6 = 3 + 6$$
$$5x = 9$$
$$\left(\tfrac{1}{5}\right)(5x) = \left(\tfrac{1}{5}\right)(9)$$
$$x = \tfrac{9}{5} \qquad \left\{\tfrac{9}{5}\right\}$$

(b) $x^2 - 1 = \dfrac{3x}{2}$

Again, we shall multiply by the denominator, 2.

$$2(x^2 - 1) = 2\left(\dfrac{3x}{2}\right)$$
$$2x^2 - 2 = 3x$$

This is a quadratic equation whose coefficients are integers.

$$2x^2 - 3x - 2 = 0$$
$$(2x + 1)(x - 2) = 0$$

$\quad 2x + 1 = 0 \qquad\qquad\qquad\qquad x - 2 = 0$
$\quad\quad 2x = -1 \qquad\qquad\qquad\qquad\quad x = 2$
$\quad\quad\quad x = -\tfrac{1}{2} \qquad\qquad\qquad\qquad\qquad \{2\}$
$\quad\quad\quad \{-\tfrac{1}{2}\} \qquad\qquad \{-\tfrac{1}{2}\} \cup \{2\} = \{-\tfrac{1}{2}, 2\}$

(c) $\dfrac{2x}{5} - 1 \leq 4 - 3x$

We multiply by the *positive* number, 5.

$$5\left(\dfrac{2x}{5} - 1\right) \leq 5(4 - 3x)$$
$$2x - 5 \leq 20 - 15x$$
$$17x - 5 \leq 20$$
$$17x \leq 25$$
$$x \leq \tfrac{25}{17} \qquad \left(-\infty, \tfrac{25}{17}\right]$$

(d) $\dfrac{7x}{4} - 1 = x - \dfrac{5}{4}$

There are two denominators, both 4.

$$4\left(\dfrac{7x}{4} - 1\right) = 4\left(x - \dfrac{5}{4}\right)$$
$$7x - 4 = 4x - 5$$

5.4 Equations and Inequalities

$$3x - 4 = -5$$
$$3x = -1$$
$$x = -\tfrac{1}{3} \qquad \{-\tfrac{1}{3}\}$$

When the numerators of fractions are not monomials the grouping nature of the fraction bar must be recalled. And when multiplying the expressions in an inequality by a negative number, the resulting inequality will not be equivalent to the original unless $<$ and $>$ are interchanged. But again, the technique is to multiply by the denominator of a fraction.

Example 2 Find the solution set of each of these.

(a) $x = 4 - \dfrac{x-4}{3}$ (b) $\dfrac{x-3}{-2} > x + \dfrac{1}{-2}$ (c) $\dfrac{x^2 + 5x}{2} = x + 5$

Solution

(a) $x = 4 - \dfrac{x-4}{3}$

$$3(x) = 3\left(4 - \dfrac{x-4}{3}\right)$$

$$3x = 3 \cdot 4 - 3 \cdot \dfrac{x-4}{3}$$

$$3x = 12 - (x-4)$$

Note that we replaced $3 \cdot \dfrac{x-4}{3}$ by $(x-4)$

$$3x = 12 - x + 4$$
$$4x = 16$$
$$x = 4 \qquad \{4\}$$

(b) $\dfrac{x-3}{-2} > x + \dfrac{1}{-2}$

$(-2)\left(\dfrac{x-3}{-2}\right) < (-2)\left(x + \dfrac{1}{-2}\right).$ Note that $>$ has been replaced by $<$ since we are multiplying by a negative number, -2.

$$x - 3 < -2x + 1$$
$$3x - 3 < 1$$
$$3x < 4$$
$$x < \tfrac{4}{3} \qquad \langle -\infty, \tfrac{4}{3} \rangle$$

Rational Algebraic Expressions

(c) $\dfrac{x^2 + 5x}{2} = x + 5$

$$2\left(\dfrac{x^2 + 5x}{2}\right) = 2(x + 5)$$

$$x^2 + 5x = 2x + 10$$

$$x^2 + 3x - 10 = 0$$

$$(x + 5)(x - 2) = 0 \qquad \{-5,\ 2\}$$

The domain of a rational expression must not include any replacements which give the value zero to any denominator. This fact must be carefully considered if an open sentence contains any denominators which are not constant.

Example 3 Find the solution set of each of these:

(a) $\dfrac{4}{3x} + 1 = \dfrac{2}{3x} - 5$ \qquad (b) $\dfrac{4}{x} - 3 = x$

(c) $\dfrac{x}{x - 2} + 9 = \dfrac{2}{x - 2} - 1$ \qquad (d) $\dfrac{x^2}{x - 1} = \dfrac{1}{x - 1} - 1$

Solution

(a) The replacement set is $\{x:\ x \neq 0\} = \overline{\{0\}}$.

$$\dfrac{4}{3x} + 1 = \dfrac{2}{3x} - 5$$

We multiply both expressions by the denominator.

$$3x\left(\dfrac{4}{3x} + 1\right) = 3x\left(\dfrac{2}{3x} - 5\right)$$

$$4 + 3x = 2 - 15x$$

$$4 + 18x = 2$$

$$18x = -2$$

$$x = -\tfrac{1}{9} \qquad \{-\tfrac{1}{9}\}$$

(b) The replacement set is $\overline{\{0\}}$.

$$\dfrac{4}{x} - 3 = x$$

5.4 Equations and Inequalities

We multiply by x.

$$x\left(\frac{4}{x} - 3\right) = x \cdot x$$

$$4 - 3x = x^2$$

$$0 = x^2 + 3x - 4$$

$$x^2 + 3x - 4 = 0$$

$$(x + 4)(x - 1) = 0 \qquad \{-4, 1\}$$

(c) The replacement set is $\{x: x - 2 \neq 0\} = \{x: x \neq 2\} = \overline{\{2\}}$

$$\frac{x}{x - 2} + 9 = \frac{2}{x - 2} - 1$$

We multiply both expressions by the denominator, $(x - 2)$.

$$(x - 2)\left(\frac{x}{x - 2} + 9\right) = (x - 2)\left(\frac{2}{x - 2} - 1\right)$$

$$(\cancel{x - 2}) \cdot \frac{x}{\cancel{x - 2}} + (x - 2)9 = (\cancel{x - 2}) \cdot \frac{2}{\cancel{x - 2}} - (x - 2) \cdot 1$$

$$x + 9x - 18 = 2 - x + 2$$

$$10x - 18 = 4 - x$$

$$11x - 18 = 4$$

$$11x = 22$$

$$x = 2$$

Now while the open sentence $x = 2$ can be true only when x is replaced by 2, this is the only number *not* allowable as a replacement for x in the original sentence. The solution set of the original equation is therefore the empty set, $\emptyset$.

(d) The replacement set is $\overline{\{1\}}$.

$$\frac{x^2}{x - 1} = \frac{1}{x - 1} - 1$$

We multiply both expressions by the denominator, $(x - 1)$.

$$(x - 1)\left(\frac{x^2}{x - 1}\right) = (x - 1)\left(\frac{1}{x - 1} - 1\right)$$

$$(\cancel{x - 1}) \cdot \frac{x^2}{\cancel{x - 1}} = (\cancel{x - 1}) \cdot \frac{1}{\cancel{x - 1}} - (x - 1) \cdot 1$$

$$x^2 = 1 - x + 1$$

$$x^2 + x - 2 = 0$$
$$(x+2)(x-1) = 0$$

There are two replacements for x which make the last open sentence a true statement: -2 and 1. But, while -2 is in the replacement set of the original equation, 1 is *not*. Hence 1 cannot be in its solution set, and so the solution set is $\{-2\}$.

It may appear strange that the equation

$$\frac{x}{x-2} + 9 = \frac{2}{x-2} - 1$$

of Example 3(c) has no members in its solution set while the equation obtained by multiplying both expressions by $(x-2)$ has a nonempty solution set, $\{2\}$. But there is no guarantee that the two equations need have the same solution set. The multiplication axiom assumes that two equations will be equivalent if one is obtained from the other by multiplying by a *nonzero* number. In fact, however, we have multiplied by $(x-2)$ which has value zero if we allow 2 as a replacement! And it may appear even more strange that the equation

$$\frac{x^2}{x-1} = \frac{1}{x-1} - 1$$

of Example 3(d) leads to a quadratic equation *one* of whose solutions is a member of the solution set of the original open sentence and the other which is *not*. But again, the discrepancy arises because the expression $(x-1)$ has value 0 if $x = 1$, and we multiplied the left and right by this expression.

When there are several denominators in an equation or inequality we can again obtain a sentence of the kind previously considered if we use the LCM of *all* denominators and the multiplication axioms.

Example 4 Find the solution set of each of the following:

(a) $\dfrac{3}{2x} - 1 = \dfrac{4}{3} + 5$ (b) $\dfrac{4x}{x+1} = \dfrac{3}{x} + 2$

(c) $3x - \frac{1}{2} \geq x + \frac{1}{3}$

Solution

(a) The replacement set is $\overline{\{0\}}$.

$$\frac{3}{2x} - 1 = \frac{4}{3} + 5$$

5.4 Equations and Inequalities

We multiply by the LCM of $2x$ and 3: $6x$.

$$6x\left(\frac{3}{2x} - 1\right) = 6x\left(\frac{4}{3} + 5\right)$$

$$6x \cdot \frac{3}{2x} - 6x \cdot 1 = 6x \cdot \frac{4}{3} + 6x \cdot 5$$

$$9 - 6x = 8x + 30x$$

$$9 - 6x = 38x$$

$$9 = 44x$$

$$\tfrac{9}{44} = x \qquad \left\{\tfrac{9}{44}\right\}$$

(b) The replacement set is $\{x: x + 1 \neq 0 \text{ and } x \neq 0\} = \overline{\{-1, 0\}}$.

$$\frac{4x}{x+1} = \frac{3}{x} + 2$$

We multiply by the LCM of $(x + 1)$ and x: $x(x + 1)$

$$x(x+1)\left(\frac{4x}{x+1}\right) = x(x+1)\left(\frac{3}{x} + 2\right)$$

$$x(\cancel{x+1}) \cdot \frac{4x}{\cancel{x+1}} = \cancel{x}(x+1) \cdot \frac{3}{\cancel{x}} + x(x+1) \cdot 2$$

$$4x^2 = 3x + 3 + 2x^2 + 2x$$

$$4x^2 = 2x^2 + 5x + 3$$

$$2x^2 - 5x - 3 = 0$$

$$(2x + 1)(x - 3) = 0 \qquad \left\{-\tfrac{1}{2}, 3\right\}$$

(c) The replacement set is R.

$$3x - \tfrac{1}{2} \geq x + \tfrac{1}{3}$$

We multiply by the *positive* number 6, the LCM of 2 and 3.

$$6(3x - \tfrac{1}{2}) \geq 6(x + \tfrac{1}{3})$$

$$18x - 3 \geq 6x + 2$$

$$12x - 3 \geq 2$$

$$12x \geq 5$$

$$x \geq \tfrac{5}{12} \qquad \left[\tfrac{5}{12}, \infty\right)$$

Exercises 5.4

For Exercises 1–40 find the solution set.

1. $4x - 1 = \dfrac{x}{2} - 3$
2. $2 - \dfrac{x}{3} = x - \dfrac{5}{3}$
3. $\dfrac{4x}{5} + 1 = \dfrac{6}{5}$
4. $x - \dfrac{2}{3} < 7$
5. $\dfrac{x}{5} \geq 3x + \dfrac{2}{5}$
6. $3 - \dfrac{2}{x-5} = \dfrac{1}{x-5}$
7. $\dfrac{4}{x} - 4 = \dfrac{1}{x} - x$
8. $2 - \dfrac{x-1}{5} = x - \dfrac{1}{5}$
9. $\dfrac{x}{x-3} - 5 = \dfrac{3}{x-3}$
10. $\dfrac{4}{7} + 2x = 1 - \dfrac{3x+1}{7}$
11. $\dfrac{3x+5}{5} - 1 \leq x - \dfrac{2x}{5}$
12. $\dfrac{4}{x} - 2 = x + \dfrac{1}{x}$
13. $\dfrac{4}{3} - x = 2 - \dfrac{x}{5}$
14. $\dfrac{5x}{4} - 1 \geq 2x - \dfrac{2}{5}$
15. $\dfrac{7}{5} - x = 1 - \dfrac{3x}{8}$
16. $\dfrac{x}{2} = \dfrac{8}{x}$
17. $\dfrac{x}{2} = \dfrac{8}{x} - 3$
18. $\dfrac{3}{x} - \dfrac{1}{x^2} - 2 = 0$
19. $\dfrac{x}{2} - 2 = \dfrac{2}{x-1}$
20. $\dfrac{4}{x} + 1 = \dfrac{x}{x-2}$
21. $\dfrac{3}{x-1} - \dfrac{2}{1-x} = 5$
22. $\dfrac{3}{2x-1} = \dfrac{1}{2x+1}$
23. $\dfrac{2}{3x+1} = \dfrac{x}{3x-1}$
24. $\dfrac{4}{1-x} + 1 = \dfrac{3}{x-1}$
25. $\dfrac{1}{2}(x-1) - \dfrac{2}{3}(x+1) = \dfrac{x}{8}$
26. $\dfrac{1}{1-2x} = \dfrac{3}{2x+1}$
27. $\dfrac{1}{x-6} = \dfrac{1}{x-2} - 1$
28. $\dfrac{x-2}{x+1} = \dfrac{x}{x+2}$
29. $\dfrac{x+1}{x+4} = \dfrac{2}{x-2}$
30. $\dfrac{1}{2} - 2x > \dfrac{x}{3} - 1$
31. $\dfrac{4}{-5} + x > 1 + \dfrac{x}{2}$
32. $\dfrac{x-1}{2} - \dfrac{x+1}{5} = \dfrac{x}{6}$
33. $\dfrac{x}{2} - \dfrac{2x-1}{4} + \dfrac{x}{3} = 0$
34. $\dfrac{2-x}{x} + 1 = \dfrac{x}{2}$

35. $\dfrac{x}{5} - 1 = \dfrac{x-5}{2x}$

36. $\dfrac{x}{x+1} = \dfrac{1}{x-1} - \dfrac{1}{x+1}$

37. $\dfrac{3}{2x} - \dfrac{5}{4x} = \dfrac{1}{x^2}$

38. $\dfrac{x+1}{(x+4)(x-2)} = \dfrac{4}{x^2+2x-8}$

39. $\dfrac{2}{x^2-2x-3} = \dfrac{x}{x^2-9}$

40. $\dfrac{x}{x^2-4x-5} = \dfrac{3}{x^2-25}$

5.5 Applications

In Section 3.6 we considered a number of problems which could be solved by finding the solution sets of certain open sentences. The general techniques discussed there are still applicable, but we now have the facility for finding solution sets of open sentences which we did not encounter then. We shall illustrate with a number of examples some additional types of problems now open to us. For problems involving decimals it will be important to remember that decimal fractions are simply fractions whose denominators are powers of 10. Also we need to remember that percent means hundredths.

Example 1 The interest earned in one year by a safe investment is 4.8% of the principal (amount invested). If an annual income (from interest) of $4200 is desired, how much money must be invested at this rate?

Solution Let x represent the number of dollars invested, where $x \in N$. Then the investment income each year is 4.8% of x or $(.048)x$. An open sentence we can use is

$$(.048)x = 4200$$

$$\tfrac{48}{1000}x = 4200 \qquad \text{[We multiply by 1000]}$$

$$48x = 4{,}200{,}000$$

$$x = 87{,}500$$

Then $87,500 must be invested.

Example 2 What number added to both the numerator and denominator of the fraction $\tfrac{3}{8}$ yields the multiplicative inverse of $\tfrac{3}{8}$ as a result?

Solution Let x represent the number sought. Then since $\tfrac{8}{3}$ is the multiplicative inverse of $\tfrac{3}{8}$ we may write

$$\dfrac{3+x}{8+x} = \dfrac{8}{3}, \qquad \text{where } x \in \{x : x \neq -8\}$$

The LCM here is $3(8 + x)$

$$3(8+x)\left(\frac{3+x}{8+x}\right) = 3(8+x)\left(\frac{8}{3}\right)$$

$$9 + 3x = 64 + 8x$$

$$-55 = 5x$$

$$-11 = x$$

The required number is -11. Note that

$$\frac{3 + (-11)}{8 + (-11)} = \frac{-8}{-3} = \frac{8}{3}$$

Example 3 Ten carat gold is actually $\frac{10}{24}$ pure gold mixed with base metal. How much pure gold must be combined with 20 grams of 10 carat gold to bring it up to 14 carat ($\frac{14}{24}$ pure gold)?

Solution There is an implied sentence in this problem which we may use to solve it: The amount of *pure* gold in the 10 carat gold plus the amount of pure gold added equals the amount of *pure* gold in the 14 carat gold. We observe that 20 grams of 10 carat gold contain $(\frac{10}{24})(20)$ grams of pure gold. Let x represent the amount of pure gold added, where $x \in \langle 0, \infty \rangle$. Now the weight of the 14 carat gold will be represented by $20 + x$, since we added x grams of pure gold to 20 grams of 10 carat gold. And the amount of *pure* gold in the 14 carat gold is $(\frac{14}{24})(20 + x)$. Then an open sentence is

$$(\tfrac{10}{24})(20) + x = (\tfrac{14}{24})(20 + x)$$

Pure gold in 10 carat gold plus pure gold added equals pure gold in 14 carat gold.
We multiply by 24.

$$(10)(20) + 24x = (14)(20 + x)$$

$$200 + 24x = 280 + 14x$$

$$10x = 80$$

$$x = 8$$

Then 8 grams of pure gold must be added.

Example 4 An 8% acid solution is to be diluted to only 6% concentration by the addition of 2% acid of the same kind. If 1.2 liters of the 8% acid are on hand, how many liters of the 2% acid must be added and how much of the 6% acid will result?

Solution Again, there is an implied sentence which we may use: The amount of pure acid in the 8% mixture plus the amount of pure acid

5.5 Applications

in the 2% mixture equals the amount of pure acid in the 6% mixture. We let x represent the amount of 2% solution which is added.

	Amount of *solution*	Amount of pure *acid*
8% solution	1.2	8% of $1.2 = (.08)(1.2)$
2% solution	x	2% of $x = (.02)x$
6% solution	$1.2 + x$	6% of $(1.2 + x) = (.06)(1.2 + x)$

Then an open sentence we may use is

$$(.08)(1.2) + (.02)x = (.06)(1.2 + x)$$
$$.096 + .02x = .072 + .06x$$
$$96 + 20x = 72 + 60x \qquad \text{[Multiplying by 1000]}$$
$$24 = 40x$$
$$x = \tfrac{24}{40} = .6$$

Hence .6 liters of 2% acid must be added. This will result in $1.2 + .6 = 1.8$ liters of 6% acid.

Here is a similar problem even though it appears quite remote in subject matter.

Example 5 A young executive is left an inheritance of $6000 and decided to invest some of it at 5% interest and the rest at 8% interest each year. During one year his investments yielded an interest return of $396. How much did he invest at each rate that year?

Solution Here we shall use the obvious fact that the interest from his 5% investment plus the interest from his 8% investment equals his total interest. Let x represent the number of dollars invested at 5%, where $x \in N$.

	Dollars invested	Interest received
5% investment	x	$(.05)x$
8% investment	$6000 - x$	$(.08)(6000 - x)$
Total investment	$x + (6000 - x) = 6000$	396

192 Rational Algebraic Expressions

$$(.05)x + (.08)(6000 - x) = 396$$
$$5x + 8(6000 - x) = 39{,}600 \qquad \text{[Multiplying by 100]}$$
$$5x + 48{,}000 - 8x = 39{,}600$$
$$-3x = -8400$$
$$x = 2800$$

The executive invested $2800 at 5% and $3200 at 8%.

Many problems involve the relationship between distance, rate of speed, and time of travel. When the appropriate units of each are used, the distance can be determined by multiplying the rate by the time. It is also possible to find the time of travel by dividing the distance by the rate, when both of these are known. And we can also, should we wish to, find the rate by dividing the distance traveled by the time.

Example 6 A man can paddle a canoe at the rate of 5 miles per hour in a lake where there is no current to contend with. But when he attempts a river trip to a point 21 miles away, the round trip takes 10 hours. How fast does the current in the river flow?

Solution We shall use several obvious facts implied here. When going downstream the current will *add* to his rate, but upstream it will *subtract* from his rate. The time it takes to paddle downstream plus the time it takes to return upstream equals the total time of the round trip. Let x represent the number of miles per hour in the rate of the current, where $x \in \langle 0, 5 \rangle$. (If the current were more than 5 miles per hour he could never travel upstream.) We will use the fact that the time is the distance divided by the rate.

	Distance	Rate	Time
Downstream	21	$5+x$	$\dfrac{21}{5+x}$
Upstream	21	$5-x$	$\dfrac{21}{5-x}$

Then an open sentence is

$$\frac{21}{5+x} + \frac{21}{5-x} = 10$$

5.5 Applications

The LCM is $(5 + x)(5 - x)$.

$$(5 + x)(5 - x)\left(\frac{21}{5 + x} + \frac{21}{5 - x}\right) = (5 + x)(5 - x)(10)$$

$$(\cancel{5 + x})(5 - x) \cdot \frac{21}{\cancel{5 + x}} + (5 + x)(\cancel{5 - x}) \cdot \frac{21}{\cancel{5 - x}} = (25 - x^2)(10)$$

$$105 - 21x + 105 + 21x = 250 - 10x^2$$

$$210 = 250 - 10x^2$$

$$10x^2 - 40 = 0$$

$$x^2 - 4 = 0 \quad \left[\text{Multiplying by } \tfrac{1}{10}\right]$$

$$(x + 2)(x - 2) = 0$$

Since the replacement set is $\langle 0, 5 \rangle$, the solution set is $\{2\}$. The current flows at a rate of 2 miles per hour.

Example 7 The regular postman completes his route normally in exactly 6 hours. With the help of an assistant he can actually finish in just 4 hours. How many hours would this assistant take to cover the route working by himself?

Solution We will find the fractional part of the route each man could cover working alone in one hour. Then the sum of these two fractions will be the fractional part they can cover working together for one hour. Since the regular postman requires 6 hours to complete the route by himself, he could complete $\tfrac{1}{6}$ of it in only one hour working alone. Let x represent the number of hours the assistant working alone would require for the complete job, where $x \in \langle 0, \infty \rangle$. Then in only one hour he could complete $1/x$ of the route. Since the two working together can complete the route in 4 hours, in only one hour they could complete $\tfrac{1}{4}$ of the route. Then an open sentence we may use here is

$$\frac{1}{6} + \frac{1}{x} = \frac{1}{4}$$

The LCM is $12x$.

$$12x\left(\frac{1}{6} + \frac{1}{x}\right) = 12x\left(\frac{1}{4}\right)$$

$$2x + 12 = 3x$$

$$12 = x$$

The assistant would require 12 hours to do the job alone!

194   Rational Algebraic Expressions

We end this section by returning to Example 15 of Section 3.6 where we encountered the equation

$$\frac{2024}{x+4} = \frac{2024}{x} - 1, \quad x \in N$$

We now see that the LCM of $x + 4$ and x is $x(x + 4)$.

$$x(\cancel{x+4}) \left(\frac{2024}{\cancel{x+4}}\right) = x(x + 4)\left(\frac{2024}{x} - 1\right)$$

$$2024x = (x + 4)(2024) - x(x + 4)$$

$$2024x = 2024x + 8096 - x^2 - 4x$$

$$x^2 + 4x - 8096 = 0$$

$$(x - 88)(x + 92) = 0$$

Since $x \in N$, we see that the solution set is $\{88\}$, and the original price of a pair of socks was 88 cents.

Exercises 5.5

For Exercises 1–36 write an open sentence and find its solution set to answer the stated problem.

1. Two-thirds of a number is 4 more than one half that number. What is it?
2. Three-fourths of 2 less than a number is only one-third of that number. What is it?
3. There are two consecutive integers such that one-third the first plus one-fourth the second is 12 more than their sum. What are they?
4. One-third the product of two consecutive negative integers is $-\frac{5}{3}$ times the larger integer. What are they?
5. One number is 3 more than another. The smaller plus two-thirds the larger is only one-half their sum. What are they?
6. Fred is 3 years older than his brother Jack. In six years Jack's age will be $\frac{5}{6}$ Fred's age then. How old is each now?
7. The sum of two numbers is 14. If one-third the smaller plus two-thirds the larger is 9, what are the numbers?
8. There are 31 employees in the Data Corporation. If one-fourth of the men and one-third of the women are single and altogether 22 of the employees are *married*, how many men and how many women are employed at Data Corporation?

5.5 Applications

9. If 17% of a certain number is 125.8, what is the number?
10. How much must be invested at 4.4% to yield an investment income of $52.80 for one year?
11. A new color TV can be financed at 11% per year at the Golden Appliance and Electronics Company. The interest charges for a set which was paid off in one year came to exactly $62.04. What was the cost (not including interest) of the set?
12. The Exchange State Bank in Collins, Iowa charges 7% per year on a personal loan and deducts all the interest before granting payment of the loan. On a one year personal loan George Manning actually received $767.25. What was the amount of his loan?
13. The number 3.66 is what percent of 73.2?
14. The interest on an auto loan for one year was $156. If the amount of the loan was $2000, what was the nominal interest rate?
15. Mr. Tom Wing owns a small corner grocery store. His monthly inventory revealed that he took in exactly $4514.40 on merchandise which cost him $4320.00. What was the average percent mark up on the cost of his merchandise?
16. Mr. Williams borrowed $500 from the Easy Pay Finance Company and repaid the loan plus interest in 12 monthly installments of $51.25 each. What was the nominal yearly rate of interest paid?
17. If a certain number is added to the numerator of the fraction $\frac{5}{8}$ and the denominator is multiplied by that same number, the resulting fraction equals $\frac{11}{8}$. Find the number.
18. If the same number is added to both the numerator and denominator of $\frac{3}{4}$, the resulting number is $\frac{3}{28}$ larger than before. What is the number which was added to the numerator and denominator of $\frac{3}{4}$?
19. The denominator of a certain fraction is 3 times its numerator. If the numerator is increased by 1 and the denominator is decreased by 1, the resulting fraction equals $\frac{1}{2}$. What is the original fraction?
20. One number is one-half the multiplicative inverse of another. If their sum is $\frac{27}{10}$, what are the numbers?
21. Working alone Fred can overhaul his car in 14 hours. When his younger brother Jim helps, he can do the job in only 10 hours. How long would it take Jim alone to do the job?
22. With the main inlet pipe wide open it takes 14 hours to fill the municipal swimming pool. An auxiliary pipe can fill the pool in 35 hours. How long will it take to fill the pool if both pipes are run simultaneously?
23. A union bricklayer can lay the bricks for a fireplace in 10 hours, but his apprentice helper takes 15 hours for the same job. How long would it take them if they worked together?

196   *Rational Algebraic Expressions*

24. With the old power mower it took 12 hours to trim the fairway grass on the last three holes of the Rancho Golf and Country Club. Using both the old mower and a new improved model the job can be finished in only 4 hours. How long would it take the new mower to trim the grass?

25. Joe Smart can hike three times as fast downhill as he can uphill. One day it took him 7 hours to hike over Muir Pass, a trail which is 6 miles uphill and then 7.2 miles down the other side. How fast can Joe hike uphill?

26. After overhauling his outboard motor, Jim can travel 4 miles per hour faster than before. Now he saves 1 hour on a 48 mile excursion trip on Lake Tahoe. How fast can his boat travel now?

27. Agatha Brownlee doesn't usually drive alone, but one day she drove 185 miles along the freeway to her sister's house in Atlanta. At first she maintained a steady safe speed, but during the last 45 miles of the trip she reduced her speed by 10 mph because the traffic became heavy. If the total trip took her 5 hours, how fast does she normally drive when there is no traffic?

28. A camper estimates the rate of the current in a stream to be 2.5 miles per hour. He travels by motor boat 15 miles upstream and returns to his starting point in a total of $4\frac{1}{2}$ hours. How fast does his boat travel in still water?

29. How much pure gold must be added to 12 grams of 8 carat gold to raise it to 10 carat gold?

30. The Daisey Dairy has on hand 490 gallons of fresh milk which contains 5% butterfat. Low fat milk must contain no more than 2% butterfat. How many gallons of pure butterfat must be removed from the fresh milk to bring it down to the low fat concentration of butterfat?

31. How many milliliters of 12% acid solution must be added to 70 milliliters of 4.5% acid solution to have a mixture which is 7% strong?

32. To make a premium blend of coffee the Kitchen Pride Coffee Company adds fine aged beans which sell for $1.99 per pound to ordinary beans which sell for 73¢ per pound. How much aged coffee must be added to 20 pounds of ordinary coffee to make a blend which should sell for $1.09 per pound?

33. An estate of $4000 is invested for one year. Part of the estate is loaned out at 6% interest per year, and the remainder is invested in common stock which yields an 8.3% return for one year. If the total investment income for one year was $297.50, how much was invested in stock?

5.6 *Complex Fractions and Division of Polynomials*

34. Acetic acid for photographic use is often sold in 28% concentration. How many ounces of this acid must be added to one quart (32 ounces) of water to provide a stop bath which is 1.3% acetic acid? (Express result correct to the nearest tenth of an ounce.)

35. To prevent freezing, the coolant in an automobile radiator must be at least 6.5% alcohol. During the summer months the concentration in Joe Freeman's coolant has been reduced to only 3% due to the addition of water. If his radiator holds 12 quarts and is full, how much must be drained off and replaced with auto antifreeze at 38% alcohol to raise the concentration to the safe level?

36. Barry Watson purchased $6000 worth of safe bonds which paid 5.5% interest for the first year, but at the end of a year he sold some of the bonds and reinvested the money in the Wild West Gold Mining Corporation which soon declared bankruptcy and was able to return only 20% of their investment capital to its stockholders. At the end of the second year the bonds again paid the same dividend and Barry was able to recover some of his losses from the gold mining adventure, but altogether he lost $1050 for the two year period. How much did he invest in the Wild West Gold Mining Corporation? (Assume that he did not reinvest the income from the bonds for the first year.)

5.6 Complex Fractions and Division of Polynomials

Expressions such as

$$\frac{\frac{x}{2}}{\frac{2x+1}{3}}, \quad \frac{\frac{4x+1}{x}}{\frac{3}{x}+2+x^2}, \quad \text{and} \quad \frac{\frac{2x+1}{x^2-1}}{5-\frac{3}{x^2-2x+1}}$$

are examples of **complex fractions**. While such expressions appear formidable we shall see that they actually are equivalent to rational expressions and we shall develop a technique we can use to find a standard form for rational expressions represented by complex fractions.

When finding the solution set of an open sentence which contains fractions, we have used the technique of first finding the LCM of all denominators present and then multiplying both expressions by this LCM. Complex fractions can be simplified by a process analogous to this. First we find the LCM of all denominators found within the fraction; then we multiply both the numerator and denominator of the complex fraction by this LCM.

Example 1 Find a standard form of each of the following:

(a) $\dfrac{x/2}{\frac{3}{5}}$

(b) $\dfrac{x^2 - \dfrac{3x}{2} - 10}{x + \frac{5}{2}}$

(c) $\dfrac{\dfrac{2x+1}{x}}{1 - \dfrac{4}{x} - \dfrac{3}{x^2}}$

(d) $\dfrac{\dfrac{x}{4} - \dfrac{1}{x}}{5}$

Solution

(a) The LCM of 2 and 5 is 10.

$$\frac{\frac{x}{2}}{\frac{3}{5}} = \frac{10\left(\frac{x}{2}\right)}{10\left(\frac{3}{5}\right)} = \frac{5x}{6}$$

(b) There is only one denominator here, 2.

$$\frac{x^2 - \frac{3x}{2} - 10}{x + \frac{5}{2}} = \frac{2\left(x^2 - \frac{3x}{2} - 10\right)}{2\left(x + \frac{5}{2}\right)} = \frac{2x^2 - 3x - 20}{2x + 5}$$

$$= \frac{(2x+5)(x-4)}{(2x+5)(1)} = x - 4$$

(c) The LCM of x and x^2 is x^2.

$$\frac{\frac{2x+1}{x}}{1 - \frac{4}{x} - \frac{3}{x^2}} = \frac{x^2\left(\frac{2x+1}{x}\right)}{x^2\left(1 - \frac{4}{x} - \frac{3}{x^2}\right)} = \frac{x(2x+1)}{x^2 - 4x - 3}$$

(d) The only denominators which are *within* the complex fraction are 4 and x. The LCM is $4x$.

$$\frac{\frac{x}{4} - \frac{1}{x}}{5} = \frac{4x\left(\frac{x}{4} - \frac{1}{x}\right)}{4x(5)} = \frac{x^2 - 4}{20x}$$

When some of the denominators within a complex fraction are not monomials, finding their LCM may be more difficult, but the process for simplifying remains the same.

Example 2 Find a standard form for these rational expressions.

5.6 Complex Fractions and Division of Polynomials

(a) $\dfrac{\dfrac{x+3}{x-4}}{\dfrac{x+2}{x+1}}$ 	(b) $\dfrac{1 - \dfrac{1}{2x-1}}{\dfrac{x}{2x-1} - \dfrac{3}{x+2}}$

Solution

(a) The LCM of $x - 4$ and $x + 1$ is $(x-4)(x+1)$.

$$\frac{\dfrac{x+3}{x-4}}{\dfrac{x+2}{x+1}} = \frac{(x-4)(x+1)\left(\dfrac{x+3}{x-4}\right)}{(x-4)(x+1)\left(\dfrac{x+2}{x+1}\right)} = \frac{(x+1)(x+3)}{(x-4)(x+2)}$$

(b) The LCM of $2x - 1$ and $x + 2$ is $(2x-1)(x+2)$.

$$\frac{1 - \dfrac{1}{2x-1}}{\dfrac{x}{2x-1} - \dfrac{3}{x+2}} = \frac{(2x-1)(x+2)\left(1 - \dfrac{1}{2x-1}\right)}{(2x-1)(x+2)\left(\dfrac{x}{2x-1} - \dfrac{3}{x+2}\right)}$$

$$= \frac{(2x-1)(x+2) - (x+2)}{(x+2)x - (2x-1)3} = \frac{2x^2 + 2x - 4}{x^2 - 4x + 3}$$

$$= \frac{2(x+2)(x-1)}{(x-3)(x-1)} = \frac{2(x+2)}{x-3}$$

In Sections 3.1 and 4.2 we have considered the problem of finding the standard form of an expression which is the sum, difference, or product of two polynomials. In each case the distributive axiom can be used (repeatedly, if necessary), and the result is always another polynomial. But we have not considered the problem of division of polynomials. Let us do so now. The quotient of $2x + 3$ and $x^2 + x + 3$ can be written $(2x + 3) \div (x^2 + x + 3)$ or $\dfrac{2x+3}{x^2+x+3}$. As the latter representation clearly shows, the quotient of two polynomials is always a rational expression. Indeed, expressing the quotient in a fractional form is a trivial matter; we merely write the dividend as the numerator and the divisor as the denominator of the fraction. Of course, if we seek a standard form of the rational expression which is the quotient of two polynomials, further simplification may be required, but we have already considered such techniques.

Example 3 Express each of the following quotients as a rational expression in standard form.

200 *Rational Algebraic Expressions*

(a) $(x - 3) \div (2x^2 + x + 1)$
(b) $(2x - 6) \div (x^2 - x - 6)$
(c) $(6x^2 + 7x - 3) \div (3x - 1)$

Solution

(a) $(x - 3) \div (2x^2 + x + 1) = \dfrac{x - 3}{2x^2 + x + 1}$

(b) $(2x - 6) \div (x^2 - x - 6) = \dfrac{2x - 6}{x^2 - x - 6}$

$$= \dfrac{2(x - 3)}{(x - 3)(x + 2)} = \dfrac{2}{x + 2}$$

(c) $(6x^2 + 7x - 3) \div (3x - 1) = \dfrac{6x^2 + 7x - 3}{3x - 1}$

$$= \dfrac{(2x + 3)(3x - 1)}{3x - 1} = 2x + 3$$

When the degree of the divisor is less than (or equal to) the degree of the dividend there is a process analogous to "long division" of arithmetic for finding the quotient of two polynomials. The next example shows this alternate method of division.

Example 4 Find a standard form for $(x^2 - x - 6) \div (x + 2)$.

Solution We shall first rewrite the problem as in long division.

$$x + 2 \overline{\smash{)}\, x^2 - x - 6}$$

To find the first term of the quotient we consider the first terms of the dividend and divisor, x^2 and x respectively. Since $x^2 \div x = x^2/x = x$, we see that x is the first term of the quotient.

$$\begin{array}{r} x \\ x + 2 \overline{\smash{)}\, x^2 - x - 6} \\ x^2 + 2x \end{array}$$

Notice that we have multiplied $x(x + 2) = x^2 + 2x$, as in long division. Next we shall subtract $x^2 + 2x$ from the dividend.

$$\begin{array}{r} x \\ x + 2 \overline{\smash{)}\, x^2 - x - 6} \\ x^2 + 2x \\ \hline -3x - 6 \end{array}$$

5.6 Complex Fractions and Division of Polynomials

To find the next term of the quotient we divide the first term of this difference $-3x - 6$ by the first term of the divisor. Since $(-3x) \div (x) = -3x/x = -3$, we see that -3 is the next term of the quotient.

$$\begin{array}{r} x - 3 \\ x + 2 \overline{\smash{)} x^2 - x - 6} \\ \underline{x^2 + 2x } \\ -3x - 6 \\ \underline{-3x - 6} \\ 0 \end{array}$$

Again, we have multiplied $-3(x + 2) = -3x - 6$ and subtracted $-3x - 6$ from $-3x - 6$ as in long division. Then we see that

$$(x^2 - x - 6) \div (x + 2) = x - 3$$

Of course, it would have been easier to find the quotient in Example 4 by factoring.

$$(x^2 - x - 6) \div (x + 2) = \frac{x^2 - x - 6}{x + 2} = \frac{(x - 3)(\cancel{x + 2})}{(\cancel{x + 2})(1)} = x - 3$$

But as the next example shows there are times when we will be unable to factor the dividend, yet the "long division" process will enable us to find the standard form of the quotient.

Example 5 Find the standard form of $(x^3 - 6x^2 + 5x + 12) \div (x - 3)$.

Solution

$$\begin{array}{r} x^2 - 3x - 4 \\ x - 3 \overline{\smash{)} x^3 - 6x^2 + 5x + 12} \\ \underline{x^3 - 3x^2 } \\ -3x^2 + 5x + 12 \\ \underline{-3x^2 + 9x } \\ -4x + 12 \\ \underline{-4x + 12} \end{array}$$

$$(x^3 - 6x^2 + 5x + 12) \div (x - 3) = x^2 - 3x - 4$$

It is interesting to observe that the result of Example 5 shows us how to factor the expression $x^3 - 6x^2 + 5x + 12$. Since

$$(x^3 - 6x^2 + 5x + 12) \div (x - 3) = x^2 - 3x - 4$$

we see that
$$x^3 - 6x^2 + 5x + 12 = (x - 3) \cdot (x^2 - 3x - 4)$$
But $x^2 - 3x - 4 = (x - 4)(x + 1)$ and so the complete factorization is $(x - 3)(x - 4)(x + 1)$.

As we have seen in Example 3 it is *not* always the case that the quotient of two polynomials is a polynomial. Let us see what happens in the "long division" process when the quotient is not a polynomial.

Example 6 Use "long division" to find an expression which is equivalent to
$$\frac{2x^2 - 4x + 1}{x + 1}$$

Solution $\quad \dfrac{2x^2 - 4x + 1}{x + 1} = (2x^2 - 4x + 1) \div (x + 1)$

$$\begin{array}{r} 2x - 6 \phantom{{}+1} \\ x+1 \overline{\smash{)}\, 2x^2 - 4x + 1} \\ \underline{2x^2 + 2x \phantom{{}+1}} \\ -6x + 1 \\ \underline{-6x - 6} \\ 7 \end{array}$$

We must stop the division since the degree of the last difference, 7, is less than the degree of the divisor, $x + 1$. As in long division of arithmetic, when there is a remainder we write this remainder as the numerator of a fraction whose denominator is the divisor. Then
$$\frac{2x^2 - 4x + 1}{x + 1} = 2x - 6 + \frac{7}{x + 1}$$

Let us verify this result. Since $2x - 6 + \dfrac{7}{x + 1}$ is a rational expression, we may find a standard form.
$$2x - 6 + \frac{7}{x + 1} = \frac{2x - 6}{1} + \frac{7}{x + 1}$$
The LCM is $x + 1$. Then
$$\frac{2x - 6}{1} + \frac{7}{x + 1} = \frac{(2x - 6)(x + 1)}{(1)(x + 1)} + \frac{7}{x + 1}$$
$$= \frac{(2x - 6)(x + 1) + 7}{x + 1}$$
$$= \frac{(2x^2 - 4x - 6) + 7}{x + 1} = \frac{2x^2 - 4x + 1}{x + 1}$$

5.6 Complex Fractions and Division of Polynomials

As Example 6 shows, when the quotient of two polynomials is *not* a polynomial (but the degree of the divisor is less than or equal to the degree of the dividend) the "long division" process can be used to express this quotient as the sum of a polynomial and a rational expression whose numerator has a smaller degree than its denominator. Let us give a name to this polynomial and the numerator of the rational expression. We shall call the polynomial the partial quotient and the numerator of the resulting rational expression the remainder. Then since $(2x^2 - 4x + 1) \div (x + 1)$
$= 2x - 6 + \dfrac{7}{x + 1}$ we shall say that when $2x^2 - 4x + 1$ is divided by $x + 1$ the partial quotient is $2x - 6$ and the remainder is 7.

Example 7 Find the partial quotient and the remainder and express each of the following as the sum of a polynomial and a rational expression, if possible.

(a) $\dfrac{4x^3 + 2x^2 - 5x - 1}{2x^2 - 1}$ (b) $\dfrac{1 - x + x^3}{x - 1}$

(c) $\dfrac{x^2 + 2x - 3}{2x - 1}$ (d) $\dfrac{x + 3}{x^2 + 1}$

Solution

(a)
$$\begin{array}{r} 2x + 1 \\ 2x^2 - 1 \overline{) 4x^3 + 2x^2 - 5x - 1} \\ \underline{4x^3 - 2x } \\ 2x^2 - 3x - 1 \\ \underline{2x^2 - 1} \\ -3x \end{array}$$

The partial quotient is $2x + 1$ and the remainder is $-3x$.

$$\dfrac{4x^3 + 2x^2 - 5x - 1}{2x^2 - 1} = 2x + 1 + \dfrac{-3x}{2x^2 - 1}$$

(b) First we shall write the dividend in standard form.

$$\begin{array}{r} x^2 + x \\ x - 1 \overline{) x^3 - x + 1} \\ \underline{x^3 - x^2 } \\ x^2 - x + 1 \\ \underline{x^2 - x } \\ 1 \end{array}$$

The partial quotient is $x^2 + x$ and the remainder is 1.

$$\dfrac{1 - x + x^3}{x - 1} = x^2 + x + \dfrac{1}{x - 1}$$

204   Rational Algebraic Expressions

(c)
$$2x - 1 \overline{\smash{\big)}\begin{array}{l}\tfrac{1}{2}x + \tfrac{5}{4} \\ x^2 + 2x - 3 \\ \underline{x^2 - \tfrac{1}{2}x} \\ \tfrac{5}{2}x - 3 \\ \underline{\tfrac{5}{2}x - \tfrac{5}{4}} \\ -\tfrac{7}{4}\end{array}}$$

The partial quotient is $\tfrac{1}{2}x + \tfrac{5}{4}$ and the remainder is $-\tfrac{7}{4}$

$$\frac{x^2 + 2x - 3}{2x - 1} = \frac{1}{2}x + \frac{5}{4} + \frac{-\tfrac{7}{4}}{2x - 1}$$

As we see, sometimes the polynomial has rational number coefficients which are not integers.

(d) Impossible, since the degree of the denominator is greater than the degree of the numerator.

Exercises 5.6

For Exercises 1–32 find a standard form for these rational expressions which are written as complex fractions.

1. $\dfrac{x/2}{\tfrac{1}{3}}$

2. $\dfrac{\tfrac{3}{4}}{x/3}$

3. $\dfrac{1/x}{\tfrac{3}{2}}$

4. $\dfrac{\tfrac{5}{3}}{\tfrac{2}{3x}}$

5. $\dfrac{\tfrac{x}{2} + 1}{\tfrac{5}{2}}$

6. $\dfrac{\tfrac{x}{3} - 2}{2}$

7. $\dfrac{\tfrac{3}{5}}{\tfrac{2x}{5} + 1}$

8. $\dfrac{x/4}{\tfrac{3x}{4} - 2}$

9. $\dfrac{\tfrac{x}{2} + 1}{x/3}$

10. $\dfrac{\tfrac{x}{5} - 2}{x}$

11. $\dfrac{3x/7}{\tfrac{x}{2} + 3}$

12. $\dfrac{\tfrac{x}{3} - \tfrac{1}{2}}{\tfrac{x}{2} + 1}$

5.6 Complex Fractions and Division of Polynomials

13. $\dfrac{\dfrac{x+1}{3}+1}{\dfrac{x-1}{3}}$

14. $\dfrac{4+\dfrac{2x-3}{5}}{\dfrac{x+3}{5}}$

15. $\dfrac{x-\dfrac{x+1}{3}}{\dfrac{x-1}{3}}$

16. $\dfrac{1-\dfrac{x+2}{5}}{x-\dfrac{x-1}{5}}$

17. $\dfrac{\dfrac{x+3}{x}}{\dfrac{x-1}{3x}}$

18. $\dfrac{\dfrac{x+4}{2x}-1}{\dfrac{x-1}{3x}}$

19. $\dfrac{\dfrac{x+5}{x}}{\dfrac{2x-1}{x^2}}$

20. $\dfrac{\dfrac{x-1}{2x^2}}{x-\dfrac{3x+1}{5x}}$

21. $\dfrac{1+\dfrac{2}{x}}{x+1-\dfrac{2}{x}}$

22. $\dfrac{x+2-\dfrac{3}{x}}{1+\dfrac{3}{x}}$

23. $\dfrac{1+\dfrac{8}{x}+\dfrac{16}{x^2}}{\dfrac{1}{x}+\dfrac{4}{x^2}}$

24. $\dfrac{1-\dfrac{1}{x}-\dfrac{2}{x^2}}{1+\dfrac{2}{x}+\dfrac{1}{x^2}}$

25. $\dfrac{\dfrac{x-1}{x+2}+1}{\dfrac{x}{x+2}}$

26. $\dfrac{\dfrac{x+3}{x-1}+x}{\dfrac{x+1}{x-1}}$

27. $\dfrac{\dfrac{x^2}{x+2}-1}{x+\dfrac{2x+3}{x+2}}$

28. $\dfrac{x-\dfrac{12}{x-1}}{1-\dfrac{3}{x-1}}$

29. $\dfrac{\dfrac{3}{x+1}-\dfrac{1}{x-1}}{\dfrac{x}{x+1}}$

30. $\dfrac{1-\dfrac{x+1}{x+2}}{\dfrac{x}{x+3}}$

31. $\dfrac{1 + \dfrac{5}{(x+2)(x-4)}}{\dfrac{x}{x-4} - \dfrac{15}{(x+2)(x-4)}}$

32. $\dfrac{1 + \dfrac{1}{x-1} + \dfrac{x-12}{2x^2 - x - 1}}{\dfrac{x}{2x+1} + \dfrac{1}{x-1} + \dfrac{5}{2x^2 - x - 1}}$

For Exercises 33–36 use "long division" to find the quotient.

33. $(x^2 + 5x + 6) \div (x + 2)$
34. $(x^3 + x^2 - 3x + 1) \div (x - 1)$
35. $(2x^3 - x^2 + 5x + 3) \div (2x + 1)$
36. $(x^5 + 3x^4 + x^3 + 4x^2 + 1) \div (x^2 + 1)$

For Exercises 37–40 factor completely the first expression, given that one of the factors is the second expression. [Hint: Divide the first expression by the second as a first step.]

37. $x^3 + 3x^2 - 13x - 15;\ x + 5$
38. $2x^3 - 3x^2 - 32x + 48;\ 2x - 3$
39. $x^4 - x^3 - x^2 - x - 2;\ x^2 + 1$
40. $x^4 + 4x^3 + 3x^2 - 4x - 4;\ x^2 + x - 2$

For Exercises 41–56 find the partial quotient and the remainder and write the given expression as the sum of a polynomial and a rational expression whose numerator is of smaller degree than its denominator.

41. $\dfrac{x^2 + 2x + 2}{x + 1}$

42. $\dfrac{x^2 - 4x + 5}{x - 1}$

43. $\dfrac{x^2 + 5x + 3}{x + 2}$

44. $\dfrac{2x^2 - x - 16}{x - 3}$

45. $\dfrac{2 + 5x + 2x^2}{2x - 1}$

46. $\dfrac{6x^2 + x - 3}{3x + 2}$

47. $\dfrac{x^3 + 2x^2 + x + 3}{2 + x}$

48. $\dfrac{2 - 3x + 4x^2 - x^3}{3 - x}$

49. $\dfrac{x^3 + x^2 + x + 4}{x^2 + 1}$

50. $\dfrac{x^3 - 2x - 1}{x^2 - 3}$

51. $\dfrac{2x^2 + 3x + 2}{2x^2 + 1}$

52. $\dfrac{x^3 + 3x^2 + 3x + 5}{x^2 + x + 1}$

53. $\dfrac{(x+2)^2 + 1}{x + 3}$

54. $\dfrac{2x^2 + 4x + 5}{2x + 1}$

55. $\dfrac{x^2}{2x^2 + 1}$

56. $\dfrac{x^2 + x + 1}{3x - 1}$

5.7 Integral Exponents

In Section 4.1 we considered expressions with *natural number* exponents and observed that there are three important theorems which we may use to simplify expressions with such exponents. Let us recall them now.

Theorem 1	If $a \in N$ and $b \in N$, then $x^a \cdot x^b = x^{a+b}$
Theorem 2	If $a \in N$, then $x^a \cdot y^a = (xy)^a$
Theorem 3	If $a \in N$ and $b \in N$, then $(x^a)^b = x^{ab}$

Example 1 Simplify each of these and indicate which of the above theorems were used.

(a) $5^7 \cdot 5^3$ (b) $(2x)^3$ (c) $(2x^3)^4$

Solution

(a) $5^7 \cdot 5^3 = 5^{7+3} = 5^{10}$ [Theorem 1]

(b) $(2x)^3 = 2^3 \cdot x^3 = 8x^3$ [Theorem 2]

(c) $(2x^3)^4 = 2^4 \cdot (x^3)^4$ [Theorem 2]
$= 16 \cdot x^{3 \cdot 4} = 16x^{12}$ [Theorem 3]

Now we shall make definitions for expressions with *integral* exponents. Since the positive integers are the natural numbers, and we already have definitions for powers with such exponents, we need only define powers whose exponents are zero or negative integers. We will deliberately choose these definitions so that Theorems 1-3 will also be true for all integral exponents (including, of course, natural number exponents). We shall begin with the exponent 0. Consider the expression 5^0 and its product with $5 = 5^1$. If Theorem 1 is true for integral exponents, then $5^1 \cdot 5^0 =$

$5^{1+0} = 5^1$. Notice that the product of 5^0 and 5 is 5. But the *only* number whose product with 5 is also 5 is the multiplicative identity, 1. Hence, we must define 5^0 to be 1. Similarly, $7 \cdot 7^0 = 7^{1+0} = 7$ (by Theorem 1), so 7^0 must also be the multiplicative identity. Therefore, we make the definition $7^0 = 1$. More generally if $x \neq 0$, then we define $x^0 = 1$, since, if we wish Theorem 1 to be true, $x \cdot x^0 = x^{1+0} = x$. Note that this argument fails when $x = 0$. To be sure, $0 \cdot 0^0 = 0^{1+0} = 0$. But $0 \cdot 15 = 0$, $0(-7) = 0$ and $0 \cdot 0 = 0$. As these examples show, there are many numbers whose product with 0 is also 0. Actually, 0^0 is ambiguous, and we will not attempt to define it at all. Let us summarize these observations about the exponent 0 with the following definition.

DEFINITION

If $x \neq 0$, then $x^0 = 1$

Next we shall consider the exponent -1. Observe that if Theorem 1 is true for integral exponents, we must have $5^1 \cdot 5^{-1} = 5^{1+(-1)} = 5^0 = 1$. But then 5^{-1} is the multiplicative inverse of 5, since its product with 5 is 1. Hence we must define $5^{-1} = \frac{1}{5}$. Similarly, since $7^1 \cdot 7^{-1} = 7^0 = 1$, we must have $7^{-1} = \frac{1}{7}$. More generally, if $x \neq 0$, then $x \cdot x^{-1} = x^{1+(-1)} = x^0 = 1$. Hence we must define $x^{-1} = 1/x$. Of course, zero does not have a multiplicative inverse, and we do not define 0^{-1} at all.

Now consider the exponent -2. First we observe that if Theorem 3 is true for integral exponents then $(x^2)^{-1} = x^{(2)(-1)} = x^{-2}$. But we have said that $(x^2)^{-1}$ is the multiplicative inverse of x^2. That is, $(x^2)^{-1} = \frac{1}{x^2}$. Then $x^{-2} = (x^2)^{-1} = \frac{1}{x^2}$. Again we require that $x \neq 0$. Also $x^{-3} = (x^3)^{-1}$ if Theorem 3 is to be true, and so we must define $x^{-3} = (x^3)^{-1} = \frac{1}{x^3}$. And we define $x^{-4} = (x^4)^{-1} = \frac{1}{x^4}$. In general we shall make the following definition for *negative* integral exponents.

DEFINITION

If $a \in N$ and $x \neq 0$, then $x^{-a} = \frac{1}{x^a}$

Example 2 Use the definitions of this section to write each of the following in simplest form.
(a) 3^0 (b) 5^{-3} (c) $(2 \cdot 5)^{-1}$

5.7 Integral Exponents

(d) $\dfrac{1}{3^{-2}}$ (e) $(-2)^{-2}$ (f) -2^{-2}

Solution

(a) $3^0 = 1$

(b) $5^{-3} = \dfrac{1}{5^3} = \dfrac{1}{125}$

(c) $(2 \cdot 5)^{-1} = 10^{-1} = \dfrac{1}{10}$

(d) $\dfrac{1}{3^{-2}} = \dfrac{1}{1/3^2} = \dfrac{1}{\frac{1}{9}} = 9$

(e) The parentheses indicate that the base is -2.

$$(-2)^{-2} = \dfrac{1}{(-2)^2} = \dfrac{1}{4}$$

(f) With no symbols of grouping the *base* must be positive; here it is 2. Then

$$-2^{-2} = -(2^{-2}) = -\left(\dfrac{1}{2^2}\right) = -\dfrac{1}{4} = \dfrac{-1}{4}$$

With these two definitions for negative and zero integral exponents, we could prove (but shall not do so) that Theorems 1–3 are also true for *all integral* exponents. We restate them in a way which shows this more general nature.

Theorem 1	If $a \in I$, and $b \in I$, then $x^a \cdot x^b = x^{a+b}$
Theorem 2	If $a \in I$, then $x^a \cdot y^a = (xy)^a$
Theorem 3	If $a \in I$ and $b \in I$, then $(x^a)^b = x^{ab}$

Example 3 Use Theorem 1 to replace each of the following by an equivalent expression which has only one base and one exponent.

(a) $(2^{-3})(2^{-5})$ (b) $(3^{-4})(3^6)$ (c) $(-2)^{-5}(-2)$
(d) $(\tfrac{1}{3})^{-2}(\tfrac{1}{3})^2$ (e) $(2x)^{-3}(2x)^5$

Solution

(a) $(2^{-3})(2^{-5}) = 2^{(-3)+(-5)} = 2^{-8}$

(b) $(3^{-4})(3^6) = 3^{-4+6} = 3^2$

(c) $(-2)^{-5}(-2) = (-2)^{-5+1} = (-2)^{-4}$ [Note that we *must* retain the parentheses to indicate that the base is -2.]

(d) $(\frac{1}{3})^{-2}(\frac{1}{3})^2 = (\frac{1}{3})^{-2+2} = (\frac{1}{3})^0$
(e) $(2x)^{-3}(2x)^5 = (2x)^{-3+5} = (2x)^2$

Example 4 Use Theorem 2 to replace each of these with an equivalent expression which has only one base and one exponent.
(a) $3^{-4} \cdot 2^{-4}$ (b) $(-2)^{-1} \cdot 5^{-1}$ (c) $(-2)^{-3} \cdot x^{-3}$

Solution

(a) $3^{-4} \cdot 2^{-4} = (3 \cdot 2)^{-4} = 6^{-4}$
(b) $(-2)^{-1} \cdot 5^{-1} = (-2 \cdot 5)^{-1} = (-10)^{-1}$ [Note the need for parentheses here to indicate that the base is -10.]
(c) $(-2)^{-3} \cdot x^{-3} = (-2x)^{-3}$ [Again, we must have parentheses to indicate the base.]

Example 5 Use Theorem 3 to replace each of the following by an equivalent expression which has only one base and one exponent.
(a) $(2^2)^{-2}$ (b) $[(-3)^{-2}]^{-3}$ (c) $(4^0)^{-1}$ (d) $[(5x)^2]^{-1}$

Solution

(a) $(2^2)^{-2} = 2^{2(-2)} = 2^{-4}$
(b) $[(-3)^{-2}]^{-3} = (-3)^{(-2)(-3)} = (-3)^6$
(c) $(4^0)^{-1} = 4^{(0)(-1)} = 4^0$
(d) $[(5x)^2]^{-1} = (5x)^{(2)(-1)} = (5x)^{-2}$

Example 6 First use the theorems to simplify and then use the definitions to find the simplest form for each of the following rational numbers.
(a) $(2^{-1})^{-3}$ (b) $3^{-7} \cdot 3^5$ (c) $(\frac{2}{5})^{-2} \cdot 5^{-2}$
(d) $3^7 \cdot 3^{-7}$ (e) $(\frac{1}{2})^{-3}$

Solution

(a) $(2^{-1})^{-3} = 2^{(-1)(-3)} = 2^3 = 8$
(b) $3^{-7} \cdot 3^5 = 3^{-7+5} = 3^{-2} = \frac{1}{3^2} = \frac{1}{9}$
(c) $(\frac{2}{5})^{-2} \cdot 5^{-2} = (\frac{2}{5} \cdot 5)^{-2} = 2^{-2} = \frac{1}{2^2} = \frac{1}{4}$
(d) $3^7 \cdot 3^{-7} = 3^{7-7} = 3^0 = 1$.
(e) $\frac{1}{2} = 2^{-1}$. Then $(\frac{1}{2})^{-3} = (2^{-1})^{-3} = 2^3 = 8$

Example 7 Find the value of the expression when the variable is replaced by the indicated number.

5.7 Integral Exponents

(a) $5x^{-2}$, $x = 2$ (b) $(-x)^{-3}$, $x = 2$
(c) x^{-2}, $x = -4$ (d) $-x^{-4}$, $x = -2$

Solution

(a) $5 \cdot 2^{-2} = 5 \cdot \frac{1}{4} = \frac{5}{4}$

(b) $(-2)^{-3} = \dfrac{1}{(-2)^3} = \dfrac{1}{-8} = \dfrac{-1}{8}$

(c) $(-4)^{-2} = \dfrac{1}{(-4)^2} = \dfrac{1}{16}$ [Note the need for parentheses to indicate that the base is -4.]

(d) $-(-2)^{-4} = -\dfrac{1}{(-2)^4} = -\dfrac{1}{16} = \dfrac{-1}{16}$ [Again, we need parentheses to indicate that the base is -2.]

Often we will want to replace expressions with negative (or zero) exponents by equivalent expressions in which the only exponents shown are positive. We may use Theorems 1, 2, and 3 and the definitions of this section together with other techniques we have previously employed for finding equivalent expressions.

Example 8   For each of the following find an equivalent expression in which all indicated exponents are positive. Write this expression in standard form.
(a) $4x^{-2}$ (b) $(4x)^{-2}$ (c) $x^{-2} + 1$ (d) $(1 - 3x^{-1} + 2x^{-2})x^2$

Solution

(a) The base here is x. Then

$$4x^{-2} = 4 \cdot x^{-2} = 4 \cdot \frac{1}{x^2} = \frac{4}{x^2}$$

(b) The parentheses indicate that the base is $4x$. Then

$$(4x)^{-2} = \frac{1}{(4x)^2} = \frac{1}{4^2 \cdot x^2} = \frac{1}{16x^2}$$

(c) $x^{-2} + 1 = \dfrac{1}{x^2} + 1 = \dfrac{1}{x^2} + \dfrac{x^2}{x^2} = \dfrac{1 + x^2}{x^2} = \dfrac{x^2 + 1}{x^2}$

(d) We shall use the distributive axiom.

$$(1 - 3x^{-1} + 2x^{-2})x^2 = 1 \cdot x^2 - (3x^{-1})x^2 + (2x^{-2})x^2$$
$$= x^2 - 3x + 2x^0 = x^2 - 3x + 2$$

As we see from Example 8, expressions with integral exponents are

212 *Rational Algebraic Expressions*

equivalent to rational expressions, and we may use the theorems and definitions to find a standard form. The expression $\dfrac{x^{-1} + 2x^{-2}}{1 - 4x^{-2}}$ is equivalent to a complex fraction, as we can see by using the definition of this section:

$$\frac{x^{-1} + 2x^{-2}}{1 - 4x^{-2}} = \frac{\dfrac{1}{x} + \dfrac{2}{x^2}}{1 - \dfrac{4}{x^2}}$$

We may find a standard form if we multiply both its numerator and denominator by x^2, the LCM of x^2 and x.

$$\frac{x^2\left(\dfrac{1}{x} + \dfrac{2}{x^2}\right)}{x^2\left(1 - \dfrac{4}{x^2}\right)} = \frac{x + 2}{x^2 - 4} = \frac{(x+2)(1)}{(x+2)(x-2)} = \frac{1}{x-2}$$

However, we need not write the original expression as a complex fraction to obtain this standard form. Instead, it is more direct to use the theorems about negative exponents:

$$\frac{x^{-1} + 2x^{-2}}{1 - 4x^{-2}} = \frac{(x^{-1} + 2x^{-2})x^2}{(1 - 4x^{-2})x^2} = \frac{x^{-1} \cdot x^2 + 2x^{-2} \cdot x^2}{1 \cdot x^2 - 4x^{-2} \cdot x^2}$$

$$= \frac{x + 2x^0}{x^2 - 4x^0} = \frac{x + 2}{x^2 - 4} = \frac{1}{x - 2}$$

Example 9 Use the theorems of this section to find a standard form for each of the following.

(a) $\dfrac{3x^{-1}}{2x^{-4}}$ (b) $\dfrac{(2x^2)^{-2}}{2x^{-5}}$ (c) $\dfrac{x - 3 - 4x^{-1}}{x + 1}$

Solution

(a) Since the negative exponent with the larger absolute value is -4, we shall multiply the numerator and denominator by x^4 to eliminate negative exponents.

$$\frac{3x^{-1}}{2x^{-4}} = \frac{(3x^{-1})(x^4)}{(2x^{-4})(x^4)} = \frac{3x^3}{2}$$

(b) First we shall use the theorems to simplify the expression.

$$\frac{(2x^2)^{-2}}{2x^{-5}} = \frac{2^{-2}(x^2)^{-2}}{2x^{-5}} = \frac{2^{-2} \cdot x^{-4}}{2x^{-5}}$$

5.7 Integral Exponents

To eliminate the negative exponents we shall multiply the numerator and denominator by $2^2 \cdot x^5$. Notice that 2^{-2} and x^{-5} are the negative powers of 2 and x whose exponents have the larger absolute value:

$$\frac{2^{-2}x^{-4}}{2x^{-5}} = \frac{(2^{-2}x^{-4})(2^2x^5)}{(2x^{-5})(2^2x^5)} = \frac{x}{2^3} = \frac{x}{8}$$

(c) $\dfrac{x-3-4x^{-1}}{x+1} = \dfrac{(x-3-4x^{-1})x}{(x+1)x} = \dfrac{x^2-3x-4}{(x+1)x}$

$= \dfrac{(x-4)(x+1)}{(x+1)x} = \dfrac{x-4}{x}$

Exercises 5.7

Each of the numerals in Exercises 1–28 represents a rational number. Use the theorems and definitions of this section to find its simplest form.

1. 5^{-1}
2. 4^{-2}
3. $3 \cdot 2^{-5}$
4. $5 \cdot 3^{-3}$
5. 8^0
6. $\left(\frac{3}{4}\right)^0$
7. $5 \cdot 4^0$
8. $(-3 \cdot 2)^0$
9. -2^{-2}
10. $(-3)^{-2}$
11. $-(-4)^{-2}$
12. $-(-2)^{-3}$
13. $2^7 \cdot 2^{-5}$
14. $(-3)^4(-3)^{-6}$
15. $\left(\frac{1}{3}\right)^{-4}\left(\frac{1}{3}\right)^4$
16. $(17^0)^{-3}$
17. $(2^{-2})^{-2}$
18. $(3^{-1})^2$
19. $\left(\frac{3}{19}\right)^{-2} \cdot 19^{-2}$
20. $(.25)^{-4}(4)^{-4}$
21. $\left(\frac{1}{4}\right)^{-2}$
22. $\left(\frac{3}{2}\right)^{-3}$
23. $(.5)^{-2}$
24. $(.25)^{-1}$
25. $\dfrac{1}{4^{-2}}$
26. $\dfrac{2^{-4}}{3^{-2}}$
27. $(-5^2)(-5)^{-2}$
28. $(-3)^{-4}(-3)^5$

For Exercises 29–40 find the value of the expression when the variable is replaced by the indicated number.

29. $3x^{-2}$, $x = 2$
30. $2x^{-2}$, $x = -2$
31. $5x^0$, $x = \frac{2}{3}$
32. $-4x^0$, $x = -3$
33. $-x^{-3}$, $x = 2$
34. $-x^{-3}$, $x = -2$
35. $(-x)^{-2}$, $x = 3$
36. $(-x)^{-2}$, $x = -3$

37. $x^{-1} + x^{-2}, x = 2$
38. $x^{-1} + x^{-2}, x = -2$
39. $x - x^{-1}, x = 1$
40. $x - x^{-1}, x = -1$

Each of the expressions in Exercises 41–68 is equivalent to a rational expression. Use the theorems and definitions of this section to find a standard form for this expression.

41. $2x^{-4}$
42. $-3x^{-1}$
43. $(3x)^{-2}$
44. $(-2x)^{-5}$
45. $x^7 \cdot x^{-5}$
46. $(-2x^{-3})(x^4)$
47. $(2x^{-5})(3x^2)$
48. $(x^{-2})(-4x^{-1})$
49. $x^2(x^{-1} + 2x^{-2})$
50. $(3 + 2x^{-1} - 2x^{-2})x^2$
51. $3x^{-1}(x^2 + 3x)$
52. $(-2x^{-3})(x^5 - x^3)$
53. $x^{-1} + 4$
54. $x^{-1} + x^{-2}$
55. $x^{-2} - 2x^{-1}$
56. $1 - x^{-1} + 2x^{-2}$
57. $\dfrac{2x^{-5}}{x^{-7}}$
58. $\dfrac{3x^{-2}}{3^{-1}x}$
59. $\dfrac{(2x)^{-1}}{2x^{-2}}$
60. $\dfrac{2x^{-3}}{(2x)^{-2}}$
61. $\dfrac{1 + x^{-1}}{1 - x^{-1}}$
62. $\dfrac{2x^{-1} + 1}{3x^{-2}}$
63. $\dfrac{1 - x^{-2}}{x + 1}$
64. $\dfrac{x^2 + x + 1}{x^{-1} + 1 + x}$
65. $(x^3 + x^2)(x^{-2} + 1)$
66. $(x + 1)^{-2}(x^2 - 1)$
67. $\dfrac{1 - 4x^{-1} - 5x^{-2}}{2 + 3x^{-1} + x^{-2}}$
68. $\dfrac{21x^{-2} - 13x^{-1} + 2}{1 - 3x^{-1}}$

Expressions with Two Variables

Chapter SIX

6.1 Sets of Ordered Pairs and Their Graphs

By an **ordered pair** of real numbers we shall mean a pair of numbers with one of them designated as the "first." Then (4, 5) represents an ordered pair in which the first number is 4. We shall often call the numbers in an ordered pair the **components** of that pair. In $(-5, 3)$ the first component is -5 and the second component is 3. The ordered pair $(3, -5)$ on the other hand is a *different* pair, for in it the first component is 3 and the second component is -5.

Let $A = \{1, 2, 3\}$ and $B = \{4, 5\}$. Let us form the set of *all* possible ordered pairs whose first component is a member of A and whose second component is a member of B: $\{(1, 4), (1, 5), (2, 4), (2, 5), (3, 4), (3, 5)\}$. We shall call this set the **Cartesian product** of sets A and B and designate it by $A \times B$. Cartesian products can be found for other sets in exactly the same way.

Example 1 Find the Cartesian products.

(a) $\{1, 2\} \times \{3\}$ (b) $\{1, 2\} \times \{0, 1, 2\}$ (c) $\{0, 1\} \times \{0, 1\}$

215

Solution

(a) {(1, 3), (2, 3)}
(b) {(1, 0), (1, 1), (1, 2), (2, 0), (2, 1), (2, 2)}
(c) {(0, 0), (0, 1), (1, 0), (1, 1)}

We say that the set {(1, 0), (1, 1), (1, 2), (2, 0), (2, 1), (2, 2)} has six members since there are exactly six ordered *pairs* in this set. Notice that this set is the Cartesian product of $C = \{1, 2\}$ and $D = \{0, 1, 2\}$ which we found in Example 1(b). It is interesting to observe that sets C and D have two and three members, respectively, and the *product* of two and three is six, the number of members in $C \times D$. As Example 1 illustrates, it is generally true that for finite sets the number of elements in their Cartesian product can be found by multiplying the number of elements in one set by the number of elements in the other. Note, too, that in Example 1(c) we have found the Cartesian product of a set with itself.

Often a Cartesian product will contain many members. If $A = \{-5, -4, -3, \ldots, 3, 4, 5\}$, then $A \times A$ will contain exactly 121 ordered pairs (since there are 11 members in A and $11 \cdot 11 = 121$)! The set

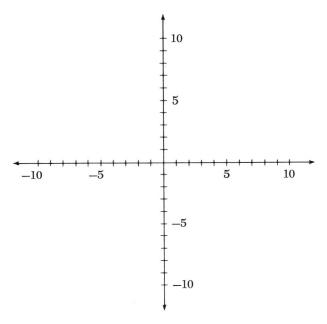

Figure 6.1

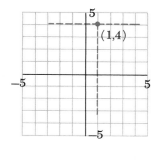
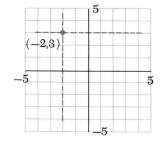
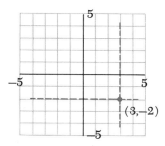
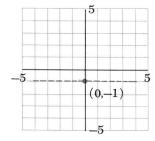

Figure 6.2

$B = \{(1, 4), (-2, 3), (3, -2), (0, -1)\}$ which contains four of these ordered pairs is a subset of $A \times A$. We shall often be interested in subsets of a Cartesian product rather than the Cartesian product itself. We note, of course, that each such subset is itself a set whose members are ordered pairs.

Let us next find the **graph** of the set $B = \{(1, 4), (-2, 3), (3, -2), (0, -1)\}$. We begin by considering two real number lines, one horizontal, the other vertical, which intersect at right angles at their origins. See Figure 6.1. Note that the positive direction on the horizontal line (called the horizontal axis) is to the right and the positive direction on the vertical line (vertical axis) is upward. We shall call the plane which these two axes determine, the coordinate plane; and we shall call their point of intersection, the origin of this plane.

Each element (ordered pair) of set B will correspond to a single point in the coordinate plane. The element $(1, 4)$ will correspond to the point determined by two intersecting lines. On the horizontal axis, at the point whose coordinate there is 1, we draw a perpendicular to this axis; on the vertical axis, at the point whose coordinate there is 4, we draw another perpendicular line. The required point is the intersection of these two lines as shown in Figure 6.2. We shall say that this *point corresponds* to the ordered pair $(1, 4)$ and that the components of this ordered pair, 1 and 4, are the coordinates of the point. And in Figure 6.2 we see that the

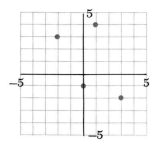

Figure 6.3

point which corresponds to (−2, 3) is *not* the same as the point which corresponds to (3, −2). These are different ordered pairs and they correspond to different points. Notice that it is always the *first* component of the ordered pair which determines the location of the perpendicular line to the *horizontal* axis; the *second* component determines the position of the perpendicular to the *vertical* axis. To find the point corresponding to the pair (0, −1) we find that point on the horizontal axis whose coordinate is 0 and observe that there is no need to draw a perpendicular there—the vertical axis already has been drawn. The point which corresponds to (0, −1) is the intersection of the vertical axis and the perpendicular to the vertical axis at −1, and it is shown in Figure 6.2. Finally, the graph of the set $B = \{(1, 4), (-2, 3), (3, -2), (0, -1)\}$ is the set of the four points which we have found, and it is shown in Figure 6.3.

The same technique can be used to find the graph of any set of ordered pairs of numbers, although this may be a tedious task if such a set contains many elements.

Example 2 Find the graph of each of these sets.
(a) $\{(-1, -1), (0, 1)\}$ (b) $\{(0, 1), (-1, -1)\}$
(c) $\{(3, 3), (-3, 3), (-3, -3), (3, -3), (0, 0)\}$
(d) $\{(1, 0), (1, 1), (1, 2), (2, 1), (2, 2), (3, 2)\}$

Solution

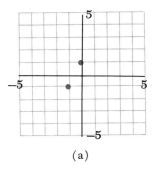

(a)

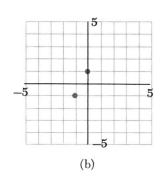

(b)

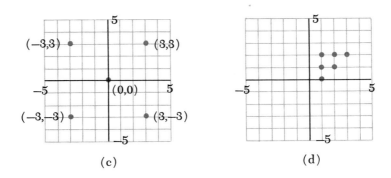

Since $\{(-1, -1), (0, 1)\} = \{(0, 1), (-1, -1)\}$ (note that they have the *same* ordered pairs as members), it is not surprising that their graphs are identical.

As Example 2 illustrates, to every set of ordered pairs of numbers corresponds a set of points in the coordinate plane. And this correspondence is reversible. To every set of points in the coordinate plane there also corresponds a set of ordered pairs.

Example 3 Find the set of ordered pairs corresponding to each of the following sets of points.

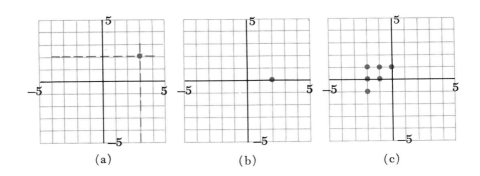

Solution

(a) $\{(3, 2)\}$ (b) $\{(2, 0)\}$
(c) $\{(0, 1), (-1, 1), (-1, 0), (-2, 1), (-2, 0), (-2, -1)\}$

Algebraic expressions such as $2x + 3y - 1$, $x^2 - 3xy + 2y^2$ and $\dfrac{x - 3y}{xy}$ contain two variables. As was true for expressions in one variable, expressions with two variables represent numbers, but unless we know the replacement for *each variable* the value of such expressions cannot be

220   *Expressions with Two Variables*

determined. Of course, if the replacements for x and y are known, the value of such expressions can be found by making the indicated replacements and performing the arithmetic operations in the resulting numerical expressions.

Example 4   Find the value of each of the following expressions in two variables for the replacements indicated.

(a) $2x + 3y - 1$ when $x = 1$ and $y = -2$
(b) $x^2 - 3xy + 2y^2$ when $x = 2$ and $y = 2$
(c) $\dfrac{x - 3y}{xy}$ when $x = -4$ and $y = -1$

Solution

(a) $2 \cdot 1 + 3(-2) - 1 = 2 - 6 - 1 = -5$
(b) $2^2 - 3(2)(2) + 2 \cdot 2^2 = 4 - 12 + 8 = 0$
(c) $\dfrac{-4 - 3(-1)}{(-4)(-1)} = \dfrac{-4 + 3}{4} = \dfrac{-1}{4}$

An algebraic expression in one variable has associated with it a set of numbers called its domain. The members of the domain are those numbers which may be replacements for the variable in the expression. For example, $3x^2 - 2x + 1$, $\{0, 1\}$ describes the expression $3x^2 - 2x + 1$ whose domain is $\{0, 1\}$. Associated with such an expression is another set called its range. The range is the set of values of the expression when all replacements of the variable by members of the domain are made. The range of $3x^2 - 2x + 1$, $\{0, 1\}$ is the set $\{1, 2\}$. When an expression in *two* variables is considered, its domain is *not* a set of numbers. As Example 4 shows, to find the value of an expression in two variables requires that we have *two* numbers—a replacement for x and a replacement for y. Let us use an ordered *pair* of numbers to indicate *one* member of the domain of an expression in two variables, with the understanding that the first component of the pair is the replacement for x and the second component, the replacement for y. Then by $x - y + 2$, $\{(1, 2), (2, 4), (3, 6)\}$ we shall mean the expression $x - y + 2$ with domain $\{(1, 2), (2, 4), (3, 6)\}$. And the range of this expression shall (again) be the set of all values it has over the given domain. In this case when $x = 1$ and $y = 2$, $x - y + 2 = 1 - 2 + 2 = 1$; when $x = 2$ and $y = 4$, $x - y + 2 = 2 - 4 + 2 = 0$; when $x = 3$ and $y = 6$, $x - y + 2 = 3 - 6 + 2 = -1$; and so the range is $\{-1, 0, 1\}$.

Example 5   For each of the following expressions in two variables find the range for the given domain.

(a) $3x + 4y - 2$, $\{(1, 1), (2, 2), (-1, 3)\}$
(b) $x^2 + 4xy - 5y^2$, $\{(-1, -1), (3, 1)\}$

6.1 Sets of Ordered Pairs and Their Graphs

Solution

(a) $3x + 4y - 2$

$(1, 1)$: $3 \cdot 1 + 4 \cdot 1 - 2 = 5$

$(2, 2)$: $3 \cdot 2 + 4 \cdot 2 - 2 = 12$

$(-1, 3)$: $3(-1) + 4 \cdot 3 - 2 = 7$

The range is $\{5, 7, 12\}$.

(b) $x^2 + 4xy - 5y^2$

$(-1, -1)$: $(-1)^2 + 4(-1)(-1) - 5(-1)^2 = 0$

$(3, 1)$: $3^2 + 4 \cdot 3 \cdot 1 - 5 \cdot 1^2 = 16$

The range is $\{0, 16\}$.

An open sentence in one variable is neither true nor untrue. After the variable in such a sentence is replaced by some number in the domain, a numerical statement results which is either true or false. That subset of the domain which results in true statements we have called the solution set of the sentence. We shall treat open sentences in two variables in exactly the same way. The domain of a sentence in two variables is a set of ordered pairs of numbers, and that subset of the domain which yields a true statement will again be the solution set.

Example 6 For each open sentence find the solution set for the given domain.

(a) $x - 2y = 5$, $\{(1, -1), (2, -2), (5, 0), (3, -1)\}$

(b) $y = x^2 - 4$, $\{(1, 3), (-2, 0), (0, 4)\}$

(c) $y \leq 2x + 1$, $\{(1, 4), (2, 5), (1, -1), (0, 0)\}$

Solution

(a) $x - 2y = 5$.

$(1, -1)$: $1 - 2(-1) = 5$

$3 = 5$ False

$(2, -2)$: $2 - 2(-2) = 5$

$6 = 5$ False

$(5, 0)$: $5 - 2(0) = 5$

$5 = 5$ True

$(3, -1)$: $3 - 2(-1) = 5$

$5 = 5$ True

The solution set is $\{(5, 0), (3, -1)\}$.

(b) $y = x^2 - 4$

$(1, 3)$: $3 = 1^2 - 4$

            $3 = -3$ False

$(-2, 0)$: $0 = (-2)^2 - 4$

            $0 = 0$ True

$(0, 4)$: $4 = 0^2 - 4$

            $4 = -4$ False.

The solution set is $\{(-2, 0)\}$.

(c) $y \leq 2x + 1$

$(1, 4)$: $4 \leq 2 \cdot 1 + 1$

            $4 \leq 3$ False

$(2, 5)$: $5 \leq 2 \cdot 2 + 1$

            $5 \leq 5$ True

$(1, -1)$: $-1 \leq 2 \cdot 1 + 1$

            $-1 \leq 3$ True

$(0, 0)$: $0 \leq 2 \cdot 0 + 1$

            $0 \leq 1$ True

The solution set is $\{(2, 5), (1, -1), (0, 0)\}$.

We can find the graph of the solution set of an open sentence in two variables as we have found the graph of solution sets of sentences in one variable.

Example 7   Find the graph of the solution set for each of the open sentences in Example 6.

Solution

(a) $\{(5, 0), (3, -1)\}$ (b) $\{(-2, 0)\}$ (c) $\{(2, 5), (1, -1), (0, 0)\}$

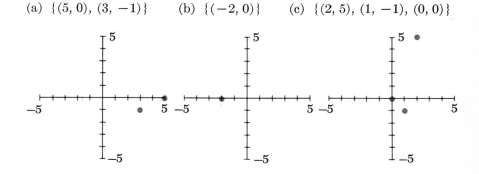

Exercises 6.1

For Exercises 1–8 let $A = \{1\}, B = \{2, 3\}, C = \{-1, 1\}, D = \{0\}$. Find each of the indicated sets.

1. $A \times B$
2. $B \times A$
3. $A \times C$
4. $C \times A$
5. $A \times A$
6. $B \times D$
7. $C \times C$
8. $D \times D$

For Exercises 9–12 let $E = \{1, 2, 3, 4, 5\}$ and $F = \{1, 2, 3, \ldots, 10\}$. How many members are there in each of these sets?

9. $E \times F$
10. $E \cup F$
11. $E \cap F$
12. $E \times E$

For Exercises 13–16 find the indicated sets.

13. $\{(1, 2), (3, 4)\} \cap \{(3, 4), (5, 6)\}$
14. $\{(2, 5)\} \cup \{(2, 6)\}$
15. $\{(3, 1)\} \cap \{(1, 3)\}$
16. $\{(1, 2\} \cap \{(1, 2)\}$

For Exercises 17–24 find the graph of the indicated set.

17. $\{(2, -1)\}$
18. $\{(0, 3)\}$
19. $\{(-3, -1), (-2, 4)\}$
20. $\{(0, 0), (2, 1)\}$
21. $\{1, 2\}$ [This is not $\{(1, 2)\}$.]
22. $\{1\} \times \{2\}$
23. $\{0\} \times \{-1, 0, 1\}$
24. $\{-3, 0, 2\} \times \{0\}$

For Exercises 25–28 find the set of ordered pairs which corresponds to the indicated set of points.

25.
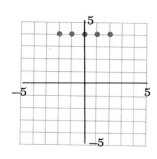

26.
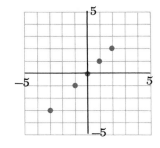

27.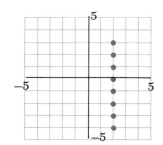
28.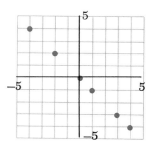

For Exercises 29–36 find the range of the given expression with indicated domain.

29. $x + y$, $\{(3, 4), (-2, 1)\}$
30. $\dfrac{y}{x}$, $\{(2, 5), (-1, 4)\}$
31. $x^2 - y$, $\{(-2, 1), (1, -2), (3, 6)\}$
32. $2x - xy$, $\{(1, 2), (3, 2), (0, 4)\}$
33. $3x - y + 2$, $\{(5, 1), (0, 1), (0, 0)\}$
34. $y - 4x + 1$, $\{(2, 1), (-1, -3)\}$
35. $y - xy^2$, $\{(2, -2)\}$
36. $2xy^3 - 3x^2y$, $\{(-3, -2)\}$

For Exercises 37–44 find the solution set.

37. $x + 2y = 5$, $\{(1, 1), (1, 2), (3, 1)\}$
38. $2x - 5y = 1$, $\{(3, 0), (3, 1), (0, 0)\}$
39. $y = 2x + 6$, $\{(1, 8), (10, 2)\}$
40. $y < x - 1$, $\{(3, 0), (2, 5)\}$
41. $x^2 = y + 4$, $\{(2, 0), (0, -4), (-2, 0)\}$
42. $x - 2y^2 = 0$, $\{(0, 0), (2, 1), (2, -1)\}$
43. $x + y = 3$, $\{(1, 1), (2, -1), (4, -3)\}$
44. $x + 2y > 1$, $\{(-1, 0), (2, -3), (0, 0)\}$

For Exercises 45–52 find the solution set and graph it.

45. $x - y = 5$, $\{(2, -3), (1, 6), (0, -5)\}$
46. $2x - y = 0$, $\{(1, 1), (2, -4), (1, 2)\}$
47. $x = -y$, $\{(0, 0), (3, -3), (-2, 2), (1, 1)\}$
48. $x = y$, $\{(3, -3), (2, 2), (0, 0), (-1, -1), (2, -2)\}$
49. $x + 0 \cdot y = 3$, $\{(3, 1), (3, -2), (3, 0), (0, 3) (0, 0)\}$
50. $x = 2$, $\{(2, 1), (2, 2), (2, 0), (0, 1), (0, 0), (2, -1)\}$ [Hint: See Problem 49.]

51. $y = -1$, $\{-1, 0, 1\} \times \{-1, 0, 1\}$
52. $x = -y$, $\{-2, -1, 0, 1, 2\} \times \{-2, -1, 0, 1, 2\}$

6.2 Equivalent Expressions

In the last section we defined the Cartesian product of two sets. When both sets are finite, the number of members in this product set is the product of the number of members in each of the original sets. But if one or both of two sets is infinite, the Cartesian product is also infinite, and we shall not be able to list its elements. Our primary interest will be with the infinite set $R \times R$. This is the set of *all* ordered pairs each of whose components is a real number.

We also considered in the last section algebraic expressions which contained two variables. Such expressions had a domain which was in every case a finite subset of $R \times R$. Let us adopt the same convention for expressions in two variables as we did for those with one variable. Unless explicitly stated otherwise, the domain of an expression in two variables is the set of all members of $R \times R$ for which the expression has a real value. Let us also agree that two expressions (in two variables) are equivalent if they have the same value for each replacement of the variables by an element in the intersection of their domains.

Example 1 For each of the following pairs of expressions decide whether or not they are equivalent:

(a) $(x - 3y)(x + 2y)$ and $x^2 - xy - 6y^2$ with domain $\{(0, 1), (1, 2)\}$

(b) $\dfrac{4x}{x - y} - \dfrac{2y}{x + y}$ and $\dfrac{2(2x^2 + xy + y^2)}{x^2 - y^2}$ with domain $\{(3, 2)\}$

(c) $(3x + 5y - 1) - (2x - 3y + 4)$ and $x + 8y - 5$

Solution

(a)

Replacement	Value of $(x - 3y)(x + 2y)$	Value of $x^2 - xy - 6y^2$
(0, 1)	-6	-6
(1, 2)	-25	-25

Equivalent over $\{(0, 1), (1, 2)\}$.

(b)

Replacement	Value of $\dfrac{4x}{x-y} - \dfrac{2y}{x+y}$	Value of $\dfrac{2(2x^2 + xy + y^2)}{x^2 - y^2}$
(3, 2)	$\dfrac{56}{5}$	$\dfrac{56}{5}$

Equivalent over $\{(3, 2)\}$.

(c) The intersection of the domains is $R \times R$. Clearly, we cannot find the value of these expressions for every element in $R \times R$. However, for each replacement of the variables the expression $(3x + 5y - 1) - (2x - 3y + 4)$ represents a real number, and we can use the definition of subtraction, the associative and commutative axioms of addition, and the distributive axiom to write the following.

$(3x + 5y - 1) - (2x - 3y + 4)$
$= (3x + 5y - 1) + (-2x + 3y - 4)$
$= [3x + (-2x)] + [5y + 3y] + [(-1) + (-4)] = x + 8y - 5$

Hence the given expressions are equivalent over $R \times R$.

Example 1(c) illustrates the fact that we shall often be able to replace a given expression in two variables by another equivalent one and this equivalence will be established by using the definitions and axioms we have made for the real numbers. Indeed, we could have used these real number properties to show that the expressions in Example 1(a) and (b) were equivalent without making any replacements and computing values. For example, using the distributive and commutative axioms we can write the following.

$(x - 3y)(x + 2y) = (x - 3y)x + (x - 3y)2y$
$\qquad = x^2 - 3xy + 2xy - 6y^2 = x^2 + (-3 + 2)xy - 6y^2$
$\qquad = x^2 - xy - 6y^2$

Or we may shorten the multiplication somewhat by the use of a schematic method similar to the one we used to find the standard form for the product of two first degree binomials in *one* variable.

$$\overset{-3xy}{(x - 3y)(x + 2y)} \qquad \overset{x^2}{(x - 3y)(x + 2y)}$$
$$\underset{2xy}{} \qquad \underset{-6y^2}{}$$

6.2 Equivalent Expressions

In this way we see at once that $(x - 3y)(x + 2y) = x^2 - xy - 6y^2$. Hence $(x - 3y)(x + 2y)$ and $x^2 - xy - 6y^2$ are equivalent over $R \times R$ and so they are certainly equivalent over the subset $\{(0, 1), (1, 2)\}$.

Example 2 Replace each of the given expressions by an equivalent expression as described.
(a) $(3x - 2y) + (2x - y) - (x - y)$, the sum of two terms
(b) $(4x - y)(3x - y)$, the sum of three terms
(c) $\dfrac{x - 2y}{4} - \dfrac{x - y}{3}$, a single fraction whose numerator has only two terms

Solution

(a) $(3x - 2y) + (2x - y) - (x - y)$
$= 3x - 2y + 2x - y - x + y$
$= (3x + 2x - x) + (-2y - y + y) = 4x - 2y$

(b) $(4x - y)(3x - y) = 12x^2 - 7xy + y^2$

(c) $\dfrac{x - 2y}{4} - \dfrac{x - y}{3} = \dfrac{3(x - 2y)}{3 \cdot 4} - \dfrac{4(x - y)}{4 \cdot 3}$

$= \dfrac{3x - 6y}{12} - \dfrac{4x - 4y}{12} = \dfrac{(3x - 6y) - (4x - 4y)}{12} = \dfrac{-x - 2y}{12}$

We see in Example 2 that the same techniques that we have developed for finding equivalent expressions in one variable can be used for treating expressions in two variables. And very similar techniques can be used to find factors of an expression in two variables.

Example 3 Replace each of the following by an expression in factored form.
(a) $6x^2 - 15xy + 3x$
(b) $4x^2 - 9y^2$
(c) $x^2 + 8xy + 16y^2$
(d) $2x^2 - 3xy - 2y^2$

Solution

(a) We remove the common factor: $3x(2x - 5y + 1)$
(b) The expression is a difference of squares: $(2x + 3y)(2x - 3y)$
(c) The trinomial is the square of a binomial: $(x + 4y)^2$
(d) We must resort to trial and error. The factors of $2x^2$ are $2x$ and x; the factors of $-2y^2$ are either $-2y$ and y, or $2y$ and $-y$. We find that $2x^2 - 3xy - 2y^2 = (2x + y)(x - 2y)$.

The technique of factoring expressions in two variables can be used effectively when fractions are encountered.

Example 4 Replace each of these by an equivalent fraction whose numerator and denominator have no common nontrivial factor.

(a) $\dfrac{x^2 + 3xy - 4y^2}{x - y} \cdot \dfrac{x + 3y}{x^2 + 9xy + 20y^2}$

(b) $\dfrac{2}{x^2 + 4xy + 4y^2} + \dfrac{1}{x^2 + xy - 2y^2}$

Solution

(a) $\dfrac{x^2 + 3xy - 4y^2}{x - y} \cdot \dfrac{x + 3y}{x^2 + 9xy + 20y^2}$

$= \dfrac{(x - y)(x + 4y)}{(x - y)(1)} \cdot \dfrac{(x + 3y)}{(x + 5y)(x + 4y)}$

$= \dfrac{(x + 4y)(x + 3y)}{(x + 5y)(x + 4y)} = \dfrac{x + 3y}{x + 5y}$

(b) $\dfrac{2}{x^2 + 4xy + 4y^2} + \dfrac{1}{x^2 + xy - 2y^2}$

$= \dfrac{2}{(x + 2y)^2} + \dfrac{1}{(x + 2y)(x - y)}$

$= \dfrac{2(x - y)}{(x + 2y)^2(x - y)} + \dfrac{(1)(x + 2y)}{(x + 2y)(x - y)(x + 2y)}$

$= \dfrac{2(x - y) + (x + 2y)}{(x + 2y)^2(x - y)} = \dfrac{3x}{(x + 2y)^2(x - y)}$

Exercises 6.2

For Exercises 1–8 find an equivalent expression in which there are as few terms as possible.

1. $(2x + y) + (x - 2y)$
2. $(-3x + 2y) + (x - 5y)$
3. $(x + 3y) - (2x + y)$
4. $(-2x + y) - (x - y)$
5. $(3x - 5y) + (x + 5y)$
6. $(4x + y + 3) + (-4x + 2y - 1)$
7. $(4x - 7y) - (-3x - 7y)$
8. $(x - 2y + 4) - (x + 2y + 4)$

For Exercises 9–16 replace each product with an equivalent sum with as few terms as possible.

6.2 Equivalent Expressions

9. $(x + 2y)(x - y)$
10. $(x + 5y)(x - 5y)$
11. $(2x + y)^2$
12. $(x - 3y)^2$
13. $(3x + y)(x - 4y)$
14. $x(x - y)^2$
15. $(3x - y)(x - 2y)$
16. $y(x + 3y)(x + y)$

For Exercises 17–28 replace each sum with an equivalent product which is completely factored.

17. $x^2 - y^2$
18. $3xy + y^2$
19. $x^2 + 2xy - 3y^2$
20. $x^3 + 2x^2y$
21. $4x^2 - 7xy - 2y^2$
22. $4x^2 - 8xy + 3y^2$
23. $2x^2y - xy^2$
24. $x^2 - 8xy + 16y^2$
25. $x^3 - 4xy^2$
26. $x^2y + 4xy^2 + 4y^3$
27. $x^3 + x^2y + 2x + 2y$
28. $x^3 - 2x^2y - xy^2 + 2y^3$

For Exercises 29–44 find a single fraction which is equivalent to the given expression and simplify this fraction if possible by canceling factors.

29. $\dfrac{-12xy^2}{3xy} \cdot \dfrac{x^2}{4y^3}$

30. $\dfrac{5x^3}{-8y^2} \div \dfrac{-10x^2y}{4y^4}$

31. $\dfrac{x^2 - y^2}{x^2 + 3xy + 2y^2}$

32. $\dfrac{3x^2 - 7xy + 2y^2}{4x^2 - 7xy - 2y^2}$

33. $\dfrac{x^2 + xy - 2y^2}{x^2 + 4xy + 4y^2} \cdot \dfrac{x^2 - xy - 6y^2}{x^2 - 4xy + 3y^2}$

34. $\dfrac{4x^2 - 9y^2}{2x^2 + xy - 3y^2} \div \dfrac{x^2 + 10xy + 25y^2}{x^2 + 4xy - 5y^2}$

35. $\dfrac{x^3 + 2x^2y + xy^2}{x^2y - 2xy^2 + y^3} \cdot \dfrac{x^2y - xy^2}{x^3 + x^2y}$

36. $\dfrac{-4x^3 - 10x^2y + 6xy^2}{16x^2y - 4y^3} \cdot \dfrac{4x^2y + 2xy^2}{x^4 + 3x^3y}$

37. $\dfrac{x - y}{3} + \dfrac{x + y}{6}$

38. $\dfrac{x + y}{x} - \dfrac{y + 1}{y}$

39. $\dfrac{x - 1}{x^2y} + \dfrac{1 - y}{xy^2}$

40. $\dfrac{2x + 1}{2xy} + \dfrac{y + 1}{y^2}$

41. $\dfrac{x - y}{x + y} + \dfrac{x + 2y}{2x + 2y}$

42. $\dfrac{2xy}{x^2 - y^2} + \dfrac{x - y}{x + y}$

43. $\dfrac{x - 1}{x^2 + 2x} + \dfrac{1 - y}{xy + 2y}$

44. $\dfrac{2}{x^2 + xy - 2y^2} + \dfrac{1}{x^2 - 2xy + y^2}$

For Exercises 45–52 use the theorems on exponents to replace each of the following with a simpler equivalent expression with positive exponents.

45. $(2xy^2)^3$

46. $\dfrac{(2xy)^2}{x^2 y^3}$

47. $xy(x^{-1} + y^{-1})$

48. $\dfrac{x^{-1} y}{xy^{-2}} \cdot \dfrac{-x^{-2} y}{xy^{-1}}$

49. $(x^{-1} y^{-3})^{-2}$

50. $\left(\dfrac{x^{-1} \cdot y}{2x^{-2} y^{-1}} \right)^{-1}$

51. $\dfrac{x^{-1} + y^{-1}}{x + y}$

52. $(x^{-1} + y^{-1})^{-1}$

For Exercises 53–54 use "long division" to find the quotient

53. $\dfrac{x^3 - x^2 y - 4xy^2 - 6y^3}{x - 3y}$

54. $\dfrac{x^3 + y^3}{x + y}$

For Exercises 55–56 factor the given expression completely.

55. $x^3 - y^3$ [Hint: One of the factors is $x - y$.]
56. $x^3 + 27y^3$ [Hint: One of the factors is $x + 3y$.]

6.3 Equations over R × R

When the domain of an equation in two variables is a *finite* set of ordered pairs, the solution set may easily be found by making all the indicated replacements from the domain. But when the domain is infinite this clearly *cannot* be done. The set $\{1\} \times R$ is an infinite set of ordered pairs each of which has for its first component the number 1, while its second component may be any real number. Then $(1, 3)$, $(1, 0)$, $(1, -5/2)$ and $(1, \sqrt{2})$ are all members of $\{1\} \times R$. Example 1 shows how to find the solution set for an equation whose domain is $\{1\} \times R$.

Example 1 Find the solution set of $x - 2y = 5$, $\{1\} \times R$.

 Solution Since each member of $\{1\} \times R$ has 1 for its *first* component, we must replace x by 1 in the given equation. Then $x - 2y = 5$ becomes

6.3 Equations over $R \times R$

$1 - 2y = 5$. But this is an open sentence in *one* variable which we may solve by the methods of Chapter 3.

$$1 - 2y = 5$$
$$-2y = 4$$
$$y = -2$$

Hence, the *only* replacement for y which makes a true statement is -2. We conclude that the solution set of the given equation is $\{(1, -2)\}$.

We may always use this technique when the domain of the equation is the Cartesian product of two sets with one of them *finite*.

Example 2   Find the solution set for each of these equations over the indicated domain.

(a) $3x + y = -2$, $R \times \{4\}$ (b) $-x + 2y = 4$, $\{0\} \times R$
(c) $x - 2y = 5$, $\{-3, 3\} \times R$

Solution

(a) Since the domain is $R \times \{4\}$, a set of ordered pairs each of whose *second* components is 4, we must replace y by 4.

$$3x + y = -2$$
$$3x + 4 = -2$$
$$3x = -6$$
$$x = -2$$

Hence the solution set we seek is $\{(-2, 4)\}$.

(b) We replace x by 0, since the domain is $\{0\} \times R$.

$$-x + 2y = 4$$
$$-0 + 2y = 4$$
$$2y = 4$$
$$y = 2$$

The solution set is $\{(0, 2)\}$.

(c) The domain is $\{-3, 3\} \times R$, a set of ordered pairs in which the first component may be *either* -3 or 3. First we shall replace x by -3.

$$x - 2y = 5$$
$$-3 - 2y = 5$$

$$-2y = 8$$
$$y = -4 \quad \{(-3, -4)\}$$

Then we shall replace x by 3.
$$x - 2y = 5$$
$$3 - 2y = 5$$
$$-2y = 2$$
$$y = -1 \quad \{(3, -1)\}$$

Finally, the solution set we seek is the *union* of these two sets: $\{(-3, -4)\} \cup \{(3, -1)\} = \{(-3, -4), (3, -1)\}$.

We have agreed that the domain of an expression in two variables will be all those members of $R \times R$ for which the expression has a real value, unless we explicitly state otherwise. Then often the domain of an equation in two variables will be the entire infinite set $R \times R$. Usually the solution set for such an equation will be an *infinite* subset of $R \times R$ and we will *not* be able to list each of its members. However, we shall use set notation similar to that in Section 3.5 to *indicate* its solution set. Thus

$$\{(x, y): x - 2y = 5\}$$

which we read "the set of all ordered pairs x, y such that $x - 2y = 5$," indicates the solution set of the equation $x - 2y = 5$ with domain $R \times R$. While we cannot list all the members of this set, it is sometimes helpful to find *some* of them as Example 3 illustrates.

Example 3 Find that *subset* of $\{(x, y): x - 2y = 5\}$ which is the solution set of $x - 2y = 5$, $\{-5, -3, -1, 1, 3, 5\} \times R$ and draw its graph.

Solution We shall make each of the replacements for x from the set $\{-5, -3, -1, 1, 3, 5\}$ in the equation $x - 2y = 5$.

$x = -5$: $-5 - 2y = 5$ $\qquad$ $x = 1$: $1 - 2y = 5$
$\qquad -2y = 10$ $\qquad\qquad\qquad -2y = 4$
$\qquad\qquad y = -5 \quad \{(-5, -5)\}$ $\qquad\qquad y = -2 \quad \{(1, -2)\}$

$x = -3$: $-3 - 2y = 5$ $\qquad$ $x = 3$: $3 - 2y = 5$
$\qquad -2y = 8$ $\qquad\qquad\qquad -2y = 2$
$\qquad\qquad y = -4 \quad \{(-3, -4)\}$ $\qquad\qquad y = -1 \quad \{(3, -1)\}$

$x = -1$: $-1 - 2y = 5$ $\qquad$ $x = 5$: $5 - 2y = 5$
$\qquad -2y = 6$ $\qquad\qquad\qquad -2y = 0$

6.3 Equations over $R \times R$

$$y = -3 \quad \{(-1, -3)\} \qquad y = 0 \quad \{(5, 0)\}$$

The set we seek is the *union* of the six sets we have found and is shown in Figure 6.4.

$$\{(-5, -5), (-3, -4), (-1, -3), (1, -2), (3, -1), (5, 0)\} \qquad \{(x, y): x - 2y = 5\}$$

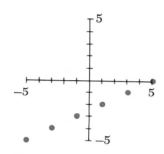

Figure 6.4

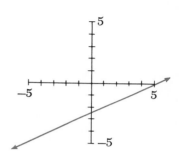

Figure 6.5

Since the graph of the *subset* of $\{(x, y): x - 2y = 5\}$ which we found in Example 3 is a set of points which all lie in a straight line (see Figure 6.4), we shall *assume* that the graph of $\{(x, y): x - 2y = 5\}$ itself is the entire line which contains these points. See Figure 6.5.

Example 4 Find six points on the graph of $\{(x, y): x + y = 3\}$ by finding the solution set of $x + y = 3$ first over the domain $\{-3, 0, 2\} \times R$ and then over the domain $R \times \{-2, 0, 4\}$.

Solution

$x = -3: \ -3 + y = 3$
$\qquad\qquad y = 6 \quad \{(-3, 6)\}$
$x = \ \ \ 0: \ \ \ 0 + y = 3$
$\qquad\qquad y = 3 \quad \{(0, 3)\}$
$x = \ \ \ 2: \ \ \ 2 + y = 3$
$\qquad\qquad y = 1 \quad \{(2, 1)\}$

$y = -2: \ x - 2 = 3$
$\qquad\qquad x = 5 \quad \{(5, -2)\}$
$y = \ \ \ 0: \ x + 0 = 3$
$\qquad\qquad x = 3 \quad \{(3, 0)\}$
$y = \ \ \ 4: \ x + 4 = 3$
$\qquad\qquad x = -1 \quad \{(-1, 4)\}$

The six points we have found are shown in Figure 6.6.

Again, we see that the points on the graph of Example 4 all lie on the same straight line. We shall assume that the graph of the entire set $\{(x, y): x + y = 3\}$ is the straight line which contains those six points. See Figure 6.7. In more advanced books it is proved that this is a generally true fact about the graph for solution sets of certain types of equations in two variables. Although we shall not try to prove it here, we shall state that result in the next theorem.

$\{(-3, 6), (-1, 4), (0, 3),$
$(2, 1), (3, 0), (5, -2)\}$

$\{(x, y): x + y = 3\}$

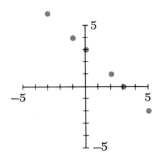

Figure 6.6

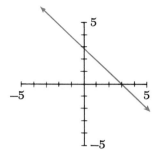

Figure 6.7

Theorem 1 If a and b are not both 0, then the graph of $\{(x, y): ax + by = c\}$ is a straight line

Example 5 Find the graph of each of these sets.
(a) $\{(x, y): x - 4y = -4\}$ (b) $\{(x, y): x = 2\}$
(c) $\{(x, y): y = -3\}$

Solution

(a) By Theorem 1 the graph will be a straight line. Then we need only find two points to determine the line. To find the solution set of $x - 4y = 4$ over $\{0\} \times R$ we replace x by 0:

$$0 - 4y = -4$$
$$y = 1 \quad \{(0, 1)\}$$

To find the solution set over $R \times \{0\}$ we replace y by 0.

$$x - 4 \cdot 0 = -4$$
$$x = -4 \quad \{(-4, 0)\}$$

The graph is shown in Figure 6.8.

(b) By Theorem 1 the graph is a line, for we may write the equation $x = 2$ as $x + 0 \cdot y = 2$. We shall first find the solution set of $x + 0 \cdot y = 2$ over $R \times \{0, 3\}$. We replace y by 0:

$$x + 0 \cdot 0 = 2$$
$$x = 2 \quad \{(2, 0)\}$$

Then we replace y by 3:

$$x + 0 \cdot 3 = 2$$
$$x = 2 \quad \{(2, 3)\}$$

Note that the graph of $\{(x, y): x = 2\}$ as shown in Figure 6.9 is a *vertical* line.

(c) The graph is a line, for we may write the equation $y = -3$ as $0 \cdot x + y = -3$. We first find the solution set over $\{0, 3\} \times R$. We replace x by 0.

$$0 \cdot 0 + y = -3$$
$$y = -3 \quad \{(0, -3)\}$$

We replace x by 3:

$$0 \cdot 3 + y = -3$$
$$y = -3 \quad \{(3, -3)\}$$

The graph of $\{(x, y): y = -3\}$ is a *horizontal* line. See Figure 6.10.

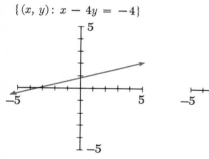

Figure 6.8

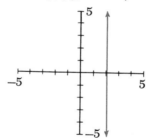

Figure 6.9

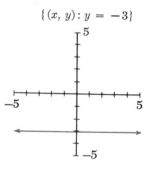

Figure 6.10

Each point in the coordinate plane corresponds to an ordered pair of real numbers. We shall often refer to the first component of this pair as the **x-coordinate** of the point and the second component as the **y-coordinate** of the point. This is a reasonable procedure since when a point is on the graph of the solution set of an open sentence its first coordinate is the replacement for x and its second coordinate is the replacement for y which result in a true statement. We shall also refer to the horizontal axis as the **x-axis** and the vertical axis as the **y-axis** and the coordinate plane as the **xy-plane**.

In Example 5(a) we showed that the graph of $\{(x, y): x - 4y = -4\}$ intersects the x-axis at $(-4, 0)$. Let us call -4 the x-intercept of this graph. Similarly, we shall call 1 the y-intercept of the graph since the graph intersects the y-axis at $(0, 1)$, a point whose y-coordinate is 1. See Figure 6.8. Often we will want to know both the x-intercept and the y-intercept of the graph and use these to draw the graph when it is a line.

Example 6 Find the x- and y-intercepts and draw the graph of the solution set of each of these equations.

(a) $x + 2y = 4$ (b) $x + 3y = 0$

Solution

(a) The x-intercept is the x-coordinate of a point on the x-axis. But *every* point on the x-axis has y-coordinate 0. Then we shall replace y by 0.

$$x + 2y = 4$$
$$x + 2 \cdot 0 = 4$$
$$x = 4$$

The x-intercept is 4. The y-intercept is the y-coordinate of a point on the y-axis. But *every* point on the y-axis has x-coordinate 0. Then we shall replace x by 0.

$$x + 2y = 4$$
$$0 + 2y = 4$$
$$y = 2$$

The y-intercept is 2. The graph is shown in Figure 6.11.

(b) To find the x-intercept we replace y by 0.

$$x + 3y = 0$$
$$x + 3 \cdot 0 = 0$$
$$x = 0$$

The x-intercept is 0. To find the y-intercept we replace x by 0.

$$x + 3y = 0$$
$$0 + 3y = 0$$
$$y = 0$$

The y-intercept is also 0. The line intersects both axes at the origin, hence, we must find another point to draw the line. We replace y by 1.

$$x + 3 \cdot 1 = 0$$
$$x = -3 \quad (-3, 1)$$

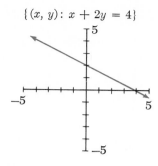

Figure 6.11

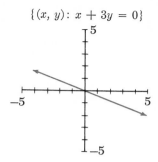

Figure 6.12

The graph is shown in Figure 6.12.

Often we will shorten the phrase "the graph of the solution set of an open sentence" to simply "the graph of the open sentence." We should observe that Theorem 1 does *not* suggest that the graph of *every* equation in two variables is a line. While the equation $2x - 3y = 5$ is of the form $ax + by = c$ and so its graph is a line, the equation $x^2 + y = 4$ is *not* of this form and its graph is *not* a line. We will study the graphs of such equations in Chapter 7.

Sometimes we will want to find the intersection of two graphs.

Example 7 Find the intersection: $\{(x, y): x + 2y = 4\} \cap \{(x, y): 2x - y = 3\}$.

Solution In Example 6(a) we found the graph of $\{(x, y): x + 2y = 4\}$. See Figure 6.11. We shall find the graph of $\{(x, y): 2x - y = 3\}$ by first finding the intercepts. Replace y by 0:

$$2x - y = 3$$
$$2x - 0 = 3$$
$$x = \tfrac{3}{2}$$

The x-intercept is $\tfrac{3}{2}$. Replace x by 0:

$$2x - y = 3$$
$$2 \cdot 0 - y = 3$$
$$y = -3$$

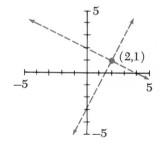

Figure 6.13

The y-intercept is -3. The graph of the intersection is shown in Figure 6.13. Note that the intersection is $\{(2, 1)\}$ and that if we replace x by 2 and y by 1 in *each* of the equations $x + 2y = 4$ and $2x - y = 3$, *true* statements will result.

Exercises 6.3

For Exercises 1–12 find the solution set of the given equation over the indicated domain.

1. $x + 3y = 7$, $\{1\} \times R$
2. $2x - y = -3$, $\{2\} \times R$
3. $2x + 3y = 1$, $R \times \{1\}$
4. $3x - y = 4$, $R \times \{2\}$
5. $5x + y = -3$, $\{0\} \times R$
6. $3x + 2y = 5$, $R \times \{0\}$
7. $2x + y = 3$, $\{0, 1\} \times R$
8. $2x - 3y = 1$, $R \times \{0, 1\}$
9. $3x - y = 0$, $R \times \{0, 2\}$
10. $x + 4y = 0$, $\{0, 4\} \times R$
11. $y = 3x + 5$, $\{-1, 0, 1\} \times R$
12. $y = 2x - 6$, $R \times \{-2, 0, 2\}$

For Exercises 13–20 find the x- and y-intercepts of the graph of the given equation with domain $R \times R$.

13. $2x + y = 4$
14. $x - 3y = 6$
15. $2x - y = 5$
16. $3x + 2y = -4$
17. $2x + 3y = 0$
18. $y = 3x - 1$
19. $x = 2y + 5$
20. $x + y = 3x - 2y + 1$

For Exercises 21–44 find the graph of the indicated set.

21. $\{(x, y): x + y = 2\}$
22. $\{(x, y): x - y = 3\}$
23. $\{(x, y): x + y = -1\}$
24. $\{(x, y): x - y = -4\}$
25. $\{(x, y): 2x + y = 6\}$
26. $\{(x, y): x - 3y = 3\}$
27. $\{(x, y): 2x + 3y = 6\}$
28. $\{(x, y): 3x - 2y = -6\}$
29. $\{(x, y): x + 3y = 2\}$
30. $\{(x, y): 3x - y = -2\}$
31. $\{(x, y): 2x - 5y = 5\}$
32. $\{(x, y): 3x + 4y = 6\}$
33. $\{(x, y): x = 4\}$
34. $\{(x, y): x = -1\}$
35. $\{(x, y): y = 2\}$
36. $\{(x, y): y = -2\}$
37. $\{(x, y): x = 0\}$
38. $\{(x, y): y = 0\}$
39. $\{(x, y): x + y = 0\}$
40. $\{(x, y): 2x + 5y = 0\}$
41. $\{2\} \times R$
42. $\{-1\} \times R$
43. $R \times \{1\}$
44. $R \times \{-2\}$

For Exercises 45–52 use graphs to find the indicated intersection.

45. $\{(x, y): x + y = 1\} \cap \{(x, y): x - y = 1\}$

46. $\{(x, y): x + 2y = 4\} \cap \{(x, y): x - y = -2\}$
47. $\{(x, y): x + 3y = 4\} \cap \{(x, y): 2x - y = 1\}$
48. $\{(x, y): x + y = -1\} \cap \{(x, y): x + 2y = 0\}$
49. $\{(x, y): 2x + y = 5\} \cap \{(x, y): y = 3\}$
50. $\{(x, y): x + 3y = 1\} \cap \{(x, y): x = 4\}$
51. $\{(x, y): 2x + y = 2 \text{ and } y = 4\}$
52. $\{(x, y): 2x + 3y = 5 \text{ and } x = 1\}$

6.4 Intersection of Solution Sets

Let us agree that two equations (in two variables) are equivalent over a domain if they have exactly the same solution set over that domain. Often we shall want to replace one equation by an equivalent one over the domain $R \times R$. We shall use axioms analogous to the substitution, addition, and multiplication axioms of Chapter 3 to accomplish this, as is illustrated by Example 1.

Example 1 Find the graph of each of the following.
(a) $y = 3x - 3$ (b) $2(x - y) = 4 - 3y$

Solution

(a)
$$y = 3x - 3$$
$$y + (-3x) = (3x - 3) + (-3x) \qquad \text{[Addition axiom]}$$
$$-3x + y = -3 \qquad \text{[Substitution axiom]}$$

This is an equation of the form $ax + by = c$, and so by Theorem 1 its graph is a line. We shall find the intercepts of this line. Replace y by 0:

$$-3x + y = -3$$
$$-3x + 0 = -3$$
$$x = 1$$

The x-intercept is 1. Replace x by 0:

$$-3x + y = -3$$
$$-3 \cdot 0 + y = -3$$
$$y = -3$$

The y-intercept is -3. Since $-3x + y = -3$ is equivalent to the original equation $y = 3x - 3$, their solution sets are equal and they have the same graph. The graph is shown in Figure 6.14.

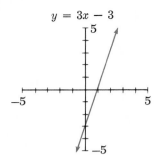

Figure 6.14   Figure 6.15

(b) $\quad 2(x - y) = 4 - 3y$
$\quad\quad 2x - 2y = 4 - 3y$ [Substitution axiom]
$\quad (2x - 2y) + 3y = (4 - 3y) + 3y$ [Addition axiom]
$\quad\quad 2x + y = 4$ [Substitution axiom]

The graph is shown in Figure 6.15.

As Example 1 shows, there are many equations which are equivalent to one of the form $ax + by = c$. Since the graph of any such equation is a line, let us agree to call them **linear** equations. And when a linear equation is written in the form $ax + by = c$, we will say that it is in standard form. We shall usually want a, b, and c to be integers.

Example 2 Find a standard form for $\dfrac{x}{3} = \dfrac{y}{7} + 1$.

Solution

$$\frac{x}{3} = \frac{y}{7} + 1$$

The LCM of 3 and 7 is 21.

$$21\left(\frac{x}{3}\right) = 21\left(\frac{y}{7} + 1\right)$$
$$7x = 3y + 21$$
$$7x - 3y = 21$$

As we saw in Section 6.3, it is sometimes possible to find the intersection of two solution sets by drawing their graphs. In Example 7 of that section we found $\{(2, 1)\}$ to be the intersection of $\{(x, y): x + 2y = 4\}$ and $\{(x, y): 2x - y = 3\}$. This result, which was obtained from Figure 6.13, may be easily verified by making the replacements $x = 2$ and $y = 1$, in

6.4 Intersection of Solution Sets

both of the equations $x + 2y = 4$ and $2x - y = 3$. However, this graphing technique will not always lead to the *exact* intersection of two solution sets.

Example 3 Use graphs to find $\{(x, y): 5x - y = -5 \text{ and } 2x + 3y = 6\}$.

Solution We shall attempt to find the intersection of the graphs of the given equations. For $5x - y = -5$ the x- and y-intercepts are found to be -1 and 5, respectively. For $2x + 3y = 6$ the intercepts are 3 and 2. The graphs are shown in Figure 6.16. It is not clear exactly which ordered pair corresponds to the intersection point, but it seems to be *approximately* $(-\frac{1}{2}, \frac{5}{2})$.

Let us determine if it is *true* that $\{(x, y): 5x - y = -5 \text{ and } 2x + 3y = 6\} = \{(-\frac{1}{2}, \frac{5}{2})\}$. We shall make the indicated replacement in *each* of the given equations.

$$5x - y = -5 \qquad\qquad 2x + 3y = 6$$
$$5(-\tfrac{1}{2}) - (\tfrac{5}{2}) = -5 \qquad\qquad 2(-\tfrac{1}{2}) + 3(\tfrac{5}{2}) = 6$$
$$(-\tfrac{5}{2}) - (\tfrac{5}{2}) = -5 \qquad\qquad -1 + (\tfrac{15}{2}) = 6$$
$$-5 = -5, \text{ True} \qquad\qquad \tfrac{13}{2} = 6, \text{ False}$$

Then we have *not* found the required set. This is not because of a mistake we have committed. Rather it is because of our difficulty in accurately determining the ordered pair corresponding to the intersection of the two lines of Figure 6.16. When the coordinates of the required point are not integers, the graphical method is often impractical. We shall develop an alternate method which does not use graphs to find these coordinates and then complete the solution to the problem of Example 3.

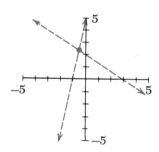

Figure 6.16

Example 4 Find $\{(x, y): 2x + 3y = 1 \text{ and } y = 2\}$.

Solution Since we seek replacements for x and y which make *both* of the equations true, it is obvious that the only desired replacement for y is 2. Then

$$2x + 3y = 1$$
$$2x + 3 \cdot 2 = 1$$
$$x = -\tfrac{5}{2}$$

The set we seek is $\{(-\tfrac{5}{2}, 2)\}$.

Example 5 Find $\{(x, y): x - y = -4 \text{ and } 2x + y = 1\}$.

Solution We seek replacements for x and y which will give the value -4 to $x - y$ and the value 1 to $2x + y$ (so that the given equations will both become true statements). Then for such a replacement the expression $(x - y) + (2x + y)$, which is the sum of the given expressions, must have value $(-4) + (1) = -3$. But $(x - y) + (2x + y)$ is equivalent to $3x$. Hence, $3x$ must have value -3, so the required replacement for x is -1. To find the replacement for y we make the replacement $x = -1$ in *either* of the original equations.

$$x - y = -4$$
$$-1 - y = -4$$
$$y = 3$$

Then the set we seek is $\{(-1, 3)\}$, as is easily verified by making the indicated replacements for x and y in both $x - y = -4$ and $2x + y = 1$.

Example 6 Find $\{(x, y): 2x = 3 - y \text{ and } 3y = 2x - 7\}$.

Solution First we shall find a standard form for each equation.

$$2x = 3 - y \qquad\qquad 3y = 2x - 7$$
$$2x + y = 3 \qquad\qquad -2x + 3y = -7$$

Then we shall add the left and right expressions.

$$2x + y = 3$$
$$\underline{-2x + 3y = -7}$$
$$4y = -4$$
$$y = -1$$

We make the replacement $y = -1$ in either of the original equations.

$$2x = 3 - y$$
$$2x = 3 - (-1)$$
$$x = 2 \qquad\qquad \{(2, -1)\}$$

Example 7 Find each of these sets:
(a) $\{(x, y): x - 4y = -3 \text{ and } 2x + y = 3\}$
(b) $\{(x, y): 3x + 2y = 8 \text{ and } 5x - 3y = 7\}$

Solution

(a) First we try the "addition" method of Example 6.

$$x - 4y = -3$$
$$\underline{2x + y = 3}$$
$$3x - 3y = 0$$

But this is not helpful, since it is still not clear what the replacement for either x or y should be. Let us replace the second equation $2x + y = 3$ by an equivalent one by multiplying each expression by 4 and then use the "addition" technique.

$$2x + y = 3$$
$$4(2x + y) = 4(3)$$
$$8x + 4y = 12$$

$$\begin{array}{ll} x - 4y = -3 & \text{[Original first equation]} \\ 8x + 4y = 12 & \text{[Equivalent to second equation]} \\ \hline 9x = 9 \\ x = 1 \end{array}$$

Note that since the coefficients, 4 and -4, are additive inverses, the y is eliminated when we add. Then we replace x by 1 in either of the original equations.

$$2x + y = 3$$
$$2 \cdot 1 + y = 3$$
$$y = 1 \qquad \{(1, 1)\}$$

(b) To eliminate the y we shall want the coefficients of y to be additive inverses. Hence we shall multiply $3x + 2y = 8$ by 3 (so that the coefficient of y will be 6), and multiply $5x - 3y = 7$ by 2 (so that the coefficient of y will be -6).

$$\begin{array}{ll} 3x + 2y = 8 & 5x - 3y = 7 \\ 3(3x + 2y) = 3(8) & 2(5x - 3y) = 2(7) \\ 9x + 6y = 24 & 10x - 6y = 14 \end{array}$$

$$\begin{array}{l} 9x + 6y = 24 \\ 10x - 6y = 14 \\ \hline 19x = 38 \\ x = 2 \end{array}$$

Finally, we shall replace x by 2 in either of the original equations.

$$3x + 2y = 8$$
$$3 \cdot 2 + 2y = 8$$
$$y = 1 \qquad \{(2, 1)\}$$

The method of finding the intersection of the solution sets which we have been considering can always be employed. We shall refer to it as the addition method. The following table summarizes its important steps.

244 *Expressions with Two Variables*

1. First use the axioms to replace each equation by an equivalent one in standard form, $ax + by = c$.
2. Use the multiplication axiom on one or both equations so that the two coefficients of x or of y are additive inverses.
3. Add, so as to obtain an equation in which only one variable appears.
4. Find the replacement for this variable.
5. Return to either of the original equations to find the replacement for the other variable.

We now can return to Example 3.
$\{(x, y): 2x + 3y = 6 \text{ and } 5x - y = -5\}$

$$5x - y = -5$$
$$3(5x - y) = 3(-5)$$
$$15x - 3y = -15$$

$$\begin{aligned} 2x + 3y &= 6 \\ 15x - 3y &= -15 \\ \hline 17x &= -9 \\ x &= -\tfrac{9}{17} \end{aligned}$$

$$5x - y = -5$$
$$5(-\tfrac{9}{17}) - y = -5$$
$$-\tfrac{45}{17} - y = -5$$
$$-y = -5 + \tfrac{45}{17} = -\tfrac{40}{17}$$
$$y = \tfrac{40}{17} \qquad \{(-\tfrac{9}{17}, \tfrac{40}{17})\}$$

Note that we obtain the (incorrect) pair $(-\tfrac{1}{2}, \tfrac{5}{2})$ from the graph (Figure 6.16) and that this is *approximately* the correct pair $(-\tfrac{9}{17}, \tfrac{40}{17})$.

Some problems can be solved by finding the intersection of two solution sets.

Example 8 The sum of the digits in a two digit natural number is 12. If the two digits are interchanged, the resulting number is 15 times the tens digit of the original number. Find the original number.

Solution We shall let x represent the tens digit and y the ones digit of the required number. Then it is given that

$$x + y = 12$$

Any two digit number is equal to 10 times its tens digit plus 1 times its ones digit. For example, $94 = 10 \cdot 9 + 1 \cdot 4$. Then our number equals

$10x + y$. But when its digits are reversed, the new number equals $10y + x$. Then
$$10y + x = 15x$$
since we are given that the resulting number is 15 times the tens digit of the original number. We shall find the set $\{(x, y): x + y = 12 \text{ and } 10y + x = 15x\}$.

We observe that the standard form of
$$10y + x = 15x$$
is
$$-7x + 5y = 0$$

We shall multiply the left and right expressions in $x + y = 12$ by 7 and then eliminate the variable x:

$$\begin{aligned} 7x + 7y &= 84 \\ -7x + 5y &= 0 \\ \hline 12y &= 84 \\ y &= 7 \end{aligned}$$

And if we replace y by 7 in $x + y = 12$, we have $x = 5$. Hence, the solution set is $\{(5, 7)\}$ and the number we seek is 57.

Exercises 6.4

For Exercises 1–8 find a standard form for the given equation and draw its graph.

1. $y = x + 2$
2. $y = 2x - 4$
3. $x = 5 - 5y$
4. $x = y + 3$
5. $2x + y = x - y + 4$
6. $3(x - y) + 1 = x - 2y - 1$
7. $\dfrac{x}{2} + \dfrac{y}{3} = 1$
8. $\dfrac{x - y}{3} = \dfrac{x + y}{2}$

For Exercises 9–28 find the indicated set.

9. $\{(x, y): 2x + 3y = 1 \text{ and } y = 1\}$
10. $\{(x, y): x - 4y = 0 \text{ and } y = -1\}$
11. $\{(x, y): 2x + y = 5 \text{ and } x = 1\}$
12. $\{(x, y): 3x - 2y = 4 \text{ and } x = 3\}$
13. $\{(x, y): x - y = 3 \text{ and } x + y = 1\}$
14. $\{(x, y): 2x + y = -3 \text{ and } x - y = 0\}$

15. $\{(x, y): x + 2y = 7 \text{ and } -x + y = 2\}$
16. $\{(x, y): -3x + y = 13 \text{ and } 3x + 8y = -4\}$
17. $\{(x, y): 3x + y = 1 \text{ and } 2x + y = 0\}$
18. $\{(x, y): 4x + 3y = 6 \text{ and } 5x + 3y = 6\}$
19. $\{(x, y): x - 2y = 7 \text{ and } -2x + 3y = -11\}$
20. $\{(x, y): 5x + y = -4 \text{ and } x - 3y = -4\}$
21. $\{(x, y): x + 3y = 1 \text{ and } 2x + y = -3\}$
22. $\{(x, y): 2x + 4y = 5 \text{ and } 3x - y = -3\}$
23. $\{(x, y): 2x - 3y = 0 \text{ and } 5x - 2y = -11\}$
24. $\{(x, y): 2x - 4y = 7 \text{ and } 3x - 5y = 9\}$
25. $\{(x, y): 4x + 5y = 2 \text{ and } -5x + 2y = 3\}$
26. $\{(x, y): 2x - 3y = 9 \text{ and } 5x + 2y = -6\}$
27. $\{(x, y): 2x + 5y = 0 \text{ and } 3x - 2y = 0\}$
28. $\{(x, y): 5x - 3y = 0 \text{ and } 4x + 5y = 0\}$

For Exercises 29–36 find a standard form for each equation and use the addition method to find the intersection of their solution sets.

29. $2x + y = 1 + x$
    $2x - y = 6 + y$

30. $x + 3y = 2x + 9$
    $2(x - y) = y - 12$

31. $3(x + 1) = 2(y + 1)$
    $y + 3 = 2(x + 2)$

32. $3(x + 2) = x$
    $3y + 2x = 0$

33. $2(x + y) + 5 = 5(x + 1)$
    $x + y - 2 = 2(y - 1)$

34. $\dfrac{x}{2} - \dfrac{3y}{4} = -2$
    $\dfrac{y}{2} + x = 4$

35. $\dfrac{x + y}{4} = \dfrac{x - y}{8}$
    $x - 3y = 6$

36. $\dfrac{2x}{3} = \dfrac{y}{4}$
    $\dfrac{x}{2} + \dfrac{y}{2} = 0$

For Exercises 37–44 use two variables and two equations to solve the stated problems.

37. The sum of two numbers is 12 and their difference is 4. What are the numbers?
38. The sum of two numbers is 9. The first plus three times the second is 3 more than the second. What are they?
39. Three more than twice one number is 4 less than a second number.

Twice their sum is 3 more than the second number. What are the numbers?

40. There are 31 students in an algebra class. If there were 3 more girls and 3 fewer boys, there would still be one more boy than girl. How many boys and how many girls are in the class?
41. The sum of the digits in a two digit natural number is 11. If the digits are reversed, the resulting number is 63 larger than before. What is the number?
42. The difference of the digits in a two digit number is one less than the ones digit. The number itself is 45 more than the sum of its digits. What is the number?
43. The sum of the ages of Mary and Bill is 18 years. In three years the sum of their ages will be twice Mary's age now. How old is each now?
44. An airplane traveled 2000 miles in 4 hours with the wind, but the return trip against the wind took 5 hours. What is the speed of the plane in still air and the speed of the wind?

6.5 Inequalities Over $R \times R$

While the addition method of the last section may always be used to find the intersection of the solution sets of two linear equations, there are times when this intersection is rather special, and we must interpret the result of our method with care.

Example 1 Find $\{(x, y): 2x - 6y = 6 \text{ and } x = 3y - 6\}$.

Solution

$x = 3y - 6$	$2x - 6y = 6$	[First equation]
$x - 3y = -6$	$-2x + 6y = 12$	[Equivalent to the second equation]
$(-2)(x - 3y) = (-2)(-6)$	$0 = 18$	
$-2x + 6y = 12$		

Since $0 = 18$ is a *false* numerical statement (and so certainly false for *any* replacement of the variables from $R \times R$), we must conclude that the set we are seeking is empty; $\{(x, y): 2x - 6y = 6 \text{ and } x = 3y - 6\} = \emptyset$. The graphs of the equations $2x - 6y = 6$ and $x = 3y - 6$ are shown in Figure 6.17. Observe that these graphs are parallel lines—they do not meet. Hence, their intersection is, indeed, empty as we had concluded from the addition method.

Example 2 Find $\{(x, y): 3x = 15 - 5y \text{ and } 6x + 10y - 30 = 0\}$.

Solution

$$3x = 15 - 5y \qquad 6x + 10y - 30 = 0$$

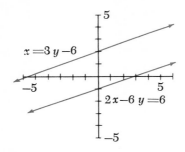

Figure 6.17	Figure 6.18

$$3x + 5y = 15 \qquad 6x + 10y = 30$$
$$\left(-\tfrac{1}{2}\right)(6x + 10y) = \left(-\tfrac{1}{2}\right)(30)$$
$$-3x - 5y = -15$$

$$\begin{aligned} 3x + 5y &= 15 \\ -3x - 5y &= -15 \\ \hline 0 &= 0 \end{aligned}$$

[Equivalent to first equation]
[Equivalent to second equation]

Since $0 = 0$ is a true numerical statement, does this imply that *every* replacement for x and y will make $3x = 15 - 5y$ and $6x + 10y - 30 = 0$ true? To see that the answer is "no" we need only try the pair $(0, 0)$. Neither $3 \cdot 0 = 15 - 5 \cdot 0$ nor $6 \cdot 0 + 10 \cdot 0 - 30 = 0$ is true. But there are infinitely many ordered pairs which do result in true statements. Actually, we may show that the original pair of equations are equivalent and so have the same solution set.

$$3x = 15 - 5y$$
$$3x + 5y - 15 = 0$$
$$6x + 10y - 30 = 0$$

[Adding $5y - 15$ to both expressions]
[Multiplying both expressions by 2]

Then $\{(x, y) : 3x = 15 - 5y\} = \{(x, y) : 6x + 10y - 30 = 0\}$, and so their intersection is either of these equal sets. That is

$$\{(x, y) : 3x = 15 - 5y \text{ and } 6x + 10y - 30 = 0\} = \{(x, y) : 3x = 15 - 5y\}$$

Of course, each set has the same graph which is shown in Figure 6.18.

While interesting, the two cases we have just considered are exceptional. In general two lines in the coordinate plane are distinct and intersect in just one point. Hence, the intersection of the solution sets of two equations of the form $ax + by = c$ (where a and b are not both zero) will usually contain only a single ordered pair, although in exceptional cases it may contain infinitely many ordered pairs or be the empty set.

6.5 *Inequalities over* $R \times R$

Next let us consider the solution set for certain inequalities. An inequality such as $2x - y < -2$ is neither true nor false until replacements are made for the variables. In Section 6.1 we learned that when the replacement set is finite, we may make the indicated replacements and find its solution set. But now we will want to find its solution set when the domain is infinite, $R \times R$. As with equations the solution set itself will be infinite, and we will indicate it by $\{(x, y): 2x - y < -2\}$. We shall find its graph. First let us consider an infinite domain which is a subset of $R \times R$.

Example 3 Find the graph of the solution set of $2x - y < -2$ over $R \times \{4, 2, 0, -2, -4\}$.

Solution We shall make the indicated replacements for y. Replace y by 4:

$$2x - y < -2$$
$$2x - 4 < -2$$
$$x < 1 \qquad \langle -\infty, 1 \rangle$$

Then over $R \times \{4\}$ the given inequality has a solution set which is a set of ordered pairs in which each pair has a second component of 4, but the first component is any member of $\langle -\infty, 1 \rangle$. Figure 6.19 shows the graph of this set. Replace y by 2:

$$2x - 2 < -2$$
$$x < 0 \qquad \langle -\infty, 0 \rangle$$

Then over $R \times \{2\}$ the solution set is a set of ordered pairs with second component 2, but the first component may be any member of $\langle -\infty, 0 \rangle$. See Figure 6.20. Similarly, we may find the solution set over the entire set $R \times \{4, 2, 0, -2, -4\}$ and this graph is shown in Figure 6.21.

Notice that all the end points of the open intervals in Figure 6.21 lie on a straight line, and that this line is the graph of the *equation* $2x - y =$

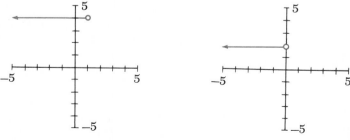

Figure 6.19   Figure 6.20

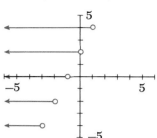

Figure 6.21

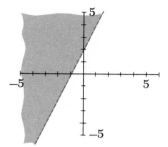

Figure 6.22

-2, as we can verify by noting from the figure that the x- and y-intercepts are -1 and 2 respectively and that both $(-1, 0)$ and $(0, 2)$ are replacements which make $2x - y = -2$ a true statement. Let us now make the assumption that over $R \times R$ the solution of $2x - y < -2$ has a graph which contains *every* point in the plane which lies to the left of the line $2x - y = -2$. It is shown in Figure 6.22. We shall call such a set of points which lie on only one side of a line a **half plane**. We see in Figure 6.22 that the boundary line is broken which indicates that this line itself is *not* in the graph of $\{(x, y): 2x - y < -2\}$ for none of the points on this line has coordinates which make the inequality $2x - y < -2$ a true statement. We shall call a graph of this sort an **open** half plane.

The assumption made in the last paragraph can actually be shown to be true. In more advanced books we can find proofs of the following theorem.

Theorem 2

If a and b are not both 0, the graphs of $\{(x, y): ax + by < c\}$ and $\{(x, y): ax + by > c\}$ are both open half planes whose boundary line is the graph of $\{(x, y): ax + by = c\}$

We may use Theorem 2 to find the graphs of many inequalities.

Example 4 Find the graph of these.
(a) $\{(x, y): 2x + 3y < 6\}$ (b) $\{(x, y): -3x + y > 3\}$

Solution

(a) By Theorem 2 the graph is an open half plane whose boundary line is the graph of $2x + 3y = 6$. See Figure 6.23. We now choose *any*

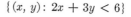
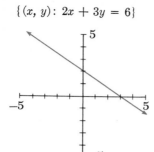
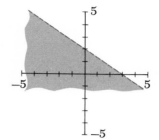

Figure 6.23

Figure 6.24

point in either half plane determined by this line. A convenient point is the one corresponding to (0, 0). When this replacement is made in $2x + 3y < 6$ we obtain $0 < 6$, a true statement. Hence, we can conclude that the graph is the open half plane which contains the origin. See Figure 6.24.

(b) We want a half plane bounded by the line $-3x + y = 3$ whose graph is shown in Figure 6.25. Since (0, 0) is a replacement which makes $-3x + y > 3$ a *false* statement, we choose the half plane which does *not* contain the origin. The graph is shown in Figure 6.26.

Example 5 Find the graph of $\{(x, y): -3x + y \geq 3\}$.

Solution Since

$$\{(x, y): -3x + y \geq 3\} = \{(x, y): -3x + y = 3 \text{ or } -3x + y > 3\}$$
$$= \{(x, y): -3x + y = 3\} \cup \{(x, y): -3x + y > 3\}$$

we want the *union* of the graphs shown in Figures 6.25 and 6.26. We shall call this union a **closed** half plane, and it is shown in Figure 6.27.

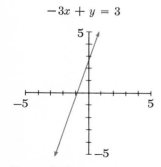
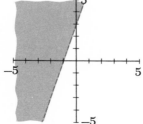

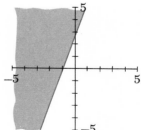

Figure 6.25

Figure 6.26

Figure 6.27

251

252 Expressions with Two Variables

Notice the solid boundary line to indicate that this line is contained in the graph.

We can also find the intersection of solution sets of inequalities, but since these intersections are usually infinite, we shall only graph them.

Example 6   Find the graphs of these sets.
(a) $\{(x, y): x > -2\} \cap \{(x, y): x - y < 0\}$
(b) $\{(x, y): x - y \geq -1 \text{ and } 2x + y \geq 4\}$

Solution

(a) The graph of $\{(x, y): x > -2\}$ is the open half plane in Figure 6.28. The graph of $\{(x, y): x - y < 0\}$ is the open half plane in Figure 6.29. The intersection of these graphs is the solution and is shown in Figure 6.30.

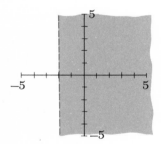

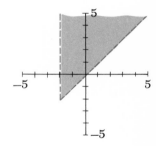

Figure 6.28   Figure 6.29   Figure 6.30

(b) Figure 6.31 shows the closed half plane which is the graph of $\{(x, y): x - y \geq -1\}$. Figure 6.32 shows the closed half plane which is the graph of $\{(x, y): 2x + y \geq 4\}$. The desired graph is the intersection of these two and is shown in Figure 6.33.

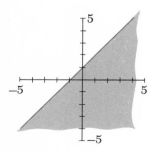

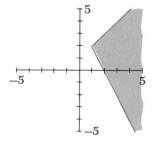

Figure 6.31   Figure 6.32   Figure 6.33

Exercises 6.5

For Exercises 1–8 find the indicated set.

1. $\{(x, y): y = 2x - 3 \text{ and } 2x - y = 5\}$
2. $\{(x, y): x = 1 + 3y \text{ and } y = x - 3\}$
3. $\{(x, y): x = 6 - 2y \text{ and } x/2 + y = 3\}$
4. $\{(x, y): y = 2x + 4 \text{ and } x = y/2 + 3\}$
5. $\{(x, y): y = 3x + 1\} \cap \{(x, y): x = 3y + 1\}$
6. $\{(x, y): 2x + y = 0\} \cap \{(x, y): x + y/2 = 1\}$
7. $\{(x, y): y = 3x + 5\} \cap \left\{(x, y): x = \dfrac{y - 5}{3}\right\}$
8. $\{(x, y): y = 5x - 1\} \cap \left\{(x, y): x = \dfrac{y + 1}{5}\right\}$

For Exercises 9–20 graph the indicated set.

9. $\{(x, y): x + y < 3\}$
10. $\{(x, y): x + y > 1\}$
11. $\{(x, y): 2x - y > 2\}$
12. $\{(x, y): x - 3y < 3\}$
13. $\{(x, y): 4x + y \geq 4\}$
14. $\{(x, y): x + 3y \leq 3\}$
15. $\{(x, y): 2x - 3y \geq 6\}$
16. $\{(x, y): 2x + 5y \leq 10\}$
17. $\{(x, y): x < 3\}$
18. $\{(x, y): y \leq 2\}$
19. $\{(x, y): x \geq -1\}$
20. $\{(x, y): y \geq 1\}$

For Exercises 21–32 graph the solution set for the given open sentences in two variables.

21. $y < x + 3$
22. $y > 2 - x$
23. $y \geq 2x - 2$
24. $y \leq 3x + 3$
25. $x > y + 4$
26. $x \geq 3y - 3$
27. $y \leq x$
28. $y > -x$
29. $2(x + y) < x + 4$
30. $x + 1 > y + 4$
31. $\dfrac{x + y}{2} \leq \dfrac{x - 2}{4}$
32. $\dfrac{x - y}{2} \geq \dfrac{x + 1}{3}$

For Exercises 33–44 find the graph of the indicated intersection.

33. $\{(x, y): x + y < 2 \text{ and } x - y < 2\}$
34. $\{(x, y): x + y \geq 2 \text{ and } x - y \geq 2\}$
35. $\{(x, y): x + y < 2 \text{ and } x - y > 2\}$

36. $\{(x, y): x + y \leq 2 \text{ and } x - y \geq 2\}$
37. $\{(x, y): x \geq -1\} \cap \{(x, y): y \leq 1\}$
38. $\{(x, y): x < 3\} \cap \{(x, y): y > -2\}$
39. $\{(x, y): y < x\} \cap \{(x, y): y < -x\}$
40. $\{(x, y): y \geq x\} \cap \{(x, y): y \geq -x\}$
41. $\{(x, y): -1 \leq x \text{ and } x \leq 3\}$
42. $\{(x, y): -1 \leq y \leq 1\}$
43. $\{(x, y): 0 \leq x \leq 2\}$
44. $\{(x, y): 0 \leq x \leq 2 \text{ and } -1 \leq y \leq 1\}$

Functions and Relations Chapter SEVEN

7.1 Linear Functions

Let us call any set of ordered pairs (of real numbers) a **relation.** We shall call the set of all first components of these pairs the **domain** of the relation and the set of all second components, the **range.** The graph of the set of ordered pairs will be called the **graph** of the relation.

Example 1 Find the domain, range, and graph of each of the following relations.
(a) $\{(-2, 1), (3, 4)\}$ (b) $\{1\} \times I$

Solution

(a) Domain = $\{-2, 3\}$. Range = $\{1, 4\}$. Graph, Figure 7.1.
(b) Domain = $\{1\}$. Range = I. Since $I = \{\ldots, -3, -2, -1, 0, 1, 2, 3, \ldots\}$, then $\{1\} \times I = \{\ldots(1, -3), (1, -2), (1, -1), (1, 0), (1, 1), (1, 2), (1, 3), \ldots\}$. The graph, which is incomplete, is shown in Figure 7.2.

Actually the concept of a relation is not new to us. In Chapter 6 we learned that the solution set for an open sentence in two variables is a

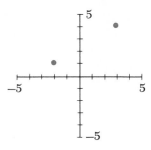

Figure 7.1   Figure 7.2

set of ordered pairs. Now we are merely naming such sets (and others) relations.

Example 2   Find the graph of each of these relations.
(a) $\{(x, y): y = 2x - 2\}$ (b) $\{(x, y): x + y < 3\}$

Solution

(a) Since $y = 2x - 2$ is equivalent to $-2x + y = -2$, the graph is a line with x-intercept 1, and y-intercept -2 (Figure 7.3).
(b) The graph is an open half plane (Figure 7.4).

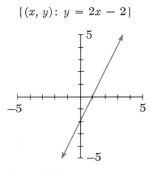
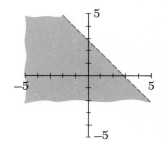

Figure 7.3   Figure 7.4

In Section 1.3 we first introduced the concept of a function and used such notation as $f(x) = x + 3$ to represent the function f which is defined by the expression $x + 3$. As we shall soon see, every such function determines a set of ordered pairs and hence is a relation.

Example 3   By pairing the numbers in the domain with those in the range, show that $f(x) = x + 3$, $\{1, 2, 3\}$ determines a relation.

Solution Since $f(1) = 1 + 3 = 4$, we have the pair $(1, 4)$. Then $f(2) = 5$ and $f(3) = 6$ also give the pairs $(2, 5)$ and $(3, 6)$. Thus the relation is the set containing these three ordered pairs, and we may indicate this by writing $f = \{(1, 4), (2, 5), (3, 6)\}$.

When the domain of a function is an infinite set, we will not be able to list all the ordered pairs which it determines, but often we will be able to find the graph of such a relation.

Example 4 Find the graph of the relation determined by $f(x) = 2x - 2$.

Solution We shall first find some of the pairs in the relation and draw their graph.

$f(-1) = -4, \quad (-1, -4) \qquad\qquad f(\tfrac{3}{2}) = 1, \ (\tfrac{3}{2}, 1)$

$f(-\tfrac{1}{2}) = -3, \ (-\tfrac{1}{2}, -3) \qquad\qquad f(2) = 2, \ (2, 2)$

$f(0) = -2, \quad\ \ (0, -2) \qquad\qquad\ \ f(\tfrac{5}{2}) = 3, \ (\tfrac{5}{2}, 3)$

$f(\tfrac{1}{2}) = -1, \quad\ \ (\tfrac{1}{2}, -1) \qquad\qquad\ \ f(3) = 4, \ (3, 4)$

$f(1) = 0, \quad\quad\ \ (1, 0)$

While the nine pairs we have found form only a *subset* of f, their graph, Figure 7.5, suggests that the graph of f is the line which contains these points. See Figure 7.6.

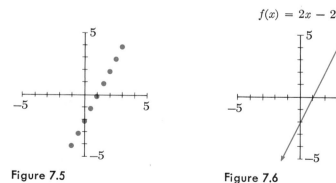

Figure 7.5   Figure 7.6

When we compare the graph of $f(x) = 2x - 2$, Figure 7.6, with that of $\{(x, y): y = 2x - 2\}$, Figure 7.3, we see that they are *exactly the same graph!* In fact every function defined by an algebraic expression determines a set of ordered pairs which is also the solution set for an equation in two variables.

Example 5 Represent the function g defined by $g(x) = 3x + 1$ as the solution set for an equation in two variables.

Solution $g = \{(x, y): y = 3x + 1\}$

Example 6 Without actually drawing its graph, show that the graph of $f(x) = 2x + 3$ is a straight line.

Solution We may also represent this function as $f = \{(x, y): y = 2x + 3\}$. And since $y = 2x + 3$ is equivalent to $-2x + y = 3$, Theorem 1 of Chapter 6 proves that its graph is a line.

As in the above example it is easy to show that every first degree polynomial expression defines a function whose graph is a straight line (if the domain is R). We shall call any such function a **linear function**.

Often (but not always) the solution set for an equation in two variables will determine a function, and we may also represent this function by an algebraic expression.

Example 7 Find an algebraic expression, $f(x)$, which represents the function determined by $f = \{(x, y): y = x^2 - 1\}$.

Solution $f(x) = x^2 - 1$. Note that since $x^2 - 1$ is not a first degree polynomial, this is not a linear function. We will find its graph in Section 7.2.

Example 8 Find $g(x)$ if

$$g = \{(x, y): 2x + y = 4\}.$$

Solution We must first find the "y-form" of the equation.

$$2x + y = 4$$
$$y = -2x + 4$$

Then $g(x) = -2x + 4$. This is a linear function, since $-2x + 4$ is a first degree polynomial.

Example 9 If $g = \{(x, y): 2x + y = 4\}$, find $g(3)$.

Solution Since we are to find the number in the range of g which is paired with 3 in the domain, we shall replace x by 3 and find the value of y.

$$2x + y = 4$$
$$2 \cdot 3 + y = 4$$
$$y = -2$$

Then $g(3) = -2$.

Alternate solution From Example 8 we have $g(x) = -2x + 4$. Hence, $g(3) = -2 \cdot 3 + 4 = -2$.

In Section 1.3 we observed that every algebraic expression (in one variable) determines a correspondence between the numbers in the domain and the numbers in the range; we agreed to call this correspondence a function. We now see that if an equation (with variables x and y) can be written in y-form, it also defines a function. Algebraic expressions and equations in y-form are the most commonly used devices for representing a function, but they are not the only ones. We shall now make a formal definition for a function which will somewhat enlarge our previous concept, but not in any way contradict it.

DEFINITION

A **function** is a set of ordered pairs in which no two distinct pairs have the same first component.

As we see from the definition, a function is merely a special type of relation. That is, some relations are functions and some are not. We shall often want to know whether or not a particular relation is also a function. When the relation contains only a few ordered pairs, it is easy to do this. For example, $\{(2, 3), (4, 5), (2, 1)\}$ is *not* a function since the pairs (2, 3) and (2, 1) have the same first component. But $\{(1, 2), (3, 4), (5, 4)\}$ is a function since each pair has a different *first* component. The fact that (3, 4) and (5, 4) have the same *second* component is irrelevant. However, it may be more difficult to tell whether or not a relation which contains infinitely many pairs is a function.

Example 10 Show that $A = \{(x, y): x + y < 3\}$ is not a function.

Solution Since both (1, 0) and (1, 1) are replacements which make $x + y < 3$ a true statement, they are both members of A. But these pairs have the same first component; hence A is not a function. (Actually we see that there will be infinitely many pairs with the same first component.)

The graph of a relation is very helpful in determining whether or not it is a function. Any two points which are in the same vertical line must have the same first component. For example (1, 2) and (1, 4) are in the same vertical line (Figure 7.7) and both have the same first component. Then if the graph of a relation has any two points in the same vertical line, this relation is *not* a function. The relation whose graph is shown in Figure 7.8 is not a function, since it has two points, A and B, in the same vertical line. But $\{(x, y): y = 2x - 2\}$ is a function since its

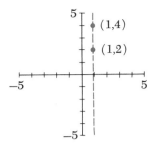

Figure 7.7

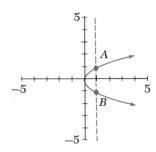

Figure 7.8

graph, Figure 7.3, has no two points in the same vertical line. In fact every vertical line will intersect this graph in just one point.

Example 11 Use its graph to show that $\{(x, y): x + y < 3\}$ is not a function.

Solution Since there is a vertical line which will intersect its graph, Figure 7.4, in more than one point (in infinitely many points), the relation is not a function.

The example suggests that a relation determined by an *inequality* will never be a function. Neither will a relation defined by an *equation* which cannot be written (uniquely) in y-form, as we shall see in Section 7.2.

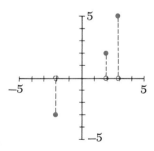

Figure 7.9

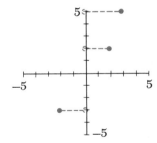

Figure 7.10

The graph of a relation is also useful in determining its domain and range. The domain of $A = \{(-2, -3), (2, 2), (3, 5)\}$ is $\{-2, 2, 3\}$. When vertical lines are drawn through the three points which are the graph of A, they intersect the x-axis at $-2, 2,$ and 3. The points $(-2, 0)$, $(2, 0)$ and $(3, 0)$ are called the projection of the graph on the x-axis (Figure 7.9). And if the graph is projected on the y-axis, Figure 7.10, we obtain $(0, -3), (0, 2)$ and $(0, 5)$, and the range of A is $\{-3, 2, 5\}$.

7.1 Linear Functions

Example 12 What are the domain and range of the relation whose graph is Figure 7.11?

Solution The projection on the x-axis consists of infinitely many points, all those in the closed interval, $[-1, 2]$, which is the domain. See Figure 7.12. The range is also infinite, the closed interval $[1, 4]$, which is the projection on the y-axis.

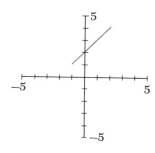

Figure 7.11

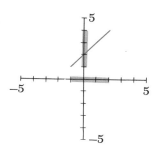

Figure 7.12

Example 13 What are the domain and range of the function $f(x) = 2x - 2$.

Solution From the graph, Figure 7.6, we see that the projections on both the x- and y-axes will consist of the entire axes. Hence both the domain and range are R, the set of all real numbers.

Exercises 7.1

For Exercises 1–8 determine the domain and range and indicate whether or not the relation is a function.

1. $\{(2, 1), (3, 0)\}$
2. $\{(5, 2), (4, 2), (-1, 2)\}$
3. $\{(1, 5)\}$
4. $\{(2, 3), (2, -1), (2, 0)\}$
5. $\{-3\} \times \{1, 2, 3\}$
6. $\{2, -2\} \times \{5\}$
7. $\{3\} \times R$
8. $I \times \{0\}$

For Exercises 9–12 represent the function as the solution set for an equation in two variables, x and y, and indicate whether or not it is a linear function.

9. $f(x) = 3x - 2$
10. $g(x) = 5 - x$
11. $h(x) = x^2 + 3$
12. $k(x) = 4 - x^2$

For Exercises 13–24 find the expression $f(x)$ which will determine the indicated function, and indicate whether or not f is a linear function.

13. $f = \{(x, y): y = 3x + 5\}$
14. $f = \{(x, y): y = |x + 1|\}$
15. $f = \{(x, y): y = 3\}$
16. $f = \{(x, y): y = x^2\}$
17. $f = \{(x, y): x + y = 5\}$
18. $f = \{(x, y): 2x - y = 1\}$
19. $f = \{(x, y): x^2 + y = 3\}$
20. $f = \{(x, y): 3x - y = -4\}$
21. $f = \{(x, y): 3y = x\}$
22. $f = \{(x, y): x + 2y = 3\}$
23. $f = \{(x, y): 3x - 4y = 5\}$
24. $f = \{(x, y): 5x + 2y = 1\}$

For Exercises 25–32 find $g(3)$.

25. $g = \{(x, y): y = 2x + 5\}$
26. $g = \{(x, y): y = x^2 - 1\}$
27. $g = \left\{(x, y): y = \dfrac{2x - 1}{5}\right\}$
28. $g = \left\{(x, y): y = \dfrac{1 - 3x}{4}\right\}$
29. $g = \{(x, y): 2x + y = 4\}$
30. $g = \{(x, y): x + 2y = 1\}$
31. $g = \{(x, y): 3x - y = 5\}$
32. $g = \{(x, y): x - 2y = -1\}$

For Exercises 33–40 draw the graph of the indicated linear function.

33. $f(x) = x + 2$
34. $g(x) = x - 3$
35. $h(x) = 4 - x$
36. $k(x) = -x - 1$
37. $F(x) = 2x - 4$
38. $G(x) = 3x$
39. $H(x) = 6 - 2x$
40. $K(x) = -2x$

For Exercises 41–52 indicate whether or not the relation whose graph is shown is a function and find its domain and range.

41.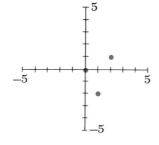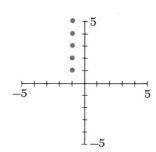

43.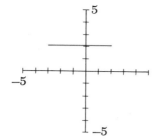
44.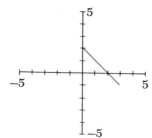
45.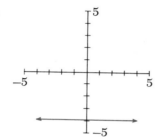
46.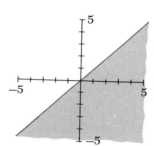
47.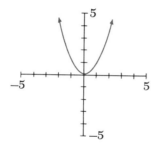
48.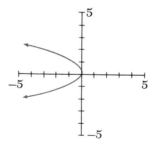
49.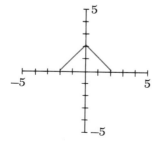
50.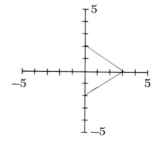

51. 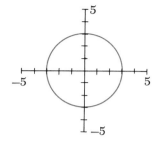   **52.**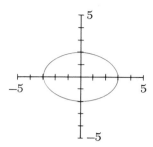

For Exercises 53–56 determine whether or not the indicated relation is a function. The graph may be helpful.

53. $\{(x, y): y = 3x - 3\}$ **54.** $\{(x, y): 2x + 3y = 6\}$
55. $\{(x, y): 2x + y < 4\}$ **56.** $\{(x, y): x = 2\}$

7.2 Nonlinear Functions

In the preceding section we observed that any function which may be represented by a first degree polynomial expression with domain R is a linear function, and its graph is a straight line. In this section we shall see that many functions are not linear and their graphs are not straight lines.

Example 1 For each of the following a function is defined by an expression and a finite domain. Express the function as a set of ordered pairs and find its graph.

(a) $f(x) = x^2$, $\{0, 1, -1, 2, -2\}$
(b) $g(x) = |x|$, $\{0, 1, -1, 2, -2\}$

Solution

(a) We shall make the indicated replacements for x.

$f(0) = 0$ $(0, 0)$ $f(-1) = (-1)^2 = 1$ $(-1, 1)$
$f(1) = 1$ $(1, 1)$ $f(2) = 4$ $(2, 4)$
                      $f(-2) = 4$ $(-2, 4)$

Then $f = \{(0, 0), (1, 1), (-1, 1), (2, 4), (-2, 4)\}$ and its graph is shown in Figure 7.13.

(b) We recall that $|\ |$ represents absolute value.

$g(0) = |0| = 0$ $g(-1) = |-1| = 1$ $g(-2) = |-2| = 2$
$g(1) = |1| = 1$ $g(2) = |2| = 2$

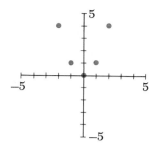
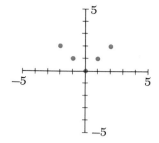

Figure 7.13   Figure 7.14

Hence, $g = \{(0, 0), (1, 1), (-1, 1), (2, 2), (-2, 2)\}$ and its graph is shown in Figure 7.14.

If we consider the function of Example 1(a) with infinite domain R, we cannot hope to represent it as a set of ordered pairs by listing all of them. We can use the notation $f = \{(x, y): y = x^2\}$, but this really conveys no more information than $f(x) = x^2$. Let us see how the graph of f can be used to further describe it. This graph certainly includes the five points shown in Figure 7.13, but these points do not lie on a straight line and so do not furnish us with enough information to complete the graph. Let us use some replacements for x which lie between the numbers already considered in Example 1(a).

$$f(\tfrac{1}{2}) = (\tfrac{1}{2})^2 = \tfrac{1}{4} \qquad (\tfrac{1}{2}, \tfrac{1}{4})$$
$$f(-\tfrac{1}{2}) = (-\tfrac{1}{2})^2 = \tfrac{1}{4} \qquad (-\tfrac{1}{2}, \tfrac{1}{4})$$
$$f(\tfrac{3}{2}) = \tfrac{9}{4} \qquad (\tfrac{3}{2}, \tfrac{9}{4})$$
$$f(-\tfrac{3}{2}) = \tfrac{9}{4} \qquad (-\tfrac{3}{2}, \tfrac{9}{4})$$

When the four points corresponding to these pairs are included with those of Figure 7.13, we obtain the graph shown in Figure 7.15. While this is not the complete graph of x^2 (with domain R), we now have nine points which do lie on the desired graph. Other points can be determined by considering replacements for x which have not already been used. Figure 7.16 shows the result of including one more point between each of those in Figure 7.15. We shall not continue to find individual points on the graph of x^2. In more advanced books it can be shown that the complete graph is a smooth, continuous curve which includes the points we have found. This curve is called a **parabola** and a portion of the graph is shown in Figure 7.17.

Notice that the function x^2 whose graph is a parabola is represented by a second degree polynomial. We shall often refer to such a function as a **quadratic** function. We cannot prove it here, but it can be shown that

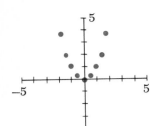
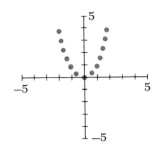
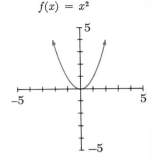

$f(x) = x^2$

Figure 7.15   Figure 7.16   Figure 7.17

the graph of every quadratic function is a parabola. We shall state this result in the following theorem.

Theorem 1 The graph of any function determined by a second degree polynomial $(ax^2 + bx + c, a \neq 0)$ is a parabola

Example 2 Find the graph of each of these functions.
(a) $F(x) = x^2 - 2$
(b) $G = \{(x, y): y = -2x^2\}$
(c) $H = \{(x, y): x^2 = 2x + y\}$

Solution

(a) Since $x^2 - 2$ is a second degree polynomial, by Theorem 1 the graph of F is a parabola. By making several replacements for x, we obtain

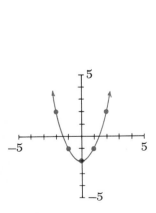
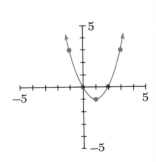

Figure 7.18   Figure 7.19   Figure 7.20

266

7.2 Nonlinear Functions

the following ordered pairs which are members of F: $(0, -2)$, $(1, -1)$, $(2, 2)$, $(-1, -1)$, $(-2, 2)$. Figure 7.18 shows the points corresponding to these pairs contained in a smooth curve which is the graph of F.

(b) Again, $G(x) = -2x^2$, so G is a quadratic function and its graph is a parabola. Some of the members of G are $(0, 0)$, $(1, -2)$, $(2, -8)$, $(-1, -2)$ and $(-2, -8)$. The graph is Figure 7.19.

(c) The y-form of $x^2 = 2x + y$ is $y = x^2 - 2x$. Hence, H is also a quadratic function and $H(x) = x^2 - 2x$. $(0, 0)$, $(1, -1)$, $(2, 0)$, $(3, 3)$ and $(-1, 3)$ are all members of H. Figure 7.20 shows the graph.

In Example 1(b) we graphed the function determined by $|x|$ with a finite domain. We now wish to consider $|x|$ with domain R. We can, of course, represent this function by $g = \{(x, y): y = |x|\}$. Certainly the five points in Figure 7.14 are included in the graph of g, and we may determine other points by making additional replacements for x. Some of these are the points corresponding to $(\frac{1}{2}, \frac{1}{2})$, $(-\frac{1}{2}, \frac{1}{2})$, $(\frac{3}{2}, \frac{3}{2})$, $(-\frac{3}{2}, \frac{3}{2})$, $(3, 3)$ and $(-3, 3)$. When these points are included with those of Figure 7.14, we obtain the graph shown in Figure 7.21. Determining additional points on the graph of $|x|$ would eventually lead us to the V-shaped graph we see in Figure 7.22.

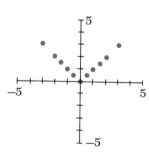

Figure 7.21

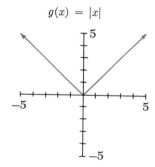

Figure 7.22

Similar treatment enables us to find the graph of other functions which involve absolute value.

Example 3 Find the graph of $f(x) = |x + 2|$.

Solution We shall find several points on the graph and draw a curve containing them.

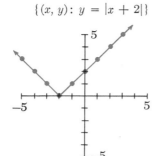

$\{(x, y): y = |x + 2|\}$

(0, 2)   (3, 5)   (−3, 1)
(1, 3)   (−1, 1)   (−4, 2)
(2, 4)   (−2, 0)   (−5, 3)

The graph is shown in Figure 7.23.

We can use techniques similar to those of Section 6.5 to find the graphs of relations which are not functions and which are defined by *inequalities*.

Figure 7.23

Example 4 Find the graph of each of the following.
(a) $\{(x, y): y < |x + 2|\}$ (b) $\{(x, y): x^2 \leq 2x + y\}$

Solution

(a) The graph we seek is the set of points on one side of the curve determined by the *equation* $y = |x + 2|$. But this is the function whose graph we found in Example 3 (Figure 7.23). Since (0, 0) is a replacement which makes $y < |x + 2|$ the true statement $0 < |0 + 2|$, we conclude that the graph is the side which includes the origin. The graph is shown in Figure 7.24.

(b) Again, we want a region on one side of the graph of $x^2 = 2x + y$ which is the parabola we found in Example 2(c) and is shown in Figure 7.20. The pair (0, 1) is a replacement which makes $x^2 \leq 2x + y$ true, so the point corresponding to (0, 1) is on the graph we seek. (Note that we get no additional information if we choose the origin, for this point is actually on the parabola which is the boundary curve.) The graph is shown in Figure 7.25.

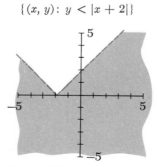

Figure 7.24

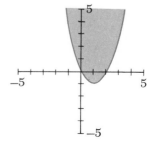

Figure 7.25

7.2 Nonlinear Functions

Some relations which are not functions have graphs which resemble those of functions we have already considered. These graphs can also be found by first determining some of the points on them and again drawing a smooth curve containing these points.

Example 5 Find the graph of $\{(x, y): x = y^2\}$.

Solution Since the equation $x = y^2$ is in x-form, we find it more convenient to make replacements for y.

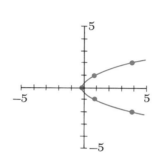

Figure 7.26

$$y = 0: x = 0^2 = 0 \quad (0, 0)$$
$$y = 1: x = 1^2 = 1 \quad (1, 1)$$
$$y = 2: x = 2^2 = 4 \quad (4, 2)$$
$$y = -1: x = (-1)^2 = 1 \quad (1, -1)$$
$$y = -2: x = (-2)^2 = 4 \quad (4, -2)$$

The graph is shown in Figure 7.26.

We note that $x = y^2$ defines a relation which is not a function, since there are (at least) two points on its graph (Figure 7.26) in the vertical line through the point (1, 0). Also note that we cannot write the equation $x = y^2$ in y-form.

Let us summarize the general technique for graphing relations which we have illustrated in this section.

1. If the relation is represented by an equation in y-form, make several replacements for x to determine points on the graph and draw a smooth curve containing them.
2. If the relation is represented by an equation in x-form, make replacements for y to determine points on the graph and then draw a smooth curve containing them.
3. If the relation is represented by an *inequality*, find the graph of the corresponding *equation*. This is the boundary curve of the desired graph. Choose a point on either side of the boundary and make the indicated replacement in the given *inequality* to determine which side of the boundary curve is desired.

We shall illustrate the general technique with an example.

Example 6 Find the graph of $\{(x, y): |x| + y = 2\}$.

Solution First we shall find the y-form of the given equation.

$$|x| + y = 2$$
$$y = 2 - |x|$$

Then we make several replacements for x.

$\{(x, y): |x| + y = 2\}$

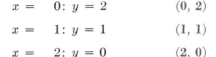

Figure 7.27

$x =$	$0: y = 2$	$(0, 2)$
$x =$	$1: y = 1$	$(1, 1)$
$x =$	$2: y = 0$	$(2, 0)$
$x =$	$3: y = -1$	$(3, -1)$
$x =$	$-1: y = 1$	$(-1, 1)$
$x =$	$-2: y = 0$	$(-2, 0)$
$x =$	$-3: y = -1$	$(-3, -1)$

The graph is shown in Figure 7.27. Note that the smooth "curve" containing the points we found is the union of two half lines.

Exercises 7.2

For Exercises 1–16 find the graph of the given quadratic function.

1. $f(x) = 2x^2$
2. $g(x) = x^2/2$
3. $h(x) = -x^2$
4. $k(x) = -x^2/4$
5. $F(x) = x^2 + 1$
6. $G(x) = x^2 - 1$
7. $H(x) = 1 - x^2$
8. $K(x) = x^2 - 4$
9. $f(x) = x^2 - 4x$
10. $g(x) = x^2 - 3x$
11. $h(x) = -x^2 + 4x$
12. $k(x) = x^2 + 2x$
13. $F(x) = x^2 - 2x + 1$
14. $G(x) = x^2 + 2x + 1$
15. $H(x) = x^2 + 4x + 3$
16. $K(x) = -x^2 + 2x + 3$

For Exercises 17–24 find the graph of the indicated nonlinear function.

17. $f(x) = -|x|$
18. $g(x) = |-x|$
19. $h(x) = |2x|$
20. $k(x) = |x| + 1$
21. $F(x) = |x + 1|$
22. $G(x) = |x - 1|$
23. $H(x) = 1 - |x|$
24. $K(x) = -|x + 2|$

For Exercises 25–32 find the graph of the indicated relation.

7.3 *Special Graphing Techniques*

25. $\{(x, y): y \geq x^2\}$
26. $\{(x, y): y < x^2\}$
27. $\{(x, y): y > x^2 - 1\}$
28. $\{(x, y): y \leq 1 - x^2\}$
29. $\{(x, y): y > |x|\}$
30. $\{(x, y): y \geq |x| - 1\}$
31. $\{(x, y): y \leq |x - 2|\}$
32. $\{(x, y): y < |x| + 1\}$

For Exercises 33–40 find either an x-form or a y-form for the given sentence, and then find its graph.

33. $\{(x, y): 4y = x^2\}$
34. $\{(x, y): y - x^2 + 3 = 0\}$
35. $\{(x, y): 2y - |x| = 0\}$
36. $\{(x, y): x^2 + 4y = 4\}$
37. $\{(x, y): x + y^2 = 0\}$
38. $\{(x, y): x - |y| = 0\}$
39. $\{(x, y): y^2 = x + 1\}$
40. $\{(x, y): x + y^2 = 2y\}$

7.3 Special Graphing Techniques

When graphing linear equations in two variables (Section 6.3) we found it helpful to first locate the x- and y-intercepts. The intercepts are also useful when making a quick sketch of the graph of certain quadratic functions.

Example 1 Find the intercepts and use them to sketch the graph of each of these functions.

(a) $f = \{(x, y): y = 4 - x^2\}$ (b) $g(x) = x^2 + x - 6$

Solution

(a) To find the x-intercepts we replace y by 0.

$$y = 4 - x^2$$
$$0 = 4 - x^2 = (2 + x)(2 - x) \qquad \{-2, 2\}$$

There are two x-intercepts, -2 and 2. To find the y-intercepts we replace x by 0.

$$y = 4 - x^2$$
$$y = 4 - 0^2 = 4 \qquad \{4\}$$

The only y-intercept is 4. Since f is defined by a second degree polynomial, Theorem 1 shows that its graph is a parabola. Using only the intercepts we shall make a rough sketch which is shown in Figure 7.28. Notice that the parabola must open downward if it is to include the three points already determined.

(b) Since $g(0) = -6$, we have the point $(0, -6)$, and the y-intercept is

272 Functions and Relations

-6. We may also represent this function by $g = \{(x, y): y = x^2 + x - 6\}$. To find the x-intercepts we replace y by 0.

$$0 = x^2 + x - 6$$
$$0 = (x + 3)(x - 2) \qquad \{-3, 2\}$$

The x-intercepts are -3 and 2. This parabola may be sketched using only the intercepts (Figure 7.29).

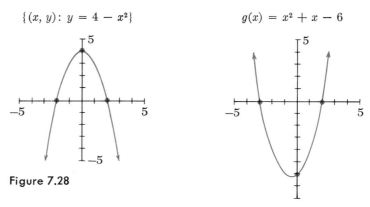

Figure 7.28

Figure 7.29

We have said that the zeros of a function are those numbers in the domain of the function which correspond to 0 in the range. There is an interesting connection between the zeros of a function and the x-intercepts of the graph of that function.

Example 2 Find the zeros of each of these functions.
(a) $f(x) = 4 - x^2$ \qquad (b) $g = \{(x, y): y = x^2 + x - 6\}$

Solution

(a) We may also represent the function by $f = \{(x, y): y = 4 - x^2\}$. Since we want the number in the range to be zero, we replace y by 0.

$$y = 4 - x^2$$
$$0 = 4 - x^2 = (2 + x)(2 - x) \qquad \{-2, 2\}$$

The zeros of f are -2 and 2. Note that these are also the x-intercepts of the graph of f (Figure 7.28).

(b)
$$y = x^2 + x - 6$$
$$0 = x^2 + x - 6 = (x+3)(x-2) \qquad \{-3, 2\}$$

Again, the zeros, -3 and 2, are also the x-intercepts of the graph of g (Figure 7.29).

The observations made in Example 2 are quite general. The zeros of any relation (whether or not it is a function) are the numbers in the domain which correspond to 0 in the range and they are also the x-intercepts of the graph of that relation.

Example 3 Find the zeros and show that they are also the x-intercepts of the graph of $r = \{(x, y): x + y < 4\}$.

Solution To find the zeros we replace y by 0.

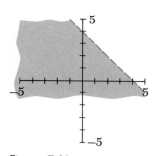

Figure 7.30

$$x + y < 4$$
$$x + 0 < 4$$
$$x < 4$$
$$\langle -\infty, 4 \rangle$$

There are infinitely many zeros—all members of $\langle -\infty, 4 \rangle$. The graph of r is the open half plane bounded by the line which is the graph of $x + y = 4$. In Figure 7.30 we see that the x-intercepts of this half plane are all the members of $\langle -\infty, 4 \rangle$.

When it is difficult to find the zeros of a relation by making the replacement $y = 0$, it may sometimes be easier to find the x-intercepts of its graph.

Example 4 Find the set of all zeros of the relation $p = \{(x, y): y \leq 4 - x^2\}$.

Solution If we replace y by 0, we obtain the second degree inequality $0 \leq 4 - x^2$. While this may be solved by the methods of Section 4.6, it will be easier to find the x-intercepts of the graph of this relation. The graph of p is the region on one side of the parabola $y = 4 - x^2$ which is shown in Figure 7.28. Since $(0, 0)$ is a replacement which makes $y \leq 4 - x^2$ a true statement, we conclude that the graph of p is the region which includes the origin. Figure 7.31 shows the graph of p, and we see from the figure that the x-intercepts are the members of $[-2, 2]$. Hence, the set of all zeros of the given relation is $[-2, 2]$.

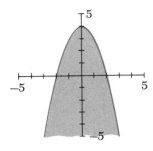

Figure 7.31

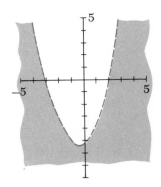

Figure 7.32

Next we shall use the graphing technique to solve Example 4 of Section 4.6.

Example 5 Find the solution set of $x^2 + x > 6$.

  Solution $x^2 + x > 6$ is equivalent to $x^2 + x - 6 > 0$. We observe that the solution set of the inequality will be the set of all zeros of the relation $q = \{(x, y): x^2 + x - 6 > y\}$. For to find the zeros of q we could replace y by 0 and $x^2 + x - 6 > y$ would become $x^2 + x - 6 > 0$. We shall use the graph of q to find its zeros. This graph is the region on one side of the parabola $x^2 + x - 6 = y$ which we sketched in Example 1 and is shown in Figure 7.29. Since $(0, 0)$ is a replacement which makes $x^2 + x - 6 > y$ false, we conclude that the region we seek is on the side of the parabola which does *not* include the origin. The graph of q is shown in Figure 7.32 and we see that the intercepts are the members of $\langle -\infty, -3 \rangle \cup \langle 2, \infty \rangle$, which is exactly the solution set for $x^2 + x > 6$. (Compare this with the solution of Example 4 of Section 4.6.)

When we examine the graphs shown in Figures 7.28 and 7.29, we see that for each there is an axis of symmetry. In Figure 7.28 we see that the the graph is symmetric about the y-axis—the portion of the graph to the right of the y-axis is the mirror image of that portion to the left of the y-axis. And in Figure 7.29 there appears to be some axis of symmetry. It is not the y-axis, but rather some vertical line to the left of the y-axis. The intercepts of this graph will help us to locate that axis exactly. Note that the intercepts are -3 and 2, that the mean of these numbers is $\dfrac{-3 + 2}{2} = \dfrac{-1}{2}$, and that the axis of symmetry intersects the x-axis at $(-\frac{1}{2}, 0)$. See Figure 7.33. Also note that this axis of symmetry intersects the parabola at its turning point V, called the **vertex** of the parabola. We may find the ordered pair which corresponds to the vertex by making the replacement $x = -\frac{1}{2}$ in the equation $y = x^2 + x - 6$ (since the parabola is the graph of this equation). Then $y = (-\frac{1}{2})^2 + (-\frac{1}{2}) - $

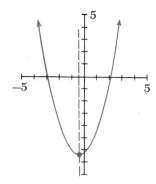

Figure 7.33

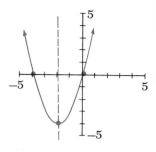

Figure 7.34

$6 = -\frac{25}{4}$, and the vertex corresponds to the pair $(-\frac{1}{2}, -\frac{25}{4})$.

Example 6 Find the ordered pair which corresponds to the vertex of the parabola which is the graph of $\{(x, y): y = x^2 + 4x\}$.

Solution
$$y = x^2 + 4x = x(x + 4)$$

Then 0 and -4 are the x-intercepts. The axis of symmetry intersects the x-axis at $(-2, 0)$, since -2 is the mean of 0 and -4. To find the y-coordinate of the vertex we shall replace x by -2.

$$y = (-2)^2 + 4(-2) = -4$$

Thus $V = (-2, -4)$. See Figure 7.34.

Example 7 Find the domain and range of $f(x) = -x^2 - 2x + 3$.

Solution Since every real replacement of x is possible in this polynomial expression, the domain is R. We shall use the graph of f to find its range. Since $f(0) = 3$, the y-intercept is 3. Then since

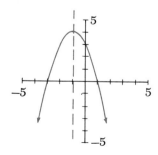

Figure 7.35

$$f(x) = -x^2 - 2x + 3$$
$$= -(x^2 + 2x - 3)$$
$$= -(x + 3)(x - 1)$$

We see that -3 and 1 are the zeros of f and also the x-intercepts. The axis of symmetry is the vertical line $\{(x, y): x = -1\}$. Next we find $f(-1) = -(-1)^2 - 2(-1) + 3 = 4$, so the vertex is at $(-1, 4)$. From the graph, Figure 7.35, we see that the range is $\langle -\infty, 4]$ since this is the projection on the y-axis.

Exercises 7.3

For Exercises 1–12 use only the intercepts to find a quick sketch of the graph of the indicated quadratic function.

1. $f = \{(x, y): y = x^2 - 1\}$
2. $g = \{(x, y): y = x^2 - 4\}$
3. $h = \{(x, y): y = 1 - x^2\}$
4. $k = \{(x, y): y = 9 - x^2\}$
5. $F = \{(x, y): y = x^2 + x - 2\}$
6. $G = \{(x, y): y = x^2 - 2x - 3\}$
7. $H(x) = x^2 + 2x - 3$
8. $K(x) = x^2 - 3x - 4$
9. $f(x) = -x^2 - x + 2$
10. $g(x) = -x^2 + 3x - 2$
11. $h(x) = x^2 - 4x + 3$
12. $k(x) = x^2 + 5x + 4$

For Exercises 13–16 find the set of all x-intercepts of the graph without actually drawing the graph.

13. $\{(x, y): x + y \leq 2\}$
14. $\{(x, y): -x + y > 4\}$
15. $\{(x, y): x \leq 1\}$
16. $\{(x, y): 2x + 3y + 4 < 0\}$

For Exercises 17–24 make a quick sketch of the graph and find the set of all x-intercepts.

17. $\{(x, y): y \geq x^2 - 1\}$
18. $\{(x, y): y \leq x^2 - 1\}$
19. $\{(x, y): y \leq x^2 - 4\}$
20. $\{(x, y): y \geq x^2 - 4\}$
21. $\{(x, y): y > x^2 + x - 2\}$
22. $\{(x, y): y < x^2 + x - 2\}$
23. $\{(x, y): y > x^2 - 4x + 3\}$
24. $\{(x, y): y < x^2 + 4x + 3\}$

For Exercises 25–32 use the x-intercepts of an appropriate graph to find the solution set.

25. $x^2 - 1 < 0$
26. $x^2 - 4 > 0$
27. $x^2 > 1$
28. $x^2 < 4$
29. $x^2 + x - 2 \leq 0$
30. $x^2 + x \geq 2$
31. $x^2 + 4x + 3 \leq 0$
32. $x^2 - 6x + 5 \geq 0$

For Exercises 33–40 find the ordered pair which corresponds to the vertex of the parabola which is the graph of the indicated relation.

33. $\{(x, y): y = x^2 - 4\}$
34. $\{(x, y): y = 4 - x^2\}$
35. $\{(x, y): y = x^2 - 2x\}$
36. $\{(x, y): y = x^2 - 4x\}$

37. $\{(x, y): y = x^2 - 4x + 3\}$
38. $\{(x, y): y = x^2 - 4x - 5\}$
39. $\{(x, y): y = x^2 + 2x - 3\}$
40. $\{(x, y): y = x^2 + 6x + 5\}$

For Exercises 41–44 find the domain and range of the function.

41. $f(x) = x^2 - 4$
42. $g(x) = 1 - x^2$
43. $h(x) = x^2 - 3x + 2$
44. $k(x) = x^2 - x - 2$

7.4 The Inverse of a Relation

We have defined a relation to be a set whose members are ordered pairs of real numbers. It is always possible to form a new relation from a given one by interchanging the components of each ordered pair. We shall call this new relation the **inverse** of the original relation.

Example 1. Find the inverse of the given relation.

(a) $\{(3, 5), (-2, 1), (0, 4)\}$
(b) $\{1, 2\} \times \{3, 4\}$

Solution

(a) $\{(5, 3), (1, -2), (4, 0)\}$
(b) $\{1, 2\} \times \{3, 4\} = \{(1, 3), (1, 4), (2, 3), (2, 4)\}$. Then the inverse is $\{(3, 1), (4, 1), (3, 2), (4, 2)\}$ which may also be represented by the Cartesian product $\{3, 4\} \times \{1, 2\}$.

If R is a given relation, then we shall use the notation R^{-1} to represent the inverse of R. Thus if $R = \{(1, 2), (3, 4)\}$, then $R^{-1} = \{(2, 1), (4, 3)\}$. Although the superscript -1 may appear to be an exponent, it is clear that R^{-1} is *not* a power of R. Only *numbers* have powers, and R is a relation, not a number.

Example 2 Let $G = \{(0, 1), (0, 2), (1, 2)\}$.

(a) Find the domain and range of both G and G^{-1}
(b) Find the graph of G, G^{-1} and $G \cup G^{-1}$

Solution

(a)

Relation	Domain	Range
$G = \{(0, 1), (0, 2), (1, 2)\}$	$\{0, 1\}$	$\{1, 2\}$
$G^{-1} = \{(1, 0), (2, 0), (2, 1)\}$	$\{1, 2\}$	$\{0, 1\}$

278   Functions and Relations

We observe that the domain of G is the range of G^{-1} and the range of G is the domain of G^{-1}.

(b) The graphs are shown in Figures 7.36, 7.37 and 7.38. We observe the axis of symmetry in Figure 7.38. This line is the graph of the equation $y = x$.

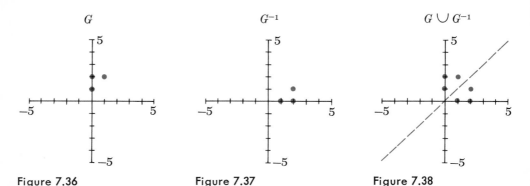

Figure 7.36   Figure 7.37   Figure 7.38

The observations made in Example 2 are quite generally true. Whenever one relation is the inverse of another, the domain of each is the range of the other and the graph of their union has the line $y = x$ as an axis of symmetry.

Example 3   Let A be the relation whose graph is shown in Figure 7.39.
(a) Find the domain and range of A^{-1}
(b) Find the graph of A^{-1}

Solution

(a) From the figure we see that the domain and range of A are $[-3, -1]$

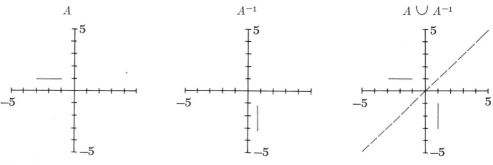

Figure 7.39   Figure 7.40   Figure 7.41

and {1}, respectively. Hence, the domain of A^{-1} is {1} (the range of A), while the range of A^{-1} is $[-3, -1]$ (the domain of A).

(b) Since the graph of $A \cup A^{-1}$ must be symmetric about the line $y = x$, the graph of A^{-1} will also be a line segment. The end points of the graph of A are at $(-3, 1)$ and $(-1, 1)$. Hence, the end points of the graph of A^{-1} are at $(1, -3)$ and $(1, -1)$. See Figure 7.40. Note that the graph of $A \cup A^{-1}$ is symmetric about $y = x$ (Figure 7.41).

Example 4 $A = \{(x, y): x + 2y = 4\}$
$B = \{(x, y): y + 2x = 4\}$

Show that B is the inverse of A.

Solution   Figures 7.42, 7.43 and 7.44 show the graphs of A, B, and $A \cup B$. Since the line $y = x$ is an axis of symmetry in the graph of $A \cup B$, we conclude that B is the inverse of A.

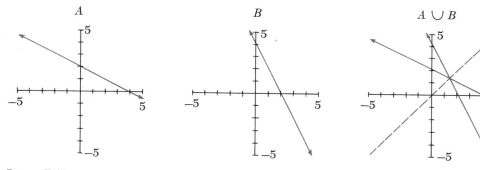

Figure 7.42   **Figure 7.43**   **Figure 7.44**

We may generalize the result of Example 4. Whenever a relation is defined by an open sentence in two variables, its inverse will be defined by the open sentence which results from interchanging the variables in the original sentence.

Example 5   Find the inverse of each of the following:
(a) $A = \{(x, y): y = 2x^2 - x - 1\}$
(b) $B = \{(x, y): |x| = y - 1\}$
(c) $C = \{(x, y): x + 2y < 5\}$
(d) $D = \{(x, y): x = 3\}$

Solution

(a) $A^{-1} = \{(x, y): x = 2y^2 - y - 1\}$

(b) $B^{-1} = \{(x, y): |y| = x - 1\}$
(c) $C^{-1} = \{(x, y): y + 2x < 5\}$
(d) $D^{-1} = \{(x, y): y = 3\}$

We have said that a function is a relation in which no two different pairs have the same first component. Then any function will have an inverse which is a relation, and it is interesting to see whether or not the inverse of a function is also a *function*.

Example 6 Find the inverse of the given function and determine whether or not it is also a function.
(a) $f = \{(4, 3), (5, 4)\}$ (b) $g = R \times \{1\}$

Solution

(a) $f^{-1} = \{(3, 4), (4, 5)\}$. Hence, f^{-1} is also a function.
(b) $g^{-1} = \{1\} \times R$. But g^{-1} is a set of many ordered pairs in which each member has the same first component. Hence, g^{-1} is *not* a function.

As we see from Example 6 the inverse of a function is always a relation, but it may or may not be a function.

Example 7 Each of these is a relation which is *not* a function. Determine whether or not the *inverse* of each is a function.
(a) $r = \{(1, 2), (1, 3), (4, 3)\}$ (b) $s = \{0\} \times R$

Solution

(a) $r^{-1} = \{(2, 1), (3, 1), (3, 4)\}$ and r^{-1} is *not* a function
(b) $s^{-1} = R \times \{0\}$ and s^{-1} *is* a function

Then Examples 6 and 7 illustrate that the inverse of a relation may, or may not, be a function regardless of whether or not the original relation is itself a function. Since a relation is a function only if each of its pairs has a different *first* component, we see that the *inverse* of a relation will be a function only if each of the pairs of the original relation has a different *second* component—for all such second components become first components of the inverse.

The graph of a relation may be used to tell whether or not its inverse is a function. We have already seen that a relation is a function if no *vertical* line intersects its graph in more than one point. If no *horizontal* line intersects its graph in more than one point, then the *inverse* of that relation is a function. For any two ordered pairs with the same second component will correspond to points which lie on the same horizontal line.

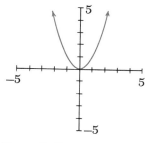

Figure 7.45

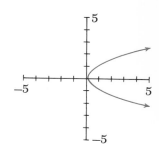

Figure 7.46

Thus Figure 7.45 shows the graph of a function whose inverse is *not* a function, and Figure 7.46 shows the graph of a relation (which is not a function) whose inverse is a function.

Example 8   Determine whether or not the inverse of each of the following functions is also a function.
(a) $f(x) = -x + 2$ (b) $g(x) = x^2 - x$

Solution

(a) Since the graph of f, Figure 7.47, does not have two points on the same horizontal line, then f^{-1} is a function.
(b) The graph of g, Figure 7.48, has two points on the same horizontal line. Hence, g^{-1} is not a function.

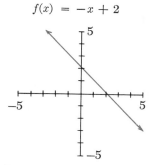

Figure 7.47

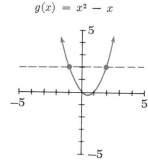

Figure 7.48

Every first degree polynomial defines a function whose graph is a (nonhorizontal) line. As in Example 8(a), it is easy to see that the inverse of such a function will also be a function. However, since the graph of a function defined by a second degree polynomial is a parabola, as in

Functions and Relations

Example 8(b), it will have two points on the same horizontal line. Thus the inverse of a quadratic function is not a function.

Example 9 Let $f(x) = 2x + 1$. Find the expression, $f^{-1}(x)$, which defines the inverse of f.

Solution $f = \{(x, y): y = 2x + 1\}$. Hence the inverse may be written as $f^{-1} = \{(x, y): x = 2y + 1\}$. We shall find the y-form.

$$x = 2y + 1$$
$$x - 1 = 2y$$
$$\frac{x - 1}{2} = y$$

Then $f^{-1} = \left\{(x, y): y = \frac{x - 1}{2}\right\}$ or $f^{-1}(x) = \frac{x - 1}{2}$.

Example 10 Let $f(x) = 2x + 1$. Find $f^{-1}(2)$.

Solution We want to find the number in the range of f^{-1} which corresponds with 2 in the domain. But this is also the number in the domain of f (the inverse of f^{-1}) which corresponds with 2 in the range of f. Now $f = \{(x, y): y = 2x + 1\}$. If 2 is the number in the range, then $y = 2$. Then

$$2 = 2x + 1$$
$$x = \tfrac{1}{2}$$

Hence, $g^{-1}(2) = \tfrac{1}{2}$.

Alternate solution In Example 9 we showed that if $f(x) = 2x + 1$, then $f^{-1}(x) = \frac{x - 1}{2}$. Thus $f^{-1}(2) = \frac{2 - 1}{2} = \frac{1}{2}$.

Exercises 7.4

For Exercises 1–12 find the inverse of the given relation.

1. $\{(1, 3)\}$
2. $\{(2, 1), (3, -1)\}$
3. $\{(1, 1), (2, 2)\}$
4. $\{(0, 3), (-2, 4)\}$
5. $\{2\} \times \{3, 4\}$
6. $\{1, 2\} \times \{3, 4\}$
7. $R \times \{3\}$
8. $\{1, 2\} \times R$
9. $\{(x, y): x + 2y = 6\}$
10. $\{(x, y): x^2 + y = 1\}$
11. $\{(x, y): x < 2y + 3\}$
12. $\{(x, y): y \geq 3x - 1\}$

7.4 The Inverse of a Relation

For Exercises 13–20 the graph of a relation is given. Sketch the graph of its inverse.

13.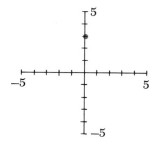

14.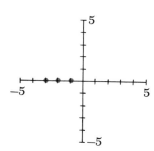

15.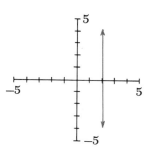

16.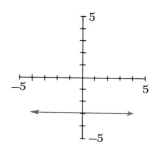

17.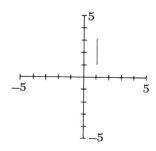

18.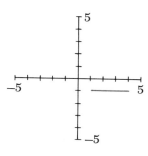

19.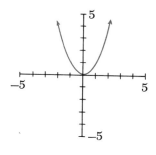

20.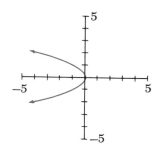

284   *Functions and Relations*

For Exercises 21–24 the graph of a relation is given. Find the domain and range of the *inverse* of this relation.

21.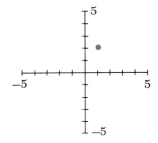

22.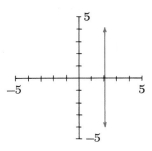

23.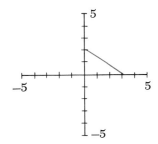

24.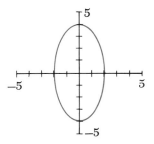

For Exercises 25–28 the graph of a relation is given. Determine whether or not its *inverse* is a function.

25.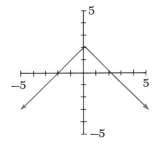

26.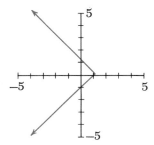

7.4 The Inverse of a Relation

27.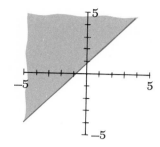

28.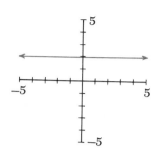

For Exercises 29–36 find the expression $f^{-1}(x)$ which defines the inverse of the given function.

29. $f(x) = 3x$
30. $f(x) = x/5$
31. $f(x) = x$
32. $f(x) = x + 1$
33. $f(x) = 2x - 3$
34. $f(x) = \dfrac{x - 5}{3}$
35. $f(x) = 4 - 3x$
36. $f(x) = 3x + 1$

For Exercises 37–40 find $f^{-1}(3)$.

37. $f(x) = 2x - 1$
38. $f(x) = 2 - x$
39. $f(x) = 3 - 2x$
40. $f(x) = 4x + 11$

$(-x)^b = (-1)^b x^{ab}$ ← sample ? sign

$(x^a)^b = x^{ab}$

Expressions with Radicals Chapter EIGHT

8.1 Roots and Radicals

When the exponent of an expression is a natural number, it indicates how many times the base is to be used as a factor. And since $3^2 = 3 \cdot 3 = 9$, we say that 9 is the second power (or square) of 3. We shall also refer to the base as a **root** and say that 3 is a second or square root of 9. Similarly, $2^3 = 2 \cdot 2 \cdot 2 = 8$, and we say that 8 is the third power (or cube) of 2, while 2 is a third root (or cube root) of 8. We observe that both -3 and 3 are square roots of 9, for $(-3)^2 = 9$ and $3^2 = 9$. But -2 is *not* a cube root of 8, since $(-2)^3 = (-2)(-2)(-2) = -8$.

Example 1 Find all real numbers which are:
(a) Square roots of 4   (b) Squares of 4
(c) Cube roots of 27   (d) Cubes of 27

Solution

(a) Both 2 and -2, since $2^2 = 4$ and $(-2)^2 = 4$
(b) Only 16, since $4^2 = 16$
(c) Only 3, since $3^3 = 27$, but $(-3)^3 = -27$
(d) Only 19,683, since $27^3 = 19,683$

While 3 is the only *real* number whose cube is 27 and hence 3 is the only *real* cube root of 27, there exist other *nonreal* numbers whose cubes are also 27. We shall confine our attention in this book to real numbers exclusively and will not further qualify our answers.

We may similarly define fourth, fifth, sixth roots, etc.

Example 2 Find all the roots indicated.

(a) Square roots of 36
(b) Cube roots of -27
(c) Fourth roots of 16
(d) Fifth roots of 32
(e) Sixth roots of 1
(f) Square roots of 0
(g) Square roots of -4

Solution

(a) 6 and -6
(b) -3, since $(-3)^3 = -27$
(c) 2 and -2, since $2^4 = 16$ and $(-2)^4 = 16$
(d) 2, since $2^5 = 32$
(e) 1 and -1, since $1^6 = (-1)^6 = 1$
(f) Only 0, since $0^2 = 0$
(g) There is no (real) square root of -4. Note that $(-2)^2 = 4$ and $2^2 = 4$, so neither -2 nor 2 is a square root of -4.

Let us agree that a square root, a fourth root, a sixth root, etc. is an *even* root, while a cube root, a fifth root, a seventh root, etc. is an *odd* root. Example 2 illustrates some general observations which we may make about roots.

1. Every positive number has two even roots, and each is the additive inverse of the other.
2. A negative number does not have an even root.
3. Every positive number has one odd root which is positive.
4. Every negative number has one odd root which is negative.
5. Zero has only one (even or odd) root and it is 0.

When a number has both a positive and a negative root, we shall say that the positive root is its principal root. Thus, both 5 and -5 are square roots of 25, but only 5 is its principal square root. For numbers which have only *one* root, we shall call this root, positive or negative, the principal root. So 2 is the principal (only) cube root of 8, and -2 is the principal cube root of -8.

The symbol $\sqrt[2]{16}$ is used to indicate the *principal* square root of 16.

Thus $\sqrt[2]{16} = 4$. We shall call $\sqrt{}$ the **radical sign** and 2 the **index** of this radical. Note carefully that 16 has *two* square roots, but $\sqrt[2]{16}$ represents only *one* of these roots, the *positive* one. The symbol $-\sqrt[2]{16}$ is used to represent the additive inverse of $\sqrt[2]{16}$, and we see that $-\sqrt[2]{16} = -4$, which is the negative square root of 16. Similarly, $\sqrt[3]{27}$ represents the principal (only) cube root of 27, and $\sqrt[3]{27} = 3$. And $-\sqrt[3]{27}$ represents the additive inverse of $\sqrt[3]{27}$, so $-\sqrt[3]{27} = -3$. In general, the index indicates the root desired. Thus $\sqrt[4]{16} = 2$, $\sqrt[5]{-32} = -2$, etc. Of course, since negative numbers do not have even roots, symbols such as $\sqrt[2]{-1}$ and $\sqrt[4]{-16}$ do not represent any *real* number. When the index is 2, it is customary to omit it. Thus, $\sqrt{25} = \sqrt[2]{25} = 5$. But we shall never omit the index when it is a natural number other than 2.

Example 3 Find the simplest form of each of these, if the symbol represents a real number.

(a) $\sqrt{49}$ (b) $\sqrt[3]{64}$ (c) $\sqrt[4]{1}$ (d) $-\sqrt{36}$
(e) $\sqrt[3]{-27}$ (f) $-\sqrt[6]{1}$ (g) $\sqrt{-25}$ (h) $-\sqrt[3]{8}$
(i) $\sqrt[3]{0}$ (j) $-\sqrt[3]{-8}$ (k) $\sqrt{\frac{1}{4}}$

Solution

(a) $\sqrt{49} = 7$ (b) $\sqrt[3]{64} = 4$
(c) $\sqrt[4]{1} = 1$ (d) $-\sqrt{36} = -6$
(e) $\sqrt[3]{-27} = -3$ (f) $-\sqrt[6]{1} = -1$
(g) $\sqrt{-25}$ is not a real number
(h) $-\sqrt[3]{8} = -2$, since $\sqrt[3]{8} = 2$
(i) $\sqrt[3]{0} = 0$ (j) $-\sqrt[3]{-8} = -(-2) = 2$
(k) $\sqrt{\frac{1}{4}} = \frac{1}{2}$, since $(\frac{1}{2})^2 = \frac{1}{4}$

Although the real numbers in Example 3 are all *rational*, many numbers which are represented by the radical sign are *irrational*. In Section 2.6 we said that $\sqrt{2}$ represents an irrational number. Recall that a rational number is one which may be represented *exactly* by a fraction whose numerator and denominator are integer symbols. We shall prove, by a rather indirect method, that there is no such fraction which represents $\sqrt{2}$. We shall assume that there *is* such a fraction and then show that this leads to an obvious contradiction. Then we must conclude that our assumption was, in fact, not true.

Before we begin the argument, however, let us observe that if *a* represents an *even* number, then 2 is a factor of a, and we may write $a = 2c$, where c is some integer. For example, 14 is even and $14 = 2 \cdot 7$. Also, we observe that the square of any even number is even ($8^2 = 64$), but the square of any odd number is odd ($7^2 = 49$). Thus, the square *root* of an even number cannot be odd.

Next we observe that when a rational number is in simplest form, its numerator and denominator cannot *both* be even. (For then we would be able to cancel a factor; for example, $\frac{2}{6} = \frac{1}{3}$.) Thus if a/b is the simplest form of a rational number, then a and b are not *both* even.

Now we are ready to make the assumption which we will show to be false. Suppose that $\sqrt{2}$ *is* a rational number. Then it may be represented *exactly* by a fraction. Let a/b be the simplest form of this rational number, $a/b = \sqrt{2}$. Then the square of a/b must be 2.

$$\left(\frac{a}{b}\right)^2 = 2$$

$$\frac{a^2}{b^2} = 2$$

(1) $$a^2 = 2b^2 \qquad \text{[Multiplying by } b^2\text{]}$$

From Equation (1) we see that a^2 is an *even* number (since 2 is one of its factors). Then a itself must *also* be even. Hence, $a = 2c$, and

(2) $$a^2 = (2c)^2 = 4c^2$$

Comparing Equations (1) and (2), we see that

$$2b^2 = 4c^2$$

$$b^2 = 2c^2$$

Hence, b^2 is even (it has the factor 2, since $b^2 = 2c^2$), and b is also even. Now we have a contradiction, for a and b cannot both be even. We must conclude that our *assumption* was not true, and hence $\sqrt{2}$ is *not* a rational number.

We shall not do so here, but it can be shown that symbols such as $\sqrt{3}$, $\sqrt[3]{2}$, and $\sqrt[4]{17}$ also represent irrational numbers. In fact, the use of radicals is the easiest way to represent many (but not all) irrational numbers. We have pointed out in Section 2.6 that it is possible to represent any irrational number by a rational number *approximation*. Thus we may replace $\sqrt{2}$ by 1.4142 and $\sqrt{3}$ by 1.7321, but these decimal forms do not *exactly* represent the irrational numbers $\sqrt{2}$ and $\sqrt{3}$. In this book we shall rarely need to make such approximations and shall prefer to represent an irrational number by means of a radical or equivalent symbol, if possible.

Example 4 A function with a finite domain is given. Find its range.
(a) $f(x) = \sqrt{3x + 4}$, $\{4\}$
(b) $g(x) = -\sqrt{8 - x^2}$, $\{-2\}$
(c) $h(x) = x^2$, $\{\sqrt{3}, -\sqrt{5}\}$

Solution

(a) We note that the bar $\overline{}$ is used here as a symbol of grouping. Then $f(4) = \sqrt{3 \cdot 4 + 4} = \sqrt{16} = 4$. The range of f is $\{4\}$.
(b) $g(-2) = -\sqrt{8 - (-2)^2} = -\sqrt{4} = -2$. The range is $\{-2\}$.
(c) $h(\sqrt{3}) = (\sqrt{3})^2 = 3$; $h(-\sqrt{5}) = (-\sqrt{5})^2 = 5$. The range is $\{3, 5\}$.

Example 5 For each equation a finite domain is given. By making all replacements find the solution set.
(a) $\sqrt{x + 3} = 4$, $\{1, -1, 13, -19\}$
(b) $x^2 = 5$, $\{5, -5, \sqrt{5}, -\sqrt{5}\}$
(c) $\sqrt{2x + 5} = -1$, $\{0, -2, -3\}$

Solution

(a) $\sqrt{x + 3} = 4$

$x = 1$: $\sqrt{1 + 3} = 4$ False

$x = -1$: $\sqrt{-1 + 3} = 4$ False

$x = 13$: $\sqrt{13 + 3} = 4$ True

$x = -19$: $\sqrt{-19 + 3} = 4$ False

The solution set is $\{13\}$
(b) $x^2 = 5$

$x = 5$: $5^2 = 5$ False

$x = -5$: $(-5)^2 = 5$ False

$x = \sqrt{5}$: $(\sqrt{5})^2 = 5$ True

$x = -\sqrt{5}$: $(-\sqrt{5})^2 = 5$ True

The solution set is $\{\sqrt{5}, -\sqrt{5}\}$
(c) $\sqrt{2x + 5} = -1$

$x = 0$: $\sqrt{5} = -1$ False

291   8.1 *Roots and Radicals*

$$x = -2: \quad \sqrt{1} = -1 \quad \text{False}$$
$$x = -3: \quad \sqrt{-1} = -1 \quad \text{False}$$

The solution set is $\emptyset$. In fact we see that the sentence $\sqrt{2x+5} = -1$ becomes a false statement for *every* replacement of x, since a radical expression with an even index always represents a nonnegative number. Then even with domain R this equation has solution set $\emptyset$.

Exercises 8.1

For each Exercise express rational numbers in simplest form and irrational numbers in radical form.

For Exercises 1–12 list all the numbers which are:

1. Square roots of 1
2. Squares of 16
3. Square roots of 5
4. Square roots of 11
5. Cubes of 8
6. Cube roots of -64
7. Cube roots of -3
8. Cube roots of 125
9. Fourth roots of 16
10. Fifth roots of -1
11. Fourth roots of 3
12. Fourth roots of 0

For Exercises 13–32 indicate whether the given number is rational, irrational or not real and find the simplest form if it is rational.

13. $\sqrt{64}$
14. $\sqrt{20}$
15. $-\sqrt{25}$
16. $\sqrt{-16}$
17. $\sqrt[3]{1}$
18. $\sqrt[3]{-1}$
19. $\sqrt[3]{4}$
20. $-\sqrt[3]{8}$
21. $\sqrt[4]{-1}$
22. $\sqrt[5]{-1}$
23. $\sqrt[5]{0}$
24. $-\sqrt[4]{1}$
25. $-\sqrt{0}$
26. $-\sqrt[3]{-1}$
27. $-\sqrt[4]{-1}$
28. $-\sqrt[5]{1}$
29. $\sqrt{\frac{1}{9}}$
30. $\sqrt{.25}$
31. $\sqrt{\frac{4}{25}}$
32. $\sqrt{.04}$

For Exercises 33–40 a function with a finite domain is given. Find its range.

33. $f(x) = \sqrt{2x - 1}$, $\{1, 2, 3\}$
34. $f(x) = \sqrt{4 - x}$, $\{2, -2, 4\}$
35. $f(x) = -\sqrt{3x + 1}$, $\{0, 1, 2\}$
36. $f(x) = -\sqrt{x + 5}$, $\{-4, -1, 0\}$
37. $f(x) = \sqrt[3]{x + 3}$, $\{-4, 5\}$
38. $f(x) = -\sqrt[3]{x^2 - 1}$, $\{0, 1, 2\}$
39. $f(x) = x^2$, $\{\sqrt{2}, -\sqrt{3}\}$
40. $f(x) = -x^2$, $\{\sqrt{7}, -\sqrt{11}\}$

For Exercises 41–48 an equation with finite domain is given. Find its solution set by making all indicated replacements.

41. $\sqrt{x} = 9$, $\{3, 9, 81\}$
42. $\sqrt{x + 3} = 2$, $\{0, 1, 2\}$
43. $\sqrt{1 - 3x} = 4$, $\{-5, -1, 2\}$
44. $\sqrt[3]{x + 3} = -1$, $\{-4, -1, 1\}$
45. $x^2 = 3$, $\{9, \sqrt{3}, -\sqrt{3}, \sqrt{-3}\}$
46. $x^2 = -2$, $\{2, -2, \sqrt{2}, -\sqrt{2}\}$
47. $x^3 = 2$, $\{8, 2, \sqrt{2}, \sqrt[3]{2}, -\sqrt[3]{2}\}$
48. $x^2 = 7$, $\{\sqrt{7}, -\sqrt{7}, \sqrt{-7}\}$

8.2 Rational Number Exponents

In Chapter 4 we stated three theorems about powers whose exponents are natural numbers. Then in Chapter 5 we defined powers whose exponents are integers in such a way that these same three theorems remained true even for negative integers. But expressions such as $5^{1/2}$, $4^{2/3}$ and $7^{-5/4}$, whose exponents are rational numbers, have not as yet been given any meaning. Let us now make definitions which will (where possible) preserve these same three theorems for powers whose exponents are rational numbers. We shall restate the theorems of Chapter 5.

Theorem 1	If $a \in I$ and $b \in I$, then $\quad x^a \cdot x^b = x^{a+b}$
Theorem 2	If $a \in I$, then $\qquad\qquad x^a \cdot y^a = (xy)^a$
Theorem 3	If $a \in I$ and $b \in I$, then $\quad (x^a)^b = x^{ab}$

8.2 Rational Number Exponents

We first consider $5^{1/2}$ and its square $(5^{1/2})^2$. If Theorem 3 is to be true for rational number exponents, $(5^{1/2})^2 = 5^{(1/2) \cdot 2} = 5^1 = 5$. Then since the square of $5^{1/2}$ is 5, we see that $5^{1/2}$ must be a square root of 5. Let us avoid any ambiguity and at the same time make use of the radical notation introduced in the preceding section. We shall define $5^{1/2}$ to be the principal or positive square root of 5: $5^{1/2} = \sqrt{5}$. Again, if Theorem 3 is to be true, $(16^{1/2})^2 = 16^{(1/2) \cdot 2} = 16^1 = 16$. Thus, since 16 is the square of $16^{1/2}$, then $16^{1/2}$ must be a square root of 16. To avoid ambiguity we again define $16^{1/2} = \sqrt{16} = 4$. Similar definitions give $8^{1/2} = \sqrt{8}$ and $17^{1/2} = \sqrt{17}$, etc. The expression $-4^{1/2}$ represents the additive inverse of $4^{1/2}$. Hence, $-4^{1/2} = -\sqrt{4} = -2$. However, $(-4)^{1/2}$ would logically represent a square root of -4. Since no such real number exists, we shall *not* define $(-4)^{1/2}$ at all. But for any $x \in [0, \infty)$ we make the general definition, $x^{1/2} = \sqrt{x}$.

Next we consider $4^{1/3}$. If Theorem 3 is to be true, $(4^{1/3})^3 = 4^{(1/3) \cdot 3} = 4^1 = 4$, and so $4^{1/3}$ is a cube root of 4. We may again use the radical notation to write $4^{1/3} = \sqrt[3]{4}$. Similarly, we define $9^{1/3} = \sqrt[3]{9}$ and $(-8)^{1/3} = \sqrt[3]{-8} = -2$. In general, we make the definition $x^{1/3} = \sqrt[3]{x}$ for all $x \in R$. We shall define $7^{1/4} = \sqrt[4]{7}$, the principal fourth root of 7, and $(-32)^{1/5} = \sqrt[5]{-32} = -2$. In fact we make the definitions $x^{1/4} = \sqrt[4]{x}$ if $x \in [0, \infty)$ and $x^{1/5} = \sqrt[5]{x}$ for all $x \in R$.

The pattern of definitions is clear.

DEFINITION 1

If b is a natural number greater than 1, then $x^{1/b} = \sqrt[b]{x}$, where $x \geq 0$, if b is even.

Example 1 Use Definition 1 to find a radical form for each of the following. Find the simplest form for any numbers which are rational.
(a) $25^{1/2}$ (b) $5^{1/3}$ (c) $16^{1/4}$ (d) $-9^{1/2}$
(e) $(-27)^{1/3}$ (f) $-1^{1/4}$ (g) $-(-1)^{1/5}$ (h) $0^{1/2}$

Solution

(a) $25^{1/2} = \sqrt{25} = 5$ (b) $5^{1/3} = \sqrt[3]{5}$
(c) $16^{1/4} = \sqrt[4]{16} = 2$ (d) $-9^{1/2} = -\sqrt{9} = -3$
(e) $(-27)^{1/3} = \sqrt[3]{-27} = -3$ (f) $-1^{1/4} = -\sqrt[4]{1} = -1$
(g) $-(-1)^{1/5} = -\sqrt[5]{-1} = -(-1) = 1$
(h) $0^{1/2} = \sqrt{0} = 0$

294 Expressions with Radicals

Definition 1 gives meaning to a power only when the exponent is a quotient number represented by a fraction whose numerator is 1. Next we shall consider expressions which have other quotient number exponents. If Theorem 3 is to be true, we must have $4^{2/3} = 4^{(1/3) \cdot 2} = (4^{1/3})^2$. But $4^{1/3}$ has been defined to be $\sqrt[3]{4}$. Then $4^{2/3} = (4^{1/3})^2 = (\sqrt[3]{4})^2$. Similarly, $27^{2/3} = (27^{1/3})^2 = (\sqrt[3]{27})^2 = 3^2 = 9$. More generally, if Theorem 3 is to be true, we must make the following definition.

DEFINITION 2

If $a \in N$ and $b \in N$ $(b > 1)$, then $x^{a/b} = (\sqrt[b]{x})^a$, where $x \geq 0$ if b is even.

Note that in Definition 2 the denominator becomes the index of the radical, and the numerator becomes the exponent of the power.

Example 2 Use Definitions 1 and 2 to find a radical form of each of the following. Find the simplest form for any number which is rational.

(a) $4^{3/2}$ (b) $(-8)^{4/3}$ (c) $7^{5/4}$ (d) $-64^{2/3}$

Solution

(a) $4^{3/2} = (\sqrt{4})^3 = 2^3 = 8$
(b) $(-8)^{4/3} = (\sqrt[3]{-8})^4 = (-2)^4 = 16$
(c) $7^{5/4} = (\sqrt[4]{7})^5$
(d) $-64^{2/3} = -(\sqrt[3]{64})^2 = -(4)^2 = -16$

Next we shall consider powers whose exponents are *negative* rational numbers. In order to preserve Theorems 1, 2, and 3, we shall make a definition which is analogous to the one we made in Section 5.7 for powers whose exponents are negative *integers*.

DEFINITION 3

If $x \neq 0$, $a \in N$ and $b \in N$ $(b > 1)$, then $x^{-a/b} = 1/x^{a/b}$, where $x > 0$ if b is even.

Example 3 Use Definitions 1, 2, and 3 to find a radical form of each of the following. Find the simplest form of any number which is rational.

(a) $5^{-1/3}$ (b) $25^{-1/2}$ (c) $4^{-3/2}$
(d) $(-27)^{-2/3}$ (e) $-16^{-1/2}$

8.2 Rational Number Exponents

Solution

(a) $5^{-1/3} = \dfrac{1}{5^{1/3}} = \dfrac{1}{\sqrt[3]{5}}$

(b) $25^{-1/2} = \dfrac{1}{25^{1/2}} = \dfrac{1}{\sqrt{25}} = \dfrac{1}{5}$

(c) $4^{-3/2} = \dfrac{1}{4^{3/2}} = \dfrac{1}{(\sqrt{4})^3} = \dfrac{1}{2^3} = \dfrac{1}{8}$

(d) $(-27)^{-2/3} = \dfrac{1}{(-27)^{2/3}} = \dfrac{1}{(\sqrt[3]{-27})^2} = \dfrac{1}{(-3)^2} = \dfrac{1}{9}$

(e) $-16^{-1/2} = -\dfrac{1}{16^{1/2}} = \dfrac{-1}{4}$

The definitions of this section have given meaning to most, but not all, expressions which are powers with rational number exponents. When we restrict them to only those expressions which we *have* defined, Theorems 1 and 2 are preserved for all rational number exponents. However, Theorem 3 will sometimes not be true when the base is a negative number even though all expressions are defined.* Then we shall restrict Theorem 3 to only those powers with a nonnegative base. We restate the theorems for rational exponents.

Theorem 1	If $a \in F$ and $b \in F$, then	$x^a \cdot x^b = x^{a+b}$
Theorem 2	If $a \in F$, then	$x^a \cdot y^a = (xy)^a$
Theorem 3	If $a \in F$ and $b \in F$, and $x \geq 0$, then	$(x^a)^b = x^{ab}$
	(All powers must represent real numbers.)	

Example 4 Find the simplest form for each of the following rational numbers.
(a) $27^{1/2} \cdot 27^{5/6}$ (b) $2^{-1/3} \cdot 4^{-1/3}$
(c) $(4^{5/6})^3$ (d) $25^{1/2} + 27^{2/3}$

Solution

(a) By Theorem 1 we have

---

*$[(-1)^2]^{1/2} \neq (-1)^{2 \cdot (1/2)}$. For $[(-1)^2]^{1/2} = [1]^{1/2} = 1$, but $(-1)^{2 \cdot (1/2)} = (-1)^1 = -1$.

296 Expressions with Radicals

$$27^{1/2} \cdot 27^{5/6} = 27^{1/2+5/6} = 27^{4/3} = (\sqrt[3]{27})^4 = 3^4 = 81$$

(b) By Theorem 2

$$2^{-1/3} \cdot 4^{-1/3} = (2 \cdot 4)^{-1/3} = 8^{-1/3} = \frac{1}{8^{1/3}} = \frac{1}{2}$$

(c) By Theorem 3

$$(4^{5/6})^3 = 4^{(5/6)\cdot 3} = 4^{5/2} = (\sqrt{4})^5 = 2^5 = 32$$

(d) Theorems 1, 2, and 3 cannot be used here. But

$$25^{1/2} + 27^{2/3} = \sqrt{25} + (\sqrt[3]{27})^2 = 5 + 3^2 = 5 + 9 = 14$$

Example 5 A function and its domain are given. Find the range.
(a) $f(x) = x^{2/3}$, $\{0, 1, -1, 8, -8\}$
(b) $g(x) = x^{-1/2}$, $\{9, \frac{1}{4}\}$
(c) $h(x) = -x^{-1/3}$, $\{8, -8, \frac{1}{8}\}$

Solution

(a) $f(0) = 0^{2/3} = 0$
$f(1) = 1^{2/3} = 1$
$f(-1) = (-1)^{2/3} = (\sqrt[3]{-1})^2 = (-1)^2 = 1$
$f(8) = 8^{2/3} = (\sqrt[3]{8})^2 = 2^2 = 4$
$f(-8) = (-8)^{2/3} = (\sqrt[3]{-8})^2 = (-2)^2 = 4$

The range is $\{0, 1, 4\}$

(b) $g(9) = 9^{-1/2} = \frac{1}{9^{1/2}} = \frac{1}{3}$. Since $\frac{1}{4} = 4^{-1}$, we may write

$$g(\tfrac{1}{4}) = (\tfrac{1}{4})^{-1/2} = (4^{-1})^{-1/2} = 4^{(-1)(-1/2)} = 4^{1/2} = 2$$

The range is $\{\frac{1}{3}, 2\}$

(c) $h(8) = -8^{-1/3} = \frac{-1}{8^{1/3}} = \frac{-1}{2}$

$h(-8) = -(-8)^{-1/3} = \frac{-1}{(-8)^{1/3}} = \frac{-1}{-2} = \frac{1}{2}$

Since $\frac{1}{8} = \frac{1}{2^3} = 2^{-3}$, then

$$h(\tfrac{1}{8}) = -(\tfrac{1}{8})^{-1/3} = -(2^{-3})^{-1/3} = -[2^{(-3)(-1/3)}]$$
$$= -[2^1] = -2.$$

The range is $\{-\frac{1}{2}, \frac{1}{2}, -2\}$

Exercises 8.2

For Exercises 1–48 find a radical form. Find the simplest form for any numbers which are rational.

1. $3^{1/2}$
2. $10^{1/2}$
3. $-5^{1/2}$
4. $-7^{1/2}$
5. $36^{1/2}$
6. $-49^{1/2}$
7. $-100^{1/2}$
8. $81^{1/2}$
9. $4^{1/3}$
10. $(-6)^{1/3}$
11. $125^{1/3}$
12. $(-64)^{1/3}$
13. $10^{1/4}$
14. $32^{1/5}$
15. $-1^{1/6}$
16. $-3^{1/4}$
17. $2^{3/2}$
18. $3^{5/2}$
19. $9^{3/2}$
20. $-4^{3/2}$
21. $2^{2/3}$
22. $-5^{2/3}$
23. $8^{4/3}$
24. $(-8)^{5/3}$
25. $1^{5/4}$
26. $(-1)^{2/3}$
27. $(-1)^{3/5}$
28. $-1^{4/3}$
29. $0^{2/3}$
30. $0^{5/4}$
31. $-0^{2/5}$
32. $(-0)^{5/3}$
33. $2^{-1/2}$
34. $3^{-1/2}$
35. $5^{-1/3}$
36. $10^{-1/3}$
37. $4^{-1/2}$
38. $9^{-1/2}$
39. $8^{-1/3}$
40. $27^{-1/3}$
41. $2^{-2/3}$
42. $4^{-3/2}$
43. $-4^{-1/2}$
44. $-4^{-3/2}$
45. $(-8)^{-2/3}$
46. $-8^{-2/3}$
47. $(-1)^{-4/3}$
48. $-1^{-2/3}$

Each of the expressions in Exercises 49–60 represents a rational number. Find its simplest form.

49. $3^{1/2} \cdot 12^{1/2}$
50. $2^{1/2} \cdot 8^{1/2}$
51. $4^{1/3} \cdot 4^{2/3}$
52. $(-3)^{1/3} \cdot (-3)^{2/3}$
53. $(4^{1/4})^2$
54. $(9^{1/4})^{-2}$
55. $(9^{-3})^{1/6}$
56. $(4^{-2})^{-1/2}$
57. $16^{1/12} \cdot 16^{1/6}$
58. $8^{-1/2} \cdot 8^{5/6}$
59. $3^{1/2}(3^{1/2} + 3^{3/2})$
60. $4^{1/8} \cdot 4^{-5/8}$

For Exercises 61–68 a function with a finite domain is given. Find its range.

61. $f(x) = x^{1/3}$, $\{0, -1, \frac{1}{27}\}$
62. $f(x) = -x^{1/2}$, $\{1, 4, \frac{1}{9}\}$
63. $f(x) = (-x)^{-1/3}$, $\{1, 8\}$
64. $f(x) = -x^{2/3}$, $\{0, -8, \frac{1}{8}\}$

65. $f(x) = 4^x$, $\{\frac{1}{2}, -\frac{1}{2}\}$
66. $f(x) = 1^x$, $\{\frac{2}{3}, -\frac{3}{4}, \frac{5}{6}\}$
67. $f(x) = (x^2)^{1/2}$, $\{2, 3, 4\}$
68. $f(x) = (x^2)^{1/2}$, $\{-2, -3, -4\}$

8.3 Simplifying Radical Expressions

When an expression contains more than one radical, it is sometimes possible to replace it with an equivalent expression with just one radical. The next two theorems show how this can be done when the expression contains the product or quotient of two square root radicals.

Theorem 4

> If $a \geq 0$ and $b \geq 0$, then $\sqrt{a} \cdot \sqrt{b} = \sqrt{ab}$

Proof If $a \geq 0$ and $b \geq 0$, then

$$\sqrt{a} \cdot \sqrt{b} = a^{1/2} \cdot b^{1/2} \qquad \text{[Definition 1]}$$
$$= (ab)^{1/2} \qquad \text{[Theorem 2]}$$
$$= \sqrt{ab} \qquad \text{[Definition 1]}$$

Theorem 5

> If $a \geq 0$ and $b > 0$, then $\dfrac{\sqrt{a}}{\sqrt{b}} = \sqrt{\dfrac{a}{b}}$

The proof is not presented but can be completed by using the definitions and theorems of the preceding section.

Example 1 Use Theorems 4 and 5 to simplify each of the following expressions.

(a) $\sqrt{3} \cdot \sqrt{2}$
(b) $\dfrac{\sqrt{80}}{\sqrt{5}}$
(c) $\sqrt{5} \cdot \sqrt{20}$
(d) $\dfrac{\sqrt{2}}{\sqrt{8}}$

Solution

(a) $\sqrt{3} \cdot \sqrt{2} = \sqrt{3 \cdot 2} = \sqrt{6}$

8.3 Simplifying Radical Expressions

(b) $\dfrac{\sqrt{80}}{\sqrt{5}} = \sqrt{\dfrac{80}{5}} = \sqrt{16} = 4$

(c) $\sqrt{5} \cdot \sqrt{20} = \sqrt{100} = 10$

(d) $\dfrac{\sqrt{2}}{\sqrt{8}} = \sqrt{\dfrac{2}{8}} = \sqrt{\dfrac{1}{4}} = \dfrac{1}{2}$

Sometimes we will want to use Theorems 4 and 5 to change the form of an expression which contains only a single radical.

Example 2 Express $\sqrt{32}$ as the product of an integer and an irrational number.

Solution There is more than one possible answer. Since $32 = 16 \cdot 2$ and $\sqrt{16} = 4$, we may write $\sqrt{32} = \sqrt{16 \cdot 2} = \sqrt{16} \cdot \sqrt{2} = 4\sqrt{2}$. But also, $\sqrt{32} = \sqrt{4 \cdot 8} = \sqrt{4} \cdot \sqrt{8} = 2\sqrt{8}$. We shall usually (but not always) want the integer to be as large as possible. Hence, we usually prefer $\sqrt{32} = 4\sqrt{2}$.

Example 3 Express $\sqrt{\tfrac{5}{8}}$ as the quotient of two numbers, one of which is an integer.

Solution Again, there is more than one solution.

Since

$$\dfrac{5}{8} = \dfrac{10}{16}, \qquad \sqrt{\dfrac{5}{8}} = \sqrt{\dfrac{10}{16}} = \dfrac{\sqrt{10}}{\sqrt{16}} = \dfrac{\sqrt{10}}{4}$$

Or since

$$\dfrac{5}{8} = \dfrac{40}{64}, \qquad \sqrt{\dfrac{5}{8}} = \sqrt{\dfrac{40}{64}} = \dfrac{\sqrt{40}}{\sqrt{64}} = \dfrac{\sqrt{40}}{8}$$

And since

$$\dfrac{5}{8} = \dfrac{25}{40}, \qquad \sqrt{\dfrac{5}{8}} = \sqrt{\dfrac{25}{40}} = \dfrac{\sqrt{25}}{\sqrt{40}} = \dfrac{5}{\sqrt{40}}$$

Indeed, there are many possible solutions.

As Examples 2 and 3 show, there may be a number of ways to alter the form of an expression which contains radicals. When the expression represents a rational number, we shall always want to find its simplest form. But when it represents an irrational number, there is no easily defined simplest form, and the form we choose will depend on the instructions in the problem, the use to which the result is to be put, or sometimes just on our personal whim.

Sometimes we may use the distributive axiom to simplify sums or differences of radicals.

Example 4 Simplify each of the following.
(a) $2\sqrt{3} + 5\sqrt{3}$
(b) $4\sqrt{5} - \sqrt{5}$
(c) $\sqrt{2} + \sqrt{8}$
(d) $\sqrt{27} - \sqrt{12}$

Solution

(a) $2\sqrt{3} + 5\sqrt{3} = (2 + 5)\sqrt{3} = 7\sqrt{3}$
(b) $4\sqrt{5} - \sqrt{5} = (4 - 1)\sqrt{5} = 3\sqrt{5}$
(c) The repeated factor is $\sqrt{2}$, since $\sqrt{8} = \sqrt{4} \cdot \sqrt{2}$.
$\sqrt{2} + \sqrt{8} = \sqrt{2} + \sqrt{4} \cdot \sqrt{2} = \sqrt{2} + 2\sqrt{2} = (1 + 2)\sqrt{2} = 3\sqrt{2}$
Note that $\sqrt{2} + \sqrt{8} \neq \sqrt{10}$.
(d) $\sqrt{27} = \sqrt{9} \cdot \sqrt{3} = 3\sqrt{3}$; $\sqrt{12} = \sqrt{4} \cdot \sqrt{3} = 2\sqrt{3}$. Then
$\sqrt{27} - \sqrt{12} = 3\sqrt{3} - 2\sqrt{3} = \sqrt{3}$.

Or we may want to use the distributive axiom to replace products of radicals with equivalent sums.

Example 5 Replace each product with an equivalent sum.
(a) $\sqrt{2}(3 + \sqrt{2})$
(b) $(\sqrt{3} + 1)(\sqrt{3} + 4)$
(c) $(\sqrt{5} - 2)(\sqrt{5} + 2)$
(d) $(1 + \sqrt{2})^2$

Solution

(a) $\sqrt{2}(3 + \sqrt{2}) = \sqrt{2} \cdot 3 + \sqrt{2} \cdot \sqrt{2} = 3\sqrt{2} + 2$
(b) $(\sqrt{3} + 1)(\sqrt{3} + 4) = \sqrt{3} \cdot \sqrt{3} + \sqrt{3} \cdot 4 + 1 \cdot \sqrt{3} + 1 \cdot 4 = 3 + 4\sqrt{3} + \sqrt{3} + 4 = 7 + 5\sqrt{3}$
(c) $(\sqrt{5} - 2)(\sqrt{5} + 2) = \sqrt{5} \cdot \sqrt{5} - 2\sqrt{5} + 2\sqrt{5} - 2 \cdot 2$
$= 5 - 4 = 1$
(d) $(1 + \sqrt{2})^2 = 1^2 + 2 \cdot 1 \cdot \sqrt{2} + (\sqrt{2})^2 = 1 + 2\sqrt{2} + 2$
$= 3 + 2\sqrt{2}$

We will always want to cancel a factor which is common to both the numerator and denominator of a fraction.

Example 6 Simplify each of these.
(a) $\dfrac{4 + \sqrt{8}}{2}$
(b) $\dfrac{2 - \sqrt{48}}{6}$

8.3 Simplifying Radical Expressions

Solution

(a) $\dfrac{4+\sqrt{8}}{2} = \dfrac{4+\sqrt{4}\cdot\sqrt{2}}{2} = \dfrac{4+2\sqrt{2}}{2} = \dfrac{2(2+\sqrt{2})}{2} = 2+\sqrt{2}$

(b) $\dfrac{2-\sqrt{48}}{6} = \dfrac{2-\sqrt{16}\cdot\sqrt{3}}{6} = \dfrac{2-4\sqrt{3}}{6} = \dfrac{2(1-2\sqrt{3})}{2\cdot 3}$

$= \dfrac{1-2\sqrt{3}}{3}$

When both the numerator and denominator of a fraction contain radicals, we may sometimes prefer to replace it with an equivalent fraction with either the numerator or the denominator a rational number symbol. We shall call this process "rationalizing the numerator" or "rationalizing the denominator."

Example 7   Rationalize the numerator.

(a) $\dfrac{\sqrt{3}}{\sqrt{2}}$ (b) $\dfrac{\sqrt{5}}{\sqrt{5}-1}$

Solution

(a) $\dfrac{\sqrt{3}}{\sqrt{2}} = \dfrac{\sqrt{3}\cdot\sqrt{3}}{\sqrt{2}\cdot\sqrt{3}} = \dfrac{3}{\sqrt{6}}$. Note that we have multiplied both the numerator and denominator by $\sqrt{3}$.

(b) $\dfrac{\sqrt{5}}{\sqrt{5}-1} = \dfrac{\sqrt{5}\cdot\sqrt{5}}{(\sqrt{5}-1)\sqrt{5}} = \dfrac{5}{\sqrt{5}\cdot\sqrt{5}-\sqrt{5}} = \dfrac{5}{5-\sqrt{5}}$

Example 8   Rationalize the denominator.

(a) $\dfrac{3}{\sqrt{6}}$ (b) $\dfrac{1+\sqrt{2}}{\sqrt{3}}$ (c) $\dfrac{1}{\sqrt{2}-1}$

Solution

(a) $\dfrac{3}{\sqrt{6}} = \dfrac{3\cdot\sqrt{6}}{\sqrt{6}\cdot\sqrt{6}} = \dfrac{3\sqrt{6}}{6} = \dfrac{\sqrt{6}}{2}$

(b) $\dfrac{1+\sqrt{2}}{\sqrt{3}} = \dfrac{(1+\sqrt{2})\sqrt{3}}{\sqrt{3}\cdot\sqrt{3}} = \dfrac{\sqrt{3}+\sqrt{6}}{3}$

(c) From Example 5(c) we see that if we multiply $\sqrt{2}-1$ by $\sqrt{2}+1$ the result will be a rational number. Hence, we shall multiply both the numerator and the denominator of the given fraction by $\sqrt{2}+1$.

Expressions with Radicals

$$\frac{1}{\sqrt{2}-1} = \frac{1(\sqrt{2}+1)}{(\sqrt{2}-1)(\sqrt{2}+1)} = \frac{\sqrt{2}+1}{(\sqrt{2})^2-1} = \frac{\sqrt{2}+1}{2-1} = \sqrt{2}+1$$

Exercises 8.3

For Exercises 1–8 use Theorem 4 to simplify.

1. $\sqrt{3} \cdot \sqrt{5}$
2. $\sqrt{2} \cdot \sqrt{7}$
3. $\sqrt{2} \cdot \sqrt{8}$
4. $\sqrt{3} \cdot \sqrt{12}$
5. $\sqrt{7} \cdot \sqrt{28}$
6. $\sqrt{5} \cdot \sqrt{45}$
7. $\sqrt{2} \cdot \sqrt{3} \cdot \sqrt{6}$
8. $\sqrt{5} \cdot \sqrt{15} \cdot \sqrt{12}$

For Exercises 9–12 use Theorem 5 to simplify.

9. $\dfrac{\sqrt{27}}{\sqrt{3}}$
10. $\dfrac{\sqrt{20}}{\sqrt{5}}$
11. $\dfrac{\sqrt{3}}{\sqrt{75}}$
12. $\dfrac{\sqrt{7}}{\sqrt{63}}$

For Exercises 13–20 replace the given expression by the product of an integer and the square root of an integer (with the first factor as large as possible).

13. $\sqrt{8}$
14. $\sqrt{12}$
15. $\sqrt{27}$
16. $\sqrt{45}$
17. $\sqrt{48}$
18. $\sqrt{72}$
19. $\sqrt{80}$
20. $\sqrt{243}$

For Exercises 21–24 replace the given expression by an equivalent fraction whose numerator represents the square root of an integer and whose denominator represents an integer (which is as small as possible).

21. $\sqrt{\frac{2}{3}}$
22. $\sqrt{\frac{3}{5}}$
23. $\sqrt{\frac{5}{8}}$
24. $\sqrt{\frac{7}{12}}$

For Exercises 25–32 replace each expression by an equivalent one with as few terms as possible.

25. $\sqrt{8} + \sqrt{18}$
26. $\sqrt{3} + \sqrt{75}$
27. $\sqrt{45} - \sqrt{20}$
28. $\sqrt{125} - \sqrt{80}$
29. $3\sqrt{8} + \sqrt{50}$
30. $2\sqrt{8} - \sqrt{32}$
31. $4\sqrt{3} - 3\sqrt{12}$
32. $5\sqrt{48} + 2\sqrt{3}$

8.3 Simplifying Radical Expressions

For Exercises 33–40 replace each product by an equivalent sum with as few terms as possible.

33. $\sqrt{3}(1 + \sqrt{3})$
34. $\sqrt{5}(\sqrt{5} - 2)$
35. $(\sqrt{2} + 1)(\sqrt{2} + 3)$
36. $(\sqrt{3} - 2)(\sqrt{3} - 5)$
37. $(\sqrt{5} + 2)(\sqrt{5} - 2)$
38. $(\sqrt{3} + 4)(\sqrt{3} - 4)$
39. $(1 + \sqrt{3})^2$
40. $(2 - \sqrt{5})^2$

For Exercises 41–48 simplify by canceling a factor which is common to both the numerator and denominator.

41. $\dfrac{2 + \sqrt{8}}{2}$
42. $\dfrac{3 + \sqrt{18}}{3}$
43. $\dfrac{6 - \sqrt{8}}{2}$
44. $\dfrac{3 - \sqrt{27}}{6}$
45. $\dfrac{-2 + \sqrt{20}}{4}$
46. $\dfrac{-4 - \sqrt{20}}{2}$
47. $\dfrac{-10 - \sqrt{75}}{5}$
48. $\dfrac{-8 + \sqrt{48}}{12}$

For Exercises 49–52 rationalize the numerator.

49. $\dfrac{\sqrt{2}}{\sqrt{5}}$
50. $\dfrac{\sqrt{5}}{\sqrt{3}}$
51. $\dfrac{\sqrt{3}}{1 + \sqrt{3}}$
52. $\dfrac{\sqrt{8}}{\sqrt{2} - 1}$

For Exercises 53–60 rationalize the denominator and simplify, if possible.

53. $\dfrac{14}{\sqrt{7}}$
54. $\dfrac{1 + \sqrt{3}}{\sqrt{3}}$
55. $\dfrac{\sqrt{2} - 1}{\sqrt{8}}$
56. $\dfrac{3 + \sqrt{3}}{\sqrt{3}}$
57. $\dfrac{1}{\sqrt{2} + 1}$
58. $\dfrac{2}{\sqrt{3} - 1}$
59. $\dfrac{1}{\sqrt{3} - \sqrt{2}}$
60. $\dfrac{4}{\sqrt{5} + 3}$

8.4 Equivalent Radical Expressions

We have said that the domain of an algebraic expression is the set of all real numbers for which the expression has a real value (unless we explicitly state otherwise). As a consequence of this we have observed that the domain of any polynomial expression (such as $x^2 - 3x + 5$) is R, but for rational expressions $\left(\text{such as } \dfrac{2x-1}{x-3}\right)$ we must exclude from the domain any replacement which will give the denominator a value of zero. Thus, the domain of $\dfrac{2x-1}{x-3}$ is $\{x: x \neq 3\}$. Let us call any expression which contains the radical sign a *radical expression*. Often the domain of a square root radical expression will *not* contain every member of R.

Example 1 Find the domain of each of these radical expressions.

(a) $\sqrt{x}$ (b) $\sqrt{-x}$ (c) $\sqrt{x^2}$ (d) $\sqrt{x-2}$

Solution

(a) Since the square root of a negative number is not real, we must have $x \geq 0$. Then the domain is $[0, \infty)$.

(b) We want $-x$ to have a nonnegative value. Then

$$-x \geq 0$$
$$(-1)(-x) \leq (-1)(0) \quad \text{[Note } \geq \text{ was replaced by } \leq \text{]}$$
$$x \leq 0$$

The domain is $(-\infty, 0]$.

(c) Since x^2 is *always* nonnegative, we conclude that the domain is R.

(d) We want $x - 2$ to have a nonnegative value,

$$x - 2 \geq 0$$
$$x \geq 2$$

The domain is $[2, \infty)$.

Example 2 Each of these radical expressions has domain $[0, \infty)$. Replace each by a simpler expression which is equivalent to it over $[0, \infty)$.

(a) $2\sqrt{x} + 3\sqrt{x}$ (b) $\sqrt{4x} - \sqrt{x}$ (c) $(\sqrt{x})^2$
(d) $(\sqrt{x} + 3)(\sqrt{x} - 3)$ (e) $(\sqrt{x} + 4)^2$

8.4 Equivalent Radical Expressions

Solution

(a) $2\sqrt{x} + 3\sqrt{x} = (2+3)\sqrt{x} = 5\sqrt{x}$

(b) Since $\sqrt{4x} = \sqrt{4} \cdot \sqrt{x} = 2\sqrt{x}$, we may write

$$\sqrt{4x} - \sqrt{x} = 2\sqrt{x} - \sqrt{x} = (2-1)\sqrt{x} = \sqrt{x}$$

(c) Since $x \geq 0$, we may use Theorem 3. Then

$$(\sqrt{x})^2 = (x^{1/2})^2 = x^{(1/2) \cdot 2} = x^1 = x$$

(d) $(\sqrt{x} + 3)(\sqrt{x} - 3) = (\sqrt{x})^2 - 3^2 = x - 9$

(e) $(\sqrt{x} + 4)^2 = (\sqrt{x})^2 + 2\sqrt{x} \cdot 4 + 4^2 = x + 8\sqrt{x} + 16$

When either of the expressions in an equation is a radical expression, we shall call that equation a *radical equation*. If the domain is finite, the solution set for a radical equation may easily be found by making all the indicated replacements.

Example 3 Find the solution set for each of the following radical equations over the domain $\{1, -1, 2, -2\}$.

(a) $\sqrt{x^2} = x$ (b) $\sqrt{x^2} = -x$

Solution

(a) $\sqrt{x^2} = x$
$x = 1$: $\sqrt{1^2} = 1$
$\sqrt{1} = 1$, True
$x = -1$: $\sqrt{(-1)^2} = -1$
$\sqrt{1} = -1$, False
$x = 2$: $\sqrt{2^2} = 2$
$\sqrt{4} = 2$, True
$x = -2$: $\sqrt{(-2)^2} = -2$
$\sqrt{4} = -2$, False
$\{1, 2\}$

(b) $\sqrt{x^2} = -x$
$x = 1$: $\sqrt{1^2} = -1$
$\sqrt{1} = -1$, False
$x = -1$: $\sqrt{(-1)^2} = -(-1)$
$\sqrt{1} = 1$, True
$x = 2$: $\sqrt{2^2} = -2$
$\sqrt{4} = -2$, False
$x = -2$: $\sqrt{(-2)^2} = -(-2)$
$\sqrt{4} = 2$, True
$\{-1, -2\}$

In Example 1(c) we observed that the expression $\sqrt{x^2}$ has domain R. We seek an expression which is equivalent to $\sqrt{x^2}$ over this domain. We may generalize the results of Example 3 to see that for all *positive* replacements $\sqrt{x^2}$ is equivalent to x, but for all *negative* replacements $\sqrt{x^2}$ is equivalent to $-x$. We state this symbolically in the next theorem.

Theorem 6

If $x > 0$, then $\sqrt{x^2} = x$
If $x < 0$, then $\sqrt{x^2} = -x$
If $x = 0$, then $\sqrt{x^2} = 0$

However, we have not as yet found a *single* expression which is equivalent to $\sqrt{x^2}$ over R.

Example 4 Find the solution set of $\sqrt{x^2} = |x|$ over the domain $\{1, -1, 2, -2, 0\}$.

Solution Recall that $|x|$ is the absolute value of x. We shall make all the indicated replacements.

$$\sqrt{x^2} = |x| \qquad\qquad \sqrt{x^2} = |x|$$
$x = 1:\ \sqrt{1^2} = |1| \qquad\qquad x = -1:\ \sqrt{(-1)^2} = |-1|$
$\qquad\quad \sqrt{1} = 1,\ \text{True} \qquad\qquad\quad \sqrt{1} = 1,\ \text{True}$
$x = 2:\ \sqrt{2^2} = |2| \qquad\qquad x = -2:\ \sqrt{(-2)^2} = |-2|$
$\qquad\quad \sqrt{4} = 2,\ \text{True} \qquad\qquad\quad \sqrt{4} = 2,\ \text{True}$
$x = 0:\ \sqrt{0^2} = 0$
$\qquad\quad \sqrt{0} = 0,\ \text{True}$

Then the solution set is the entire domain $\{1, -1, 2, -2, 0\}$.

We see by generalizing Example 4 that $|x|$ is an expression which is equivalent to $\sqrt{x^2}$ over R.

Theorem 7

For all $x \in R$, $\sqrt{x^2} = |x|$

Example 5 Determine the domain of the given expression and replace it by an equivalent expression which is simpler.

(a) $\sqrt{2x} \cdot \sqrt{8x}$ \qquad (b) $\dfrac{\sqrt{12x^4}}{\sqrt{3x^2}}$ \qquad (c) $(\sqrt{2x+1})^2$

Solution

(a) Since both $2x$ and $8x$ will have a nonnegative value only when $x \geq 0$, we see that the domain is $[0, \infty)$. Then

$$\sqrt{2x} \cdot \sqrt{8x} = \sqrt{(2x)(8x)} = \sqrt{16x^2} = \sqrt{16} \cdot \sqrt{x^2} = 4x$$

Note that since $x \geq 0$ we may use Theorem 6 to replace $\sqrt{x^2}$ by x.

(b) Since both x^4 and x^2 are nonnegative for either positive or negative replacements of x, we need exclude only 0 from the domain (for when $x = 0$, the denominator of the expression has value zero).

$$\frac{\sqrt{12x^4}}{\sqrt{3x^2}} = \sqrt{\frac{12x^4}{3x^2}} = \sqrt{4x^2} = \sqrt{4} \cdot \sqrt{x^2} = 2|x|$$

Note that we must use Theorem 7 here to replace $\sqrt{x^2}$ by $|x|$, since the domain includes both positive and negative numbers.

(c) To find the domain we want $2x + 1$ to have a nonnegative value.

$$2x + 1 \geq 0$$
$$2x \geq -1$$
$$x \geq -\tfrac{1}{2}$$

The domain is $[-\tfrac{1}{2}, \infty)$, and over this domain $2x + 1$ is nonnegative, so we may use Theorem 3 to conclude that

$$(\sqrt{2x + 1})^2 = [(2x + 1)^{1/2}]^2 = (2x + 1)^{(1/2) \cdot 2} = 2x + 1$$

We shall also want to find the solution sets for radical equations whose domains are *infinite* subsets of R. Sometimes this will be an easy task. The solution set for $\sqrt{x} = 5$ is obviously $\{25\}$ for it is true that $\sqrt{25} = 5$ and there is no other real number whose principal square root is 5. It is interesting, however, to see what happens if we square both the left and right expressions in the equation.

$$\sqrt{x} = 5$$
$$(\sqrt{x})^2 = 5^2$$
$$x = 25$$

Note that the resulting equation $x = 25$ is equivalent to $\sqrt{x} = 5$, since they both have $\{25\}$ as their solution set. It is tempting to assume that squaring both the left and right expressions will always result in an equivalent equation, but unfortunately such is *not* the case.

Consider the equation $x = 2$ whose solution set is $\{2\}$. If we square both expressions we obtain $x^2 = 4$, whose solution set is $\{2, -2\}$, and so the two equations are *not* equivalent. However, we observe that the original solution set $\{2\}$ is a *subset* of the resulting set $\{2, -2\}$. And this is generally true, as we state in the following axiom.

308   *Expressions with Radicals*

> If both the left and right expressions of an equation are squared, the solution set of the original equation is a subset of the solution set of the resulting equation.
>
> [Squaring axiom]

Let us see how we can use the squaring axiom to find the solution set for equations whose expressions contain the square root radical.

Example 6   Find the solution set.
(a) $\sqrt{2x-3} = 3$
(b) $\sqrt{x+2} = -2$
(c) $\sqrt{2x+3} = x$
(d) $\sqrt{x-1} = x - 3$

Solution
(a)
$$\sqrt{2x-3} = 3$$
$$(\sqrt{2x-3})^2 = 3^2 \qquad \text{[Squaring axiom]}$$
$$2x - 3 = 9$$
$$2x = 12$$
$$x = 6 \qquad \{6\}$$

Then the solution set is a subset of $\{6\}$. Since $\sqrt{2x-3} = 3$ becomes $\sqrt{9} = 3$, a true statement, when x is replaced by 6, we conclude that the solution set we seek is the entire set $\{6\}$.

(b)
$$\sqrt{x+2} = -2$$
$$(\sqrt{x+2})^2 = (-2)^2$$
$$x + 2 = 4$$
$$x = 2 \qquad \{2\}$$

But $\sqrt{x+2} = -2$ becomes $\sqrt{4} = -2$, a false statement, when x is replaced by 2. Hence, the solution set is $\emptyset$, the empty set (which is a subset of $\{2\}$).

(c)
$$\sqrt{2x+3} = x$$
$$(\sqrt{2x+3})^2 = x^2$$
$$2x + 3 = x^2$$
$$0 = x^2 - 2x - 3$$
$$0 = (x-3)(x+1) \qquad \{3, -1\}$$

8.4 Equivalent Radical Expressions

The solution set we seek is a subset of $\{3, -1\}$. We shall make each of the replacements in $\sqrt{2x+3} = x$.

$x = 3$: $\sqrt{2 \cdot 3 + 3} = 3$ $\qquad$ $x = -1$: $\sqrt{2(-1)+3} = -1$

$\sqrt{9} = 3$, True $\qquad\qquad\qquad$ $\sqrt{1} = -1$, False

The solution set is $\{3\}$.

(d) We first observe that squaring both expressions will not help here. For $(\sqrt{x}-1)^2 = (\sqrt{x})^2 - 2\sqrt{x} + 1 = x - 2\sqrt{x} + 1$. However, we may write the equation in radical form—with one of its expressions consisting of a single radical with no other terms.

$$\sqrt{x} - 1 = x - 3$$
$$\sqrt{x} = x - 2$$
$$(\sqrt{x})^2 = (x-2)^2$$
$$x = x^2 - 4x + 4$$
$$0 = x^2 - 5x + 4$$
$$0 = (x-4)(x-1) \quad \{1, 4\}$$

$x = 4$: $\sqrt{4} - 1 = 4 - 3$ $\qquad$ $x = 1$: $\sqrt{1} - 1 = 1 - 3$

$2 - 1 = 1$, True $\qquad\qquad\qquad$ $0 = -2$, False

The solution set is $\{4\}$.

Let us summarize the technique for finding the solution set of a square root radical equation.

1. Write the equation in radical form.
2. Square both expressions.
3. Find the solution set of the resulting equation.
4. Make all the indicated replacements from this set to find the solution set of the radical equation.

Exercises 8.4

For Exercises 1–12 determine the domain of the given expression.

1. $\sqrt{2x}$ $\qquad\qquad$ 2. $\sqrt{-3x}$
3. $\sqrt{x-1}$ $\qquad\qquad$ 4. $\sqrt{x+1}$

5. $\sqrt{x^3}$
7. $\sqrt{x^4}$
9. $\sqrt{(x+2)^2}$
11. $\sqrt{(1-x)^2}$
6. $\sqrt{-x^3}$
8. $\sqrt{x^5}$
10. $(\sqrt{x+2})^2$
12. $(\sqrt{1-x})^2$

Each of the expressions in Exercises 13–32 has domain $[0, \infty)$. Replace each by an equivalent expression which is simpler.

13. $5\sqrt{x} - 3\sqrt{x}$
15. $\sqrt{9x} + \sqrt{4x}$
17. $(\sqrt{x})^4$
19. $(\sqrt{x^3})^4$
21. $\sqrt{2} \cdot \sqrt{8x}$
23. $\sqrt{x} \cdot \sqrt{x^3}$
25. $\sqrt{x}(\sqrt{x} + 1)$
27. $(\sqrt{x} + 3)(\sqrt{x} - 1)$
29. $(\sqrt{x} + 2)(\sqrt{x} - 2)$
31. $(\sqrt{x} + 3)^2$
14. $\sqrt{x} + 2\sqrt{x}$
16. $\sqrt{8x} - \sqrt{2x}$
18. $(\sqrt{x^3})^2$
20. $(\sqrt{x})^6$
22. $\sqrt{3x} \cdot \sqrt{12x}$
24. $\sqrt{x} \cdot \sqrt{x^5}$
26. $(\sqrt{x} + 1)(\sqrt{x} + 2)$
28. $(\sqrt{x} + 1)(\sqrt{x} - 1)$
30. $(\sqrt{2x} + 1)(\sqrt{2x} - 1)$
32. $(\sqrt{x} - 1)^2$

Each of the expressions in Exercises 33–40 has a domain which includes both positive and negative numbers. Replace the given expression by an equivalent one which is simpler.

33. $\sqrt{(x+1)^2}$
35. $\sqrt{x^4}$
37. $\sqrt{x^6}$
39. $\sqrt{\dfrac{x^6}{x^4}}$
34. $\sqrt{(1-x)^2}$
36. $\sqrt{(x+1)^4}$
38. $(\sqrt{|x|})^2$
40. $\dfrac{\sqrt{x^4}}{\sqrt{x^2}}$

For Exercises 41–56 find the solution set.

41. $\sqrt{x-1} = 3$
43. $\sqrt{x^2 + x - 1} = x$
45. $\sqrt{x^2 + 3x} = x + 1$
47. $\sqrt{x+2} = x$
42. $\sqrt{2x+1} = -1$
44. $\sqrt{x^2 + 2x + 3} = -x$
46. $\sqrt{x^2 - x + 2} = x - 1$
48. $\sqrt{x+5} = x + 3$

49. $\sqrt{x+6} = x+4$
50. $\sqrt{x-1} = x-7$
51. $\sqrt{x} - x = 6 - 2x$
52. $\sqrt{x+1} - x = x - 4$

Quadratic Equations Chapter NINE

9.1 Completing the Square

We can use the factoring technique of Section 4.5 to find the solution set of the quadratic equation $x^2 - 36 = 0$.

$$x^2 - 36 = 0$$
$$(x + 6)(x - 6) = 0$$

$x + 6 = 0 \qquad\qquad x - 6 = 0$

$x = -6 \qquad\qquad x = 6 \qquad \{-6, 6\}$

But the same technique fails when we alter the equation slightly, $x^2 - 35 = 0$, since we cannot factor this equation in the same way. But it is possible to find the solution sets of these equations by another technique. We shall replace $x^2 - 36 = 0$ with the equivalent equation $x^2 = 36$ and $x^2 - 35 = 0$ with the equivalent equation $x^2 = 35$ and proceed as in Example 1.

Example 1 Find the solution set of each of the following. The replacement set is, in every case, R.

9.1 *Completing the Square*

(a) $x^2 = 36$ (b) $x^2 = 35$ (c) $x^2 = -36$ (d) $x^2 = 0$

Solution

(a) A replacement for x which makes $x^2 = 36$ true is any number whose square is 36. Then either of the square roots of 36, -6 and 6, will yield a true statement and the solution set is $\{-6, 6\}$.

(b) Either of the square roots of 35, $-\sqrt{35}$ and $\sqrt{35}$, will result in a true statement. $\{-\sqrt{35}, \sqrt{35}\}$.

(c) Since no real number replacement for x has a square of -36, the solution set is $\emptyset$.

(d) The only square root of 0 is 0, so $\{0\}$ is the solution set.

This example illustrates how to find the solution set for any quadratic equation of the form $x^2 = a$. If $a < 0$ the solution set is the empty set; if $a > 0$ the solution set includes both square roots of a; if $a = 0$ the solution set contains only 0.

There is another way to view this result. The solution set of $x^2 = 36$, $\{-6, 6\}$, is the *union* of the solution sets of $x = -6$ and $x = 6$, two equations whose right expressions are the square roots of 36, and whose left expression is that polynomial whose square appeared originally on the left. Similarly, $\{-\sqrt{35}, \sqrt{35}\}$, the solution set of $x^2 = 35$, can be considered the union of the solution sets of $x = -\sqrt{35}$ and $x = \sqrt{35}$. With this point of view we can apply the same technique for finding solution sets of more complicated quadratic equations in which some algebraic expression replaces x.

Example 2 Find the solution set of each of these.

(a) $(x - 3)^2 = 49$ (b) $(2x + 1)^2 = 64$
(c) $(x - 2)^2 = 35$

Solution

(a) We first form two equations whose right expressions are the square roots of 49, namely -7 and 7, and whose left expression is the polynomial whose square appeared on the left.

$$x - 3 = -7 \qquad x - 3 = 7$$
$$x = -4 \qquad x = 10$$

The union of the solution sets of these is $\{-4, 10\}$ and is the desired solution set of $(x - 3)^2 = 49$.

(b) $(2x + 1)^2 = 64$. We form the two equations:

$$2x + 1 = -8 \quad \text{and} \quad 2x + 1 = 8$$

314 Quadratic Equations

$$2x = -9 \qquad\qquad 2x = 7$$
$$x = -\tfrac{9}{2} \qquad\qquad x = \tfrac{7}{2}$$

The solution set is $\{-\tfrac{9}{2}, \tfrac{7}{2}\}$.

(c) $(x-2)^2 = 35$

$$x - 2 = -\sqrt{35} \qquad\qquad x - 2 = \sqrt{35}$$
$$(x-2) + 2 = -\sqrt{35} + 2 \qquad\qquad (x-2) + 2 = \sqrt{35} + 2$$
$$x = 2 - \sqrt{35} \qquad\qquad x = 2 + \sqrt{35}$$

The solution set of $(x-2)^2 = 35$ is $\{2 - \sqrt{35},\ 2 + \sqrt{35}\}$.

Examples 1 and 2 show an effective method for finding the solution set of certain quadratic equations. Regrettably, many equations whose solution set we seek are not in a form to apply this technique until we replace the equation by an equivalent one which is in the desired form.

Example 3 Find the solution set of each of the following.
(a) $x^2 + 8x + 16 = 7$ \qquad (b) $x^2 + 8x = 9$

Solution

(a) We can replace $x^2 + 8x + 16$ with the equivalent square expression, $(x+4)^2$.

$$x^2 + 8x + 16 = 7$$
$$(x+4)^2 = 7$$
$$x + 4 = -\sqrt{7} \qquad\qquad x + 4 = \sqrt{7}$$
$$(x+4) + (-4) = -\sqrt{7} + (-4) \qquad (x+4) + (-4) = \sqrt{7} + (-4)$$
$$x = -4 - \sqrt{7} \qquad\qquad x = -4 + \sqrt{7}$$

The solution set is $\{-4 - \sqrt{7},\ -4 + \sqrt{7}\}$.

(b) The left expression $x^2 + 8x$ cannot be the square of a first degree polynomial since it contains only two terms. However, $x^2 + 8x + 16$ is the square of $(x+4)$. Hence, we may use the addition axiom to find an equivalent equation whose left expression is $x^2 + 8x + 16$.

$$x^2 + 8x = 9$$
$$x^2 + 8x + 16 = 9 + 16$$
$$(x+4)^2 = 25$$
$$x + 4 = -5 \qquad\qquad x + 4 = 5$$

9.1 Completing the Square

$$x = -9 \qquad\qquad x = 1 \qquad \{-9,\ 1\}$$

It may have been clear in Example 3(b) that adding 16 to $x^2 + 8x$ would result in a trinomial which was a square since in Example 3(a) we had just considered that trinomial. But this idea can be used even when it is not obvious what number must be added to form a square. Consider the binomial $x + a$ where x is a variable whose replacement set is R, but a is a constant (which we shall not specify). Squaring $x + a$ gives $(x + a)^2 = x^2 + 2ax + a^2$. Note that for any binomial $x + a$ its square must have this form, $x^2 + 2ax + a^2$, in which the first degree term is $2ax$ and the constant term is a^2. Suppose we consider the expression $x^2 + 12x$ and ask what constant must be added if the resulting trinomial is to be the square of a binomial. In effect we are seeking the constant term a^2, knowing that the first degree term is $12x$. But the first degree term is $2ax$ and so we see that $2a$ must represent 12, or that a is 6. Then a^2 is 36 and the required square is $x^2 + 12x + 36 = (x + 6)^2$. In general, by taking the coefficient of x, dividing it by two, and squaring the result, we obtain the constant which must be added to **complete the square.**

Example 4   For each of the following determine the constant which must be added to complete the square.
 (a) $x^2 + 10x$ (b) $x^2 - 14x$ (c) $x^2 + 5x$

Solution

(a) The coefficient of x is 10. Dividing by 2 gives 5. Squaring 5 gives 25. The completed square is $x^2 + 10x + 25 = (x + 5)^2$.
(b) The coefficient of x is -14. Dividing by 2 gives -7. Squaring gives 49. The completed square is $x^2 - 14x + 49 = (x - 7)^2$.
(c) The coefficient of x is 5. Dividing by 2 gives $\frac{5}{2}$. The fact that this is not an integer does not create any problems. Squaring $\frac{5}{2}$ gives $\frac{25}{4}$. The completed square is $x^2 + 5x + \frac{25}{4} = (x + \frac{5}{2})^2$.

To see how this principal can be used in finding solution sets of quadratic equations we consider Example 5.

Example 5   Find the solution set of each of these.
 (a) $x^2 + 6x = 5$ (b) $x^2 - 10x = -4$
 (c) $x^2 + 2x - 5 = 0$

Solution

(a) $x^2 + 6x = 5$. We shall complete the square on the left by adding 9. Of course, since we want an equivalent equation we add 9 to both expressions.

Quadratic Equations

$$x^2 + 6x + 9 = 5 + 9$$
$$(x + 3)^2 = 14$$

$$x + 3 = -\sqrt{14} \qquad\qquad x + 3 = \sqrt{14}$$
$$x = -3 - \sqrt{14} \qquad\qquad x = -3 + \sqrt{14}$$
$$\{-3 - \sqrt{14},\ -3 + \sqrt{14}\}$$

(b) We complete the square by adding 25.

$$x^2 - 10x = -4$$
$$x^2 - 10x + 25 = -4 + 25$$
$$(x - 5)^2 = 21$$

$$x - 5 = -\sqrt{21} \qquad\qquad x - 5 = \sqrt{21}$$
$$x = 5 - \sqrt{21} \qquad\qquad x = 5 + \sqrt{21}$$
$$\{5 - \sqrt{21},\ 5 + \sqrt{21}\}$$

(c) We first replace this equation by an equivalent one of the form we have been considering by adding 5 to both expressions.

$$x^2 + 2x - 5 = 0$$
$$x^2 + 2x = 5$$

Now we complete the square by adding 1.

$$x^2 + 2x + 1 = 5 + 1$$
$$(x + 1)^2 = 6$$

$$x + 1 = -\sqrt{6} \qquad\qquad x + 1 = \sqrt{6}$$
$$x = -1 - \sqrt{6} \qquad\qquad x = -1 + \sqrt{6}$$
$$\{-1 - \sqrt{6},\ -1 + \sqrt{6}\}$$

The technique for completing the square we have considered is only valid when the coefficient of x^2 is 1. When the coefficient of x^2 in a quadratic equation in which we wish to use this method is some other number, we may replace the equation by an appropriate equivalent one.

Example 6   Find the solution set of each of these.
(a) $3x^2 + 12x = -2$ (b) $4x^2 + 20x + 5 = 0$
(c) $2x^2 + x = 5$ (d) $x^2 + 6x + 10 = 0$

9.1 Completing the Square

Solution

(a) We first multiply both expressions by $\frac{1}{3}$ (the reciprocal of the coefficient of x^2).

$$\left(\tfrac{1}{3}\right)(3x^2 + 12x) = \left(\tfrac{1}{3}\right)(-2)$$

$$x^2 + 4x = -\tfrac{2}{3}$$

Now we complete the square by adding 4.

$$x^2 + 4x + 4 = -\tfrac{2}{3} + 4$$

$$(x + 2)^2 = \tfrac{10}{3}$$

$$x + 2 = -\sqrt{\tfrac{10}{3}} \qquad\qquad x + 2 = \sqrt{\tfrac{10}{3}}$$

$$x = -2 - \sqrt{\tfrac{10}{3}} \qquad\qquad x = -2 + \sqrt{\tfrac{10}{3}}$$

$$\{-2 - \sqrt{\tfrac{10}{3}},\ -2 + \sqrt{\tfrac{10}{3}}\}$$

(b) First we multiply by $\frac{1}{4}$.

$$4x^2 + 20x + 5 = 0$$

$$x^2 + 5x + \tfrac{5}{4} = 0$$

$$x^2 + 5x = -\tfrac{5}{4}$$

To complete the square, we next add $\frac{25}{4}$.

$$x^2 + 5x + \tfrac{25}{4} = -\tfrac{5}{4} + \tfrac{25}{4}$$

$$(x + \tfrac{5}{2})^2 = 5$$

$$x + \tfrac{5}{2} = -\sqrt{5} \qquad\qquad x + \tfrac{5}{2} = \sqrt{5}$$

$$x = -\tfrac{5}{2} - \sqrt{5} \qquad\qquad x = -\tfrac{5}{2} + \sqrt{5}$$

$$\{-\tfrac{5}{2} - \sqrt{5},\ -\tfrac{5}{2} + \sqrt{5}\}$$

(c)
$$2x^2 + x = 5$$

$$x^2 + \left(\tfrac{1}{2}\right)x = \tfrac{5}{2}$$

The coefficient of x is $\frac{1}{2}$. Dividing by 2 and squaring, we obtain $\left(\tfrac{1}{4}\right)^2 = \tfrac{1}{16}$, which we add to complete the square.

$$x^2 + \left(\tfrac{1}{2}\right)x + \left(\tfrac{1}{16}\right) = \tfrac{5}{2} + \left(\tfrac{1}{16}\right)$$

$$(x + \tfrac{1}{4})^2 = \tfrac{41}{16} \qquad \left[\text{Note that } \sqrt{\tfrac{41}{16}} = \tfrac{\sqrt{41}}{4}\right]$$

$$x + \tfrac{1}{4} = -\tfrac{\sqrt{41}}{4} \qquad\qquad x + \tfrac{1}{4} = \tfrac{\sqrt{41}}{4}$$

$$x = \frac{-1}{4} - \frac{\sqrt{41}}{4} = \frac{-1-\sqrt{41}}{4} \qquad x = \frac{-1}{4} + \frac{\sqrt{41}}{4} = \frac{-1+\sqrt{41}}{4}$$

$$\left\{ \frac{-1-\sqrt{41}}{4}, \frac{-1+\sqrt{41}}{4} \right\}$$

(d)
$$x^2 + 6x + 10 = 0$$
$$x^2 + 6x = -10$$
$$x^2 + 6x + 9 = -10 + 9$$
$$(x+3)^2 = -1$$
$$x + 3 = -\sqrt{-1} \qquad x + 3 = \sqrt{-1}$$

But since $\sqrt{-1}$ is not a real number, we see that there are no real numbers in the solution set. Hence it is the empty set, $\emptyset$.

Let us summarize the technique for finding the solution of quadratic equations by the method of completing the square.

1. If the coefficient of x^2 is not 1, multiply both expressions by its reciprocal.
2. Replace the equation, if necessary, by an equivalent one in which the first and second degree terms are on one side and the constant is on the other.
3. Complete the square by taking half the coefficient of x, squaring, and adding this square to both expressions. Simplify.
4. Write two first degree equations whose right expressions are the positive and negative square roots of the resulting constant and whose left expressions are polynomials whose squares appear on the left of the completed square.
5. Form the union of the solution sets of these equations.

It is important to observe that while the solution sets of every quadratic equation *cannot* be found by the factoring technique of Section 4.5, the method described here is completely general—it *always* yields the solution set.

Exercises 9.1

For Exercises 1–8 find the number which must be added to the given expression to complete the square.

1. $x^2 + 6x$
2. $x^2 - 10x$
3. $x^2 - 12x$
4. $x^2 + 2x$
5. $x^2 + 3x$
6. $x^2 + x$
7. $x^2 - 5x$
8. $x^2 - 11x$

For Exercises 9–40 find the solution set by completing the square.

9. $x^2 = 5$
10. $x^2 - 2 = 0$
11. $x^2 + 3 = 0$
12. $x(x + 2) = 2x + 3$
13. $x^2 + 4x = 5$
14. $x^2 - 6x = 7$
15. $x^2 - 2x = 1$
16. $x^2 + 2x = 2$
17. $x^2 + 3x = 4$
18. $x^2 - x = 2$
19. $x^2 + 5x = 1$
20. $x^2 - 3x = 2$
21. $x^2 - 4x + 1 = 0$
22. $x^2 - 2x + 3 = 0$
23. $x^2 + 12x + 33 = 0$
24. $x^2 - 4x + 5 = 0$
25. $x^2 - 8x + 16 = 0$
26. $x^2 - x - 1 = 0$
27. $x^2 + 3x + 1 = 0$
28. $x^2 + 7x + 11 = 0$
29. $(x + 1)(x + 3) = 2$
30. $(x - 3)(x + 1) = 1$
31. $(x + 1)^2 = 4x + 3$
32. $x(2x + 3) = (x + 2)(x + 3)$
33. $2x^2 + 8x = 1$
34. $2x^2 - 4x = 5$
35. $3x^2 + 6x = 2$
36. $3x^2 - 2x = 1$
37. $3x^2 - 11x = 4$
38. $6x^2 + 6 = 13x$
39. $-x^2 + x + 3 = 0$
40. $-x^2 + 2 = 3x$

9.2 The Quadratic Theorem

We shall call

$$ax^2 + bx + c = 0$$

the general quadratic equation in standard form. Note that x is the variable here; a, b, and c are not variables, but the coefficients in the equation. We use letters to represent these constants so that we need not specify them. We observe that if b and c are replaced by any constants and a is replaced by any constant except 0, then the result is a quadratic equation in standard form.

Example 1 Make the indicated replacements in the general quadratic equation and write the resulting equation.

(a) $a = 2$, $b = -3$, $c = 4$
(b) $a = 1$, $b = 2$, $c = 0$
(c) $a = 0$, $b = 3$, $c = -4$

Solution

(a) $2x^2 - 3x + 4 = 0$ (b) $x^2 + 2x = 0$
(c) $3x - 4 = 0$. Note that this is *not* a quadratic equation, and that $a = 0$.

Example 2 Determine the replacements for a, b, and c in the general quadratic equation which will result in the indicated quadratic equation.
(a) $x^2 + 3x - 2 = 0$ (b) $3x^2 - 1 = 0$
(c) $x^2 = 2x - 5$

Solution

(a) $a = 1, b = 3, c = -2$
(b) Since $3x^2 - 1$ is equivalent to $3x^2 + 0 \cdot x - 1$, we see that $a = 3$, $b = 0, c = -1$
(c) First we must find a standard form.

$$x^2 = 2x - 5$$
$$x^2 - 2x + 5 = 0$$

Hence $a = 1, b = -2, c = 5$.

In the last section we considered a method for finding the solution set of any quadratic equation. Since we can always complete the square to find the solution set of a quadratic equation, we shall do so for the general equation $ax^2 + bx + c = 0$. Any results we obtain will then apply to every quadratic equation.

$$ax^2 + bx + c = 0$$

We first obtain an equivalent equation where the coefficient of x^2 is 1. To do so we multiply by $1/a$. Note that $a \neq 0$, so $1/a$ is a real number.

$$\frac{1}{a}(ax^2 + bx + c) = \frac{1}{a} \cdot 0$$

$$x^2 + \frac{b}{a} \cdot x + \frac{c}{a} = 0$$

Next we add $-c/a$ to both the left and right expressions.

$$x^2 + \frac{b}{a} \cdot x = \frac{-c}{a}$$

Now we complete the square. The coefficient of x is b/a. Dividing by 2 gives $b/2a$ and squaring this gives $b^2/4a^2$. We add this to both expressions.

9.2 The Quadratic Theorem

$$x^2 + \frac{b}{a} \cdot x + \frac{b^2}{4a^2} = \frac{-c}{a} + \frac{b^2}{4a^2}$$

On the left we have a square since

$$x^2 + \frac{b}{a} \cdot x + \frac{b^2}{4a^2} = \left(x + \frac{b}{2a}\right)^2$$

Thus,

$$\left(x + \frac{b}{2a}\right)^2 = \frac{-c}{a} + \frac{b^2}{4a^2}$$

We shall simplify the expression on the right.

$$\frac{-c}{a} = \frac{-c \cdot 4a}{a \cdot 4a} = \frac{-4ac}{4a^2}$$

Hence,

$$\frac{-c}{a} + \frac{b^2}{4a^2} = \frac{-4ac}{4a^2} + \frac{b^2}{4a^2} = \frac{b^2 - 4ac}{4a^2}$$

Then we have

$$\left(x + \frac{b}{2a}\right)^2 = \frac{b^2 - 4ac}{4a^2}$$

The square roots of the right expression are

$$-\sqrt{\frac{b^2 - 4ac}{4a^2}} \quad \text{and} \quad \sqrt{\frac{b^2 - 4ac}{4a^2}}$$

which we may simplify:*

$$\frac{-\sqrt{b^2 - 4ac}}{2a} \quad \text{and} \quad \frac{\sqrt{b^2 - 4ac}}{2a}$$

We then obtain the two first degree equations

$$x + \frac{b}{2a} = \frac{-\sqrt{b^2 - 4ac}}{2a} \quad \text{and} \quad x + \frac{b}{2a} = \frac{\sqrt{b^2 - 4ac}}{2a}$$

We add $-b/2a$ to both expressions.

$$x = \frac{-b}{2a} - \frac{\sqrt{b^2 - 4ac}}{2a}, \quad x = \frac{-b}{2a} + \frac{\sqrt{b^2 - 4ac}}{2a}$$

*We may assume that $a > 0$, for if $a < 0$, then we may multiply the left and right expressions in $ax^2 + bx + c = 0$ by -1.

322 Quadratic Equations

$$= \frac{-b - \sqrt{b^2 - 4ac}}{2a}, \qquad = \frac{-b + \sqrt{b^2 - 4ac}}{2a}$$

We have established an important theorem.

Theorem 1   The **quadratic theorem.** If $a \neq 0$ and if $b^2 - 4ac \geq 0$, the solution set of $ax^2 + bx + c = 0$ is

$$\left\{ \frac{-b - \sqrt{b^2 - 4ac}}{2a}, \frac{-b + \sqrt{b^2 - 4ac}}{2a} \right\}$$

Example 3   For each of the following quadratic equations determine the constants a, b, and c and use the quadratic theorem to find the solution set. The replacement set is R.

(a) $x^2 + 7x + 2 = 0$ (b) $2x^2 - x - 2 = 0$
(c) $x^2 + 4x + 4 = 0$ (d) $x^2 + x + 2 = 0$

Solution

(a) $x^2 + 7x + 2 = 0$; $a = 1$, $b = 7$, $c = 2$, $-b = -7$;
$b^2 - 4ac = 7^2 - 4(1)(2) = 49 - 8 = 41$; $2a = 2$.

$$\frac{-b - \sqrt{b^2 - 4ac}}{2a} = \frac{-7 - \sqrt{41}}{2}$$

and

$$\frac{-b + \sqrt{b^2 - 4ac}}{2a} = \frac{-7 + \sqrt{41}}{2}$$

$$\left\{ \frac{-7 - \sqrt{41}}{2}, \frac{-7 + \sqrt{41}}{2} \right\}$$

(b) $2x^2 - x - 2 = 0$; $a = 2$, $b = -1$, $c = -2$.

$$\frac{-b - \sqrt{b^2 - 4ac}}{2a} = \frac{-(-1) - \sqrt{(-1)^2 - 4(2)(-2)}}{2(2)} = \frac{1 - \sqrt{17}}{4}$$

$$\frac{-b + \sqrt{b^2 - 4ac}}{2a} = \frac{1 + \sqrt{17}}{4}$$

$$\left\{ \frac{1 - \sqrt{17}}{4}, \frac{1 + \sqrt{17}}{4} \right\}$$

(c) $x^2 + 4x + 4 = 0$; $a = 1$, $b = 4$, $c = 4$.

$$\frac{-b - \sqrt{b^2 - 4ac}}{2a} = \frac{-4 - \sqrt{4^2 - 4 \cdot 1 \cdot 4}}{2}$$

$$= \frac{-4 - \sqrt{0}}{2} = \frac{-4}{2} = -2$$

$$\frac{-b + \sqrt{b^2 - 4ac}}{2a} = \frac{-4 + \sqrt{0}}{2} = -2$$

Then the solution set is $\{-2\}$, a set with only *one* member.

(d) $x^2 + x + 2 = 0$; $a = 1$, $b = 1$, $c = 2$.

$$\frac{-b - \sqrt{b^2 - 4ac}}{2a} = \frac{-1 - \sqrt{1^2 - 4 \cdot 1 \cdot 2}}{2} = \frac{-1 - \sqrt{-7}}{2}$$

Since $\sqrt{-7}$ is *not* a real number, we conclude that the solution set is $\varnothing$.

If possible, we will always want to simplify the members of the solution set.

Example 4 Find the solution set of each of these.

(a) $2x^2 - 4x = 3$ (b) $4x = 15 - 3x^2$

Solution

(a) $2x^2 - 4x = 3$
$2x^2 - 4x - 3 = 0$; $a = 2$, $b = -4$, $c = -3$.

$$\frac{-b - \sqrt{b^2 - 4ac}}{2a} = \frac{-(-4) - \sqrt{(-4)^2 - 4(2)(-3)}}{2(2)} = \frac{4 - \sqrt{40}}{4}$$

$$= \frac{4 - 2\sqrt{10}}{4} = \frac{2(2 - \sqrt{10})}{4} = \frac{2 - \sqrt{10}}{2}$$

$$\left\{ \frac{2 - \sqrt{10}}{2}, \frac{2 + \sqrt{10}}{2} \right\}$$

(b) $4x = 15 - 3x^2$
$3x^2 + 4x - 15 = 0$; $a = 3$, $b = 4$, $c = -15$.

$$\frac{-b - \sqrt{b^2 - 4ac}}{2a} = \frac{-4 - \sqrt{4^2 - 4(3)(-15)}}{2(3)} = \frac{-4 - \sqrt{196}}{6}$$

$$= \frac{-4 - 14}{6}$$

$$= \frac{-18}{6} = -3$$

$$\frac{-b + \sqrt{b^2 - 4ac}}{2a} = \frac{-4 + \sqrt{196}}{6} = \frac{-4 + 14}{6} = \frac{5}{3}$$

$$\{-3, \tfrac{5}{3}\}$$

324 *Quadratic Equations*

Note that in Example 4(b) the members of the solution set are rational numbers. We might also observe that we could have used the method of factoring with the equation.

$$3x^2 + 4x - 15 = 0$$
$$(3x - 5)(x + 3) = 0$$

$$3x - 5 = 0 \qquad\qquad x + 3 = 0$$
$$x = \tfrac{5}{3} \qquad\qquad x = -3$$

This is generally true. When it is possible to use factoring to find the solution set of a quadratic equation (with integral coefficients) the members of this set are rational numbers. When factoring is not possible, the solution set members are irrational numbers. But note that the quadratic theorem can be used in either case.

Actually, while it is a powerful device, the quadratic theorem need not be used to find the solution set of an equation which can obviously be factored, should not be used when $b = 0$ or $c = 0$ (since other techniques are easier in these cases), and may not be needed if we wish to use the method of completing the square.

Example 5 Use any convenient method to find the solution set of each of the following.
(a) $x^2 + x - 20 = 0$ (b) $3x^2 - 5x = 0$
(c) $2x^2 - 19 = 0$

Solution

(a) $x^2 + x - 20 = 0$. We observe that the left expression can be factored.

$$(x + 5)(x - 4) = 0$$

$$x + 5 = 0 \qquad\qquad x - 4 = 0$$
$$x = -5 \qquad\qquad x = 4 \qquad \{-5, 4\}$$

(b) $3x^2 - 5x = 0$. Here $a = 3$ and $b = -5$, but $c = 0$. We can always factor a quadratic polynomial when $c = 0$.

$$x(3x - 5) = 0$$

$$x = 0 \qquad\qquad 3x - 5 = 0$$
$$\qquad\qquad\qquad 3x = 5$$
$$\qquad\qquad\qquad x = \tfrac{5}{3} \qquad \{0, \tfrac{5}{3}\}$$

9.2 The Quadratic Theorem

(c) $2x^2 - 19 = 0$. Here $a = 2$ and $c = -19$, but $b = 0$. We shall use the technique of Section 9.1.

$$2x^2 = 19$$
$$x^2 = \tfrac{19}{2}$$
$$x = -\sqrt{\tfrac{19}{2}};\ x = \sqrt{\tfrac{19}{2}} \quad \{-\sqrt{\tfrac{19}{2}},\ \sqrt{\tfrac{19}{2}}\}$$

Exercises 9.2

For Exercises 1–8 if necessary find an equivalent equation which is in the form $ax^2 + bx + c = 0$ and identify the coefficients a, b, and c.

1. $2x^2 + 5x + 4 = 0$
2. $x^2 - 3x + 2 = 0$
3. $4x^2 + 5 = 0$
4. $x^2 + 4x = 0$
5. $x^2 = 2x + 5$
6. $x(x + 3) = 1$
7. $(x + 2)(x - 1) = x$
8. $(2x - 1)(x - 3) = (x + 3)^2$

For Exercises 9–28 use the quadratic theorem to find the solution set.

9. $x^2 + 3x + 1 = 0$
10. $x^2 + 5x + 5 = 0$
11. $x^2 + x - 3 = 0$
12. $2x^2 + x - 2 = 0$
13. $2x^2 - x - 1 = 0$
14. $x^2 - 3x + 1 = 0$
15. $3x^2 - 5x + 1 = 0$
16. $x^2 - 7x + 11 = 0$
17. $2x^2 + 5x - 3 = 0$
18. $x^2 + 3x + 4 = 0$
19. $2x^2 - x - 6 = 0$
20. $2x^2 - x + 1 = 0$
21. $x^2 + 2x - 4 = 0$
22. $x^2 - 2x - 1 = 0$
23. $x^2 + 4x + 2 = 0$
24. $x^2 - 6x + 6 = 0$
25. $x^2 + 4x = 4$
26. $x^2 = 5x - 6$
27. $x(x + 1) = 4$
28. $(x + 2)^2 = 2x + 7$

For Exercises 29–40 use any convenient method to find the solution set.

29. $x^2 - 3 = 0$
30. $x^2 + 2 = 0$
31. $4x^2 - 5 = 0$
32. $9x^2 - 2 = 0$
33. $x^2 + 5x = 0$
34. $x^2 = 2x$
35. $2x^2 + 3x = 8x$
36. $x^2 = x$
37. $x^2 + 4 = 4x$
38. $4x(x - 3) = -9$
39. $2x(x + 2) = (x + 1)^2$
40. $x^2 + 200x = -9998$

9.3 The Discriminant

We shall call the expression $b^2 - 4ac$ the **discriminant** of the quadratic equation $ax^2 + bx + c = 0$. As we shall see, the value of the discriminant is easy to find and can be used to partially describe the solution set for the equation.

Example 1 Find the value of the discriminant and the solution set for each of the following quadratic equations.

(a) $x^2 + 3x + 4 = 0$ (b) $4x^2 + 12x + 9 = 0$
(c) $2x^2 + 7x + 3 = 0$ (d) $x^2 - 5x + 5 = 0$

Solution

(a) $x^2 + 3x + 4 = 0$; $a = 1$, $b = 3$, $c = 4$.
$b^2 - 4ac = 3^2 - 4 \cdot 1 \cdot 4 = -7$. We shall use the Quadratic Theorem.

$$\frac{-b - \sqrt{b^2 - 4ac}}{2a} = \frac{-3 - \sqrt{-7}}{2}$$

Since $\sqrt{-7}$ is not a real number, we conclude that the solution set is $\emptyset$. We note that the discriminant is -7, a *negative* number and the solution set is empty.

(b) $4x^2 + 12x + 9 = 0$. $b^2 - 4ac = 12^2 - 4 \cdot 4 \cdot 9 = 144 - 144 = 0$.

$$\frac{-b - \sqrt{b^2 - 4ac}}{2a} = \frac{-12 - \sqrt{0}}{8} = \frac{-12}{8} = \frac{-3}{2}$$

$$\frac{-b + \sqrt{b^2 - 4ac}}{2a} = \frac{-12 + \sqrt{0}}{8} = \frac{-12}{8} = \frac{-3}{2}$$

We note that the discriminant is 0 and that the solution set is $\{-\frac{3}{2}\}$, a set with only *one* member.

(c) $2x^2 + 7x + 3 = 0$. $b^2 - 4ac = 7^2 - 4 \cdot 2 \cdot 3 = 49 - 24 = 25$.

$$\frac{-b - \sqrt{b^2 - 4ac}}{2a} = \frac{-7 - \sqrt{25}}{4} = \frac{-7 - 5}{4} = -3$$

$$\frac{-b + \sqrt{b^2 - 4ac}}{2a} = \frac{-7 + \sqrt{25}}{4} = \frac{-7 + 5}{4} = \frac{-1}{2}$$

We note that the discriminant is 25, the square of an integer, and the solution set is $\{-3, -\frac{1}{2}\}$, a set which contains *two rational* numbers.

(d) $x^2 - 5x + 5 = 0$. $b^2 - 4ac = (-5)^2 - 4 \cdot 1 \cdot 5 = 25 - 20 = 5$.

9.3 The Discriminant

$$\frac{-b - \sqrt{b^2 - 4ac}}{2a} = \frac{-(-5) - \sqrt{5}}{2} = \frac{5 - \sqrt{5}}{2}$$

Then the solution set is $\left\{ \dfrac{5 - \sqrt{5}}{2}, \dfrac{5 + \sqrt{5}}{2} \right\}$, a set which contains *two irrational* numbers. We note that the discriminant is 5, a *positive* number which is *not* the square of an integer.

Each of the equations in Example 1 has coefficients which are *integers*. We may generalize the results of this example for any quadratic equation with *integral* coefficients.

For the quadratic equation $ax^2 + bx + c = 0$, where a, b and c are integers with $a \neq 0$,

1. If $b^2 - 4ac < 0$, the solution set is empty.
2. If $b^2 - 4ac = 0$, the solution set contains one rational number.
3. If $b^2 - 4ac > 0$, the solution set contains two members. These numbers are rational if $b^2 - 4ac$ is the square of an integer; otherwise, they are irrational.

Example 2 Without finding the solution set, determine how many members it contains and whether or not they are rational.
(a) $-x^2 - 2x + 6 = 0$ (b) $2x^2 + 1 = 2x$
(c) $x^2 - (\tfrac{9}{2})x + 2 = 0$

Solution

(a) $b^2 - 4ac = (-2)^2 - 4(-1)(6) = 4 + 24 = 28$.
The solution set contains two irrational numbers.

(b)
$$2x^2 + 1 = 2x$$
$$2x^2 - 2x + 1 = 0$$
$$b^2 - 4ac = (-2)^2 - 4 \cdot 2 \cdot 1 = -4$$

The solution set contains no members.

(c) Since the coefficients are not integers, we shall multiply by 2 to obtain an equivalent equation:
$$x^2 - (\tfrac{9}{2})x + 2 = 0$$
$$2x^2 - 9x + 4 = 0$$
$$b^2 - 4ac = (-9)^2 - 4 \cdot 2 \cdot 4 = 49$$

The solution set contains two rational numbers.

In the preceding section we observed that whenever the solution set of a quadratic equation with integral coefficients contains only rational numbers, then the method of factoring may be used to find these members, but when the solution set is empty or contains irrational numbers, then factoring cannot be used. We now see how the *discriminant* can be used to determine whether or not a second degree polynomial expression can be factored. Since the discriminant of $6x^2 + x - 2 = 0$ is $1^2 - 4(6)(-2) = 49$, we see that the solution set contains two rational numbers. Hence, the expression $6x^2 + x - 2$ can be factored. We can find its factors by trial and error: $6x^2 + x - 2 = (2x - 1)(3x + 2)$.

Example 3 Determine which of these polynomials can be factored, and find the factors of those that can.

(a) $2x^2 + 6x + 3$
(b) $x^2 + 4$
(c) $6x^2 + x - 12$
(d) $18x^3 - 24x^2 + 8x$

Solution

(a) We shall consider the corresponding quadratic equation, $2x^2 + 6x + 3 = 0$ and its discriminant. Since $b^2 - 4ac = 12$, *not* the square of an integer, we conclude that the expression $2x^2 + 6x + 3$ cannot be factored.

(b) $x^2 + 4$. The discriminant is $0^2 - 4 \cdot 1 \cdot 4 = -16$, *not* the square of an integer. Hence, the expression cannot be factored.

(c) $6x^2 + x - 12$. The discriminant is $1^2 - 4(6)(-12) = 289 = 17^2$. Hence, the expression can be factored. By trial and error we see that $6x^2 + x - 12 = (3x - 4)(2x + 3)$.

(d) First we remove the repeated factor. $18x^3 - 24x^2 + 8x = 2x(9x^2 - 12x + 4)$. The discriminant of $9x^2 - 12x + 4 = 0$, is 0, the square of the integer 0. Then this expression is the square of a binomial. $9x^2 - 12x + 4 = (3x - 2)^2$. Thus, the complete factorization of $18x^3 - 24x^2 + 8x$ is $2x(3x - 2)^2$.

In Section 9.1 we learned that the solution set for any quadratic equation may be found by the method of completing the square. Although the quadratic theorem can also be used for this purpose and is often easier to use than to complete the square, the technique of completing the square is an important one which has uses other than finding solution sets. We shall see how this technique can be used to determine the range of a quadratic function and to find the ordered pair which corresponds to the vertex of the parabola which is the graph of this function. First, we shall consider some functions in which it is not necessary to complete the square.

9.3 The Discriminant

Example 4 Find the range of the indicated function and the ordered pair which corresponds to the vertex of the parabola which is its graph. Sketch the graph.

(a) $f(x) = x^2 + 1$ (b) $g(x) = (x - 2)^2 + 1$

(c) $h(x) = 3 - (2x + 1)^2$

Solution

(a) We shall make several replacements for the variable.

$$f(2) = 2^2 + 1 = 5 \qquad (2, 5)$$
$$f(1) = 2 \qquad (1, 2)$$
$$f(0) = 1 \qquad (0, 1)$$
$$f(-1) = (-1)^2 + 1 = 2 \qquad (-1, 2)$$
$$f(-2) = 5 \qquad (-2, 5)$$

It is easily seen that the smallest number in the range is 1. For, in fact, $f(0) = 1$, and furthermore for any replacement of x except $x = 0$, the value of $x^2 + 1$ will be larger than 1. We know that the graph of f is a parabola, since $x^2 + 1$ is a second degree polynomial. Then the vertex of this parabola must be $(0, 1)$, the point at which the smallest number in the range is found. The graph is shown in Figure 9.1. Note that the range is $[1, \infty)$.

(b) We note that $g(2) = 0^2 + 1 = 1$, and, as in the above example, the smallest number in the range is 1. For if x is replaced by any number other than 2, then the expression $(x - 2)^2$ will have a *positive* value and thus $(x - 2)^2 + 1$ will be *larger* than 1. Hence the range is $[1, \infty)$, and the vertex is $(2, 1)$, the point at which the smallest number in the range is found (Figure 9.2).

(c) First we seek a replacement for x which will give the value 0 to the expression $2x + 1$. We see that $x = -\tfrac{1}{2}$ is such a replacement, and that $h(-\tfrac{1}{2}) = 3 - 0^2 = 3$. We next observe that 3 is the *largest*

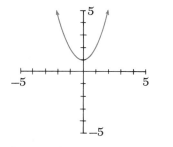

Figure 9.1

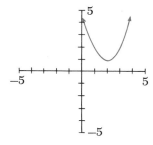

Figure 9.2

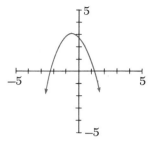
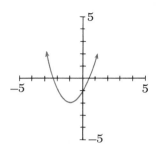

Figure 9.3 Figure 9.4

number in the range, for if x is replaced by any number other than $-\frac{1}{2}$, then $(2x + 1)^2$ will have some positive value and hence the value of $3 - (2x + 1)^2$ will be *less* than 3. For example, if $x = 0$, then $3 - (2x + 1)^2 = 3 - 1^2 = 2$. Then $\langle -\infty, 3]$ is the range, and $(-\frac{1}{2}, 3)$ is the vertex, since this is the point where the largest number in the range is found. See Figure 9.3.

Often we must replace the given expression by an equivalent one in the form of those of Example 4 by completing the square.

Example 5 Find the range of the function and the ordered pair which corresponds to the vertex of the parabola which is its graph. Sketch the graph.
(a) $f(x) = x^2 + 2x - 1$ (b) $g(x) = -x^2 - 6x - 8$
(c) $h(x) = 2x^2 - 4x + 3$

Solution

(a) We see that $x^2 + 2x + 1 = (x + 1)^2$. We may express -1 as $1 - 2$ and write

$$x^2 + 2x - 1 = x^2 + 2x + 1 - 2$$
$$= (x^2 + 2x + 1) - 2 = (x + 1)^2 - 2$$

Since $(x + 1)^2 = 0$ if $x = -1$, we have $f(-1) = 0^2 - 2 = -2$. Then the range of f is $[-2, \infty)$, and the vertex of its graph is $(-1, -2)$. The graph is shown in Figure 9.4.

(b) First, we shall remove the repeated factor -1. Then

$$-x^2 - 6x - 8 = -(x^2 + 6x + 8)$$

To complete the square with $x^2 + 6x$ we must have $x^2 + 6x + 9$. Then we shall express 8 as $9 - 1$ and have

$$g(x) = -(x^2 + 6x + 8) = -[(x^2 + 6x + 9) - 1]$$
$$= -[(x + 3)^2 - 1] = 1 - (x + 3)^2$$

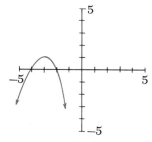

Figure 9.5

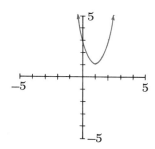

Figure 9.6

As in Example 4(c) we see that the *largest* number in the range is 1 and hence the range is $\langle-\infty, 1]$. Since $g(-3) = 1$, the vertex of the graph is $(-3, 1)$. The graph is shown in Figure 9.5.

(c) We shall remove the factor 2 and complete the square.

$$h(x) = 2x^2 - 4x + 3 = 2[x^2 - 2x + \tfrac{3}{2}]$$
$$= 2[(x^2 - 2x + 1) + \tfrac{1}{2}] = 2[(x - 1)^2 + \tfrac{1}{2}]$$
$$= 2(x - 1)^2 + 1$$

Hence, the range is $[1, \infty\rangle$ and the vertex is $(1, 1)$. The graph is shown in Figure 9.6.

Exercises 9.3

For Exercises 1–4 find the value of the discriminant.

1. $a = 2, b = 3, c = 4$
2. $a = 1, b = 0, c = -2$
3. $2x^2 - 3x + 1 = 0$
4. $x^2 + 5x + 7 = 0$

For Exercises 5–16 without actually finding the solution set determine how many members are in the solution set and whether such members are rational or irrational numbers.

5. $x^2 + 3x + 2 = 0$
6. $2x^2 - x - 1 = 0$
7. $3x^2 + 2x + 1 = 0$
8. $x^2 + 5x + 2 = 0$
9. $2x^2 - x - 2 = 0$
10. $x^2 + x + \tfrac{1}{4} = 0$
11. $9x^2 - 30x + 25 = 0$
12. $x^2 + \tfrac{5}{2}x - 6 = 0$
13. $x^2 + 2x + 5 = 0$
14. $x^2 + 10x + 20 = 0$
15. $2x^2 + \tfrac{11}{3}x + 1 = 0$
16. $5x^2 + 2x + 1 = 0$

For Exercises 17–24 use the discriminant to determine whether or not the given expression can be factored, and if so factor it.

Quadratic Equations

17. $x^2 + 5x + 3$
18. $2x^2 - x - 3$
19. $4x^2 - 4x + 1$
20. $6x^2 + x - 3$
21. $8x^2 + 10x - 3$
22. $5x^2 + 2x - 4$
23. $6x^2 - 5x - 6$
24. $2x^2 + 3x - 5$

For Exercises 25–40 find the range of the function, the ordered pair which corresponds to the vertex of the parabola which is its graph, and sketch the graph.

25. $f(x) = (x + 3)^2$
26. $f(x) = (x - 2)^2$
27. $f(x) = -(x + 1)^2$
28. $f(x) = -(x - 1)^2$
29. $f(x) = x^2 + 3$
30. $f(x) = (x + 1)^2 - 3$
31. $f(x) = 4 - (x - 1)^2$
32. $f(x) = (x + 2)^2 + 1$
33. $f(x) = x^2 + 2x + 3$
34. $f(x) = x^2 + 2x - 4$
35. $f(x) = x^2 - 4x + 3$
36. $f(x) = x^2 - 4x - 1$
37. $f(x) = -x^2 - 2x + 2$
38. $f(x) = -x^2 + 2x - 3$
39. $f(x) = 2x^2 - 4x + 1$
40. $f(x) = x^2/2 + x + 2$

Answers to Odd-Numbered Exercises

Answers 1.1

1. 23
3. 13
5. 54
7. 20
9. 51
11. 53
13. 23
15. 7
17. $\frac{1}{6}$
19. $\frac{7}{6}$
21. $\frac{3}{2}$
23. $\frac{23}{5}$
25. $\{11\}$
27. $\{3\}$
29. $\{15\}$
31. $\{12\}$
33. $\{5, 12, 21\}$
35. $\{10, 26\}$
37. $\{\frac{11}{2}, \frac{23}{5}\}$
39. $\{\frac{17}{4}, \frac{33}{16}\}$
41. $\{\frac{1}{2}\}$
43. $\{\frac{3}{10}\}$
45. $\{4\}$
47. $\{\frac{1}{2}\}$
49. $\{7\}$
51. $\{2, 3\}$
53. $(x + 3)4$
55. $5(x + 3)$
57. $x(4 + 2x)$
59. $x(x - 6)$
61. $15 + 8x$
63. $7 + 5(x - 1)$ or $2 + 5x$

Answers 1.2

1. Statement: False
3. Statement: True
5. Expression: 29
7. Statement: True
9. Statement: True
11. Statement: True
13. Statement: True
15. Statement: True
17. Sentence: {2}
19. Sentence: {0, 2}
21. Expression: {0, 32}
23. Sentence: { }
25. Sentence: {$\frac{1}{2}$}
27. Sentence: {1, 2}
29. Equivalent
31. Equivalent
33. Equivalent
35. Equivalent
37. $4 + 3x = 5x$
39. $(2x - 1)(x + 3)$
41. $\dfrac{x}{x-3} + 4$
43. $x + (2x + 5) = 29$
 (x represents the number of girls.)

Answers 1.3

1. $3x + 2$: $0 \to 2$
             $1 \to 5$
             $2 \to 8$
3. $\dfrac{5x}{3}$: $0 \to 0$
        $1 \to \frac{5}{3}$
        $3 \to 5$
5. $\dfrac{3x-1}{x+1}$: $1 \to 1$
             $2 \to \frac{5}{3}$
             $3 \to 2$
7. $\dfrac{z}{3} + \dfrac{z}{2}$: $0 \to 0$
          $2 \to \frac{5}{3}$
          $6 \to 5$
9. 7
11. 20
13. 11
15. $\frac{3}{2}$
17. 3
19. Does not exist
21. 2
23. Does not exist
25. 11
27. 1
29. 2
31. {5, 8, 17}
33. 2
35. Does not exist
37. 12
39. 7

Answers 1.4

1. True
3. True
5. True
7. False

9.	True	11.	True
13.	True	15.	True
17.	True	19.	False
21.	False	23.	False
25.	{2, 3, 4, 5}	27.	{ }
29.	{4, 5, 6, 7}	31.	{7, 8, 9, 10}
33.	{3}	35.	∅
37.	A	39.	∅
41.	B	43.	{1, 2}
45.	∅	47.	B
49.	{1, 2, 4, 5}	51.	∅
53.	{1, 2, 3, 4}	55.	V
57.	A	59.	∅
61.	Meaningless	63.	Meaningless
65.	{∅, 0}	67.	{∅}

Answers 1.5

1.	Q, F	3.	I, F
5.	F	7.	N, I, Q, F
9.	N, I, Q, F	11.	Q, F
13.	2	15.	0, 3.6, -1, 2, -5/2
17.	I, F	19.	F
21.	N, I, F, Q	23.	F
25.	N	27.	{-1, 0, 1}
29.	{-1, 0, 1, 2, 3, ...}	31.	{.5}
33.	N	35.	F
37.	{3}	39.	{1, 2, 3, 4}
41.	{$\frac{2}{3}$}	43.	∅
45.	Equivalent	47.	Equivalent
49.	Equivalent	51.	Not equivalent
53.	{2, 4, 6, 8, 10, ...}	55.	N

Answers 1.6

1.

3.

336 *Answers to Odd-Numbered Exercises*

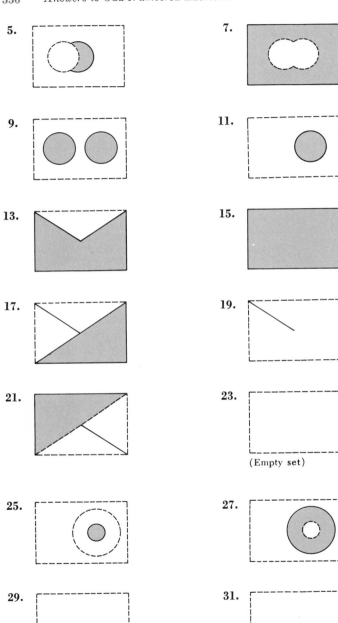

337   *Answers to Odd-Numbered Exercises*

33. 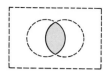   **35.**

37.    **39.**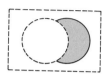
(Empty set)

41. 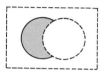   **43.**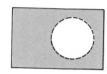

45. 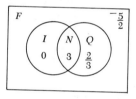   **47.** (a) 21; (b) 69; (c) 3

Answers 2.1

1. Commutative axiom of multiplication
3. Associative axiom of multiplication
5. Commutative axiom of multiplication
7. Commutative axiom of addition
9. Associative axiom of multiplication
11. Associative axiom of addition
13. Commutative axiom of addition
15. Commutative axiom of addition
17. Distributive axiom
19. Distributive axiom
21. $x + (3 + 9) = x + 12$
23. $3x + 3 \cdot 5 = 3x + 15$
25. $7(x + 4)$
27. $7x + 2$
29. $3(x + 2)$
31. $6 \cdot 7 + 6x = 42 + 6x$
33. $(2 + 5)x = 7x$
35. $(5 + 1)x = 6x$

338 Answers to Odd-Numbered Exercises

37. $(5 + 4) + 3x = 9 + 3x$
39. $[3x](x + 1)$
41. Equivalent
43. Not equivalent
45. Equivalent
47. Not equivalent
49. Equivalent
51. Not equivalent
53. Not equivalent
55. Not equivalent
57. Equivalent
59. Not equivalent

Answers 2.2

1. Not closed with respect to addition
   Closed with respect to multiplication
3. Not closed with respect to addition
   Not closed with respect to multiplication
5. Closed with respect to addition
   Closed with respect to multiplication
7. Closed with respect to addition
   Closed with respect to multiplication
9. Closed
11. Not closed
13. Not closed
15. Not closed
17. $4x + 4 \cdot \frac{1}{2} = 4x + 2$
19. $(\frac{2}{3} \cdot 6)x = 4x$
21. $(3)(x + \frac{1}{3})$
23. $(x + \frac{1}{2})[(\frac{3}{8})(16)] = (x + \frac{1}{2})(6)$
25. $\frac{7}{3}$
27. $\frac{1}{4}$
29. $\frac{5}{22}$
31. $\frac{10}{17}$
33. 0
35. -11
37. 0
39. 0
41. -6
43. 3
45. 0
47. 0
49. 8
51. 8

Answers 2.3

1. (a) Added (b) Negative
   (c) -15
3. (a) Subtracted (b) Positive
   (c) 4
5. (a) Added (b) Negative
   (c) -42
7. (a) Added (b) Negative
   (c) -15
9. 40
11. -16
13. -13
15. 0
17. -1
19. -37
21. 20
23. -24

Answers to Odd-Numbered Exercises

25. 12
27. -8
29. -13
31. -6
33. {-4}
35. {13}
37. {0, -6}
39. {0, 12}
41. {-3}
43. {-1}
45. {-1}
47. {-1, -2}
49. -1 and -2
51. -1 and -2
53. -5
55. -5

Answers 2.4

1. (a) -8 (b) $(-3) + (8)$ (c) 5
3. (a) -2 (b) $(8) + (2)$ (c) 10
5. (a) 12 (b) $8 + (-12)$ (c) -4
7. (a) -4 (b) $-6 + (4)$ (c) -2
9. -6
11. 12
13. 17
15. -6
17. -17
19. -34
21. -14
23. -5
25. -2
27. -3
29. 3
31. -3
33. $\{-13\}$
35. $\{14\}$
37. $\{-10, 2\}$
39. $\{-40, -56\}$
41. $\{10, 21, 36\}$
43. $\{5, 0, -7\}$
45. $2x$
47. $3x - 1$
49. $-3x + 1$
51. $4x - 4$
53. Not closed
55. Closed
57. Not closed
59. Closed

Answers 2.5

1. $-\frac{2}{3}$
3. 3
5. $\frac{6}{25}$
7. $-\frac{2}{9}$
9. $-\frac{25}{17}$
11. $\frac{1}{16}$
13. $\frac{3}{20}$
15. $-\frac{13}{30}$
17. $-\frac{27}{28}$
19. $\frac{10}{3}$
21. $\frac{1}{6}$
23. $-\frac{2}{5}$
25. $-\frac{23}{20}$
27. $-\frac{1}{9}$
29. $\frac{5}{4}$
31. $-\frac{10}{3}$

33. $-\frac{2}{3}$
35. $\frac{3}{4}$
37. $\frac{5}{2}$
39. $\frac{1}{3}$
41. $-\frac{3}{4}$
43. $\frac{5}{9}$
45. $-\frac{3}{4}$
47. $\frac{21}{8}$
49. 3
51. $-\frac{3}{11}$
53. $-\frac{1}{2}$
55. $-\frac{16}{3}$
57. $\{\frac{9}{2}\}$
59. $\{-\frac{6}{5}\}$
61. $\{\frac{1}{3}\}$
63. $\{\frac{2}{3}, -\frac{2}{5}\}$
65. Not closed
67. Not closed
69. Closed
71. Closed
73. $.\overline{2}\ldots$
75. $-.\overline{714285}\ldots$

Answers 2.6

1. $-6 > -10$
3. $-4 < 0$
5. $-\frac{3}{8} > -\frac{5}{13}$
7. $\frac{9}{5} < \frac{11}{6}$
9. $-2.65 < -2.64$
11. $-\frac{5}{8} > -.63$
13. $4, \frac{9}{2}, \frac{14}{3}$
15. $-\frac{7}{2}, -\frac{10}{3}, -3$
17. 3, 4
19. $-13, -12$
21. $-7, -6, -5, -4$
23. None
25. 4.3
27. $-.9$
29. R, F
31. R, F
33. R
35. R, F, I, N
37. None
39. a, b, d, e
41. b, e
43. a, b, d, e
45. c
47. e
49. g
51. g
53. b
55. a
57. b
59. e

Answers 3.1

1. 29
3. 50
5. 4
7. -4
9. 1
11. 1
13. -1
15. 1
17. -32
19. -16
21. -11
23. -18
25. 25
27. -8

Answers to Odd-Numbered Exercises

29. 16
33. 128
37. -24
41. 2
45. -2
49. -1
53. 1
57. $\{3\}$
61. $\{-1, 2\}$
65. $3x + 5$
69. $5x - 4$
73. $x + 4$
77. $4x^2 + 2x + 1$
81. $3x - 12$
85. $-5x + 2$
89. x
93. x
97. $4x - 4$

31. 16
35. 16
39. 3
43. 8
47. -13
51. -5
55. -30
59. $\{-1, 0\}$
63. $\{-2, 2\}$
67. $-5x + 4$
71. -3
75. $-8x + 7$
79. $2x^2 - 2x - 2$
83. $3x - 2$
87. $2x + 9$
91. $-x$
95. $8x$
99. $-6x + 6$

Answers 3.2

1. $\{3\}$
5. $\{\frac{5}{3}\}$
9. $\{2\}$
13. $\{2\}$
17. $\{1\}$
21. $\{\frac{2}{3}\}$
25. $\{\frac{1}{2}\}$
29. $\{\frac{4}{3}\}$
33. $\{\frac{3}{5}\}$
37. $\{-\frac{3}{10}\}$
41. $x + 3 = -5, \{-8\}$
45. $4x + 5 = 33, \{7\}$

3. $\{-8\}$
7. $\{-\frac{7}{4}\}$
11. $\{-2\}$
15. $\{2\}$
19. $\{-2\}$
23. $\{\frac{4}{3}\}$
27. $\{0\}$
31. $\{-1\}$
35. $\{\frac{1}{3}\}$
39. $\{\frac{23}{3}\}$
43. $3(x + 2) = 11, \{\frac{5}{3}\}$
47. $8x + 15 = 27, \{\frac{3}{2}\}$

Answers 3.3

1. $\{5\}$
5. $\{-2\}$
9. $\{\frac{7}{3}\}$

3. $\{-2\}$
7. $\{\frac{11}{5}\}$
11. $\{\frac{2}{3}\}$

342   Answers to Odd-Numbered Exercises

13. $\{0\}$
15. $\{0\}$
17. $\emptyset$
19. $\emptyset$
21. R
23. R
25. $\{2\}$
27. $\{\frac{5}{2}\}$
29. $2x + 3 = x - 7,\ \{-10\}$
31. $x + 6 = 3x - 2,\ \{4\}$

Answers 3.4

1. True
3. True
5. False
7. False
9. False
11. False
13. False
15. True
17. $\langle -\infty, -2 \rangle$
19. $\langle -5, \infty \rangle$
21. $\langle -\infty, 3 \rangle$
23. $\langle -4, \infty \rangle$
25. $\langle -1, \infty \rangle$
27. $\langle -\infty, -\frac{5}{2} \rangle$
29. $\langle -\infty, -2 \rangle$
31. $\langle \frac{27}{2}, \infty \rangle$
33. $\langle 0, \infty \rangle$
35. $\langle \frac{7}{3}, \infty \rangle$
37. $\emptyset$
39. R
41. $\emptyset$
43. $\langle 0, \infty \rangle$
45. $2x + 6 > 3x,\ \langle -\infty, 6 \rangle$
47. $x + 6 < 4x,\ \langle 2, \infty \rangle$

Answers 3.5

1. $\langle -\infty, 1 \rangle$

3. $[2, \infty \rangle$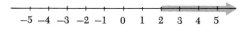

5. $\langle 2, \infty \rangle$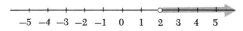

7. $[2, \infty \rangle$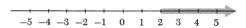

9. $\{3\}$

11. $\{0\}$

343 Answers to Odd-Numbered Exercises

13. $\langle 1, \infty \rangle$

15. $[2, \infty \rangle$

17. $\varnothing$

19. R

21. $[1, 4]$

23. $\langle -3, 3 \rangle$

25. $\varnothing$

27. $\langle -1, 2 \rangle$

29. $\langle -3, -1 \rangle$

31. $\langle -\infty, -2] \cup [2, \infty \rangle$

33. False   35. True
37. False   39. False

41. $\langle 1, \infty \rangle$

43. $\langle -\infty, -1]$

45. $\langle -1, 3 \rangle$

47. $\langle 0, 5]$

49. $\langle -1, \infty \rangle$

51. $\langle -\infty, -2 \rangle \cup \langle 4, \infty \rangle$

53. $\langle -\infty, 2 \rangle$

55. $\langle -\infty, -2]$

57. $\langle -1, \infty \rangle$

59. $\{2\}$

Answers 3.6

1. 2 miles per hour
3. 9, 10, and 11
5. 18
7. 21 acres
9. 4 and 7
11. 8 girls; 21 boys
13. 4
15. $405
17. $-12, -11,$ and -10
19. 10, 12, and 14
21. $\frac{11}{2}$ inches
23. $\frac{13}{2}$ inches by $\frac{3}{2}$ inches
25. 50 cents
27. 5 quarters, 2 dimes, and 3 nickels
29. 40 miles per hour
31. 50 miles per hour
33. John is 9 and Bill is 18
35. The man is 24; the boy is 8.
37. 7 miles per hour
39. Any three consecutive integers have this property

Answers 4.1

1. x^{10}
3. $-8x^3$
5. $-2x^6$
7. $10x^6$
9. $3x^3 + 2x^2$
11. $-12x^2 - 15x$
13. $-8x^3 + 18x$
15. $10x^4 - 2x^3$
17. $8x^3$
19. $4x^2$
21. x^4
23. x^{150}
25. x^{30}
27. x^{60}
29. $8x^6$
31. $-27x^6$
33. $4x^5$
35. $-81x^4$

37. $-128x^5$
39. $-25x^7$
41. $3x^{10}$
43. $-27x^{11}$
45. $12x^3 - 20x^2$
47. $-40x^8 + 8x^6$
49. $-18x^4$
51. $-3x^5 - x^4$
53. 7^{13}
55. 2^{15}
57. 10^5
59. 21^{12}
61. 2^7
63. 14^3

Answers 4.2

1. $x^2 + 5x + 6$
3. $x^2 + 7x + 10$
5. $x^2 - 6x + 8$
7. $x^2 - 8x + 15$
9. $x^2 + x - 12$
11. $x^2 - 2x - 35$
13. $6x^2 + 7x + 2$
15. $8x^2 - 10x + 3$
17. $12x^2 - 11x + 2$
19. $12x^2 - 32x + 5$
21. $x^2 - 25$
23. $x^2 - 49$
25. $4x^2 - 1$
27. $16x^2 - 9$
29. $x^2 + 6x + 9$
31. $4x^2 + 4x + 1$
33. $x^2 - 10x + 25$
35. $9x^2 - 6x + 1$
37. $4x^2 + 12x + 9$
39. $25x^2 + 30x + 9$
41. $4x^2 - 12x + 9$
43. $49x^2 - 28x + 4$
45. $2x^3 + 5x^2 + 7x + 6$
47. $4x^3 - 6x^2 - 8x - 2$
49. $6x^3 + x^2 + 8x - 3$
51. $4x^3 - 16x^2 + 13x + 3$
53. $5x^2 + 6x - 7$
55. $10x^2 - 12x$
57. $8x^2 - 8x - 32$
59. $5x^2 - 5$
61. $2x^3 + 11x^2 + 17x + 6$
63. $4x^3 + 28x^2 - 9x - 63$
65. $x^4 + 3x^3 + 6x^2 + 7x + 3$
67. $x^4 + 10x^3 + 16x^2 - 90x - 225$
69. $x^4 + 2x^3 + x^2$
71. $x^4 - 4x^3 + 10x^2 - 12x + 9$

Answers 4.3

1. $3(x + 3)$
3. $2(3x + 4)$
5. $x(x + 3)$
7. $x(5x^2 - 3)$
9. $3(x^2 - x + 2)$
11. $x(5x^2 + 2x - 2)$
13. $(x + 2)(x - 2)$
15. $(3x + 2)(3x - 2)$
17. $(x + 2)^2$
19. $(2x + 3)^2$
21. $(x + 2)(x + 1)$
23. $(x + 7)(x + 1)$
25. $(x - 2)(x - 1)$
27. $(x - 5)(x - 1)$
29. $(x + 3)(x + 2)$
31. $(x + 12)(x + 1)$

346   Answers to Odd-Numbered Exercises

33. $(x-4)(x-3)$
35. $(x-6)(x-3)$
37. $(x+3)(x-1)$
39. $(x+5)(x-1)$
41. $(x+3)(x-2)$
43. $(x+6)(x-3)$
45. $(x+1)(2x-1)$
47. $(3x-1)(x-2)$
49. $(2x-3)(2x-1)$
51. $(5x-1)(x-5)$
53. $(x+1)(6x-5)$
55. $4(x^2+2)$
57. $(x+3)(x-1)$
59. $(2x+1)(5x+1)$
61. $(5x+3)(2x-5)$
63. $(8x+3)(2x-9)$

Answers 4.4

1. $2x(x+3)$
3. $4x(3x-2)$
5. $2(x+1)(x-1)$
7. $2(2x+1)(2x-1)$
9. $x(x+2)(x-2)$
11. $2x(3x+2)(3x-2)$
13. $3(x+1)^2$
15. $2(2x+5)^2$
17. $x(x-1)^2$
19. $3x(x-3)^2$
21. $3(x+1)(x+2)$
23. $x(5x-1)(x-2)$
25. $-(x+3)(x-3)$
27. $-(x-3)^2$
29. $-2(x+5)(2x-1)$
31. $-2x(x+1)(3x-1)$
33. $(x+3)(x-5)$
35. $(3x+1)^2$
37. $(3x-1)^2$
39. $(3x^2+1)(2x+3)$
41. $(x^2+3)(x+2)$
43. $(x^2+1)(3x+1)$
45. $(x+3)(x-3)(x+1)$
47. $2(x^2+1)(x+3)$
49. $(\frac{1}{2})(4x+1)$
51. $(\frac{1}{2})(x+6)$
53. $(\frac{1}{2})x(x+2)$
55. $(\frac{1}{2})(2x+1)^2$
57. $3(x^2+8x+2)$
59. $x(x^2+4)$
61. Impossible
63. $x(x^2+4)(x+2)(x-2)$

Answers 4.5

1. $\{1, 2\}$
3. $\{-1, \frac{1}{2}\}$
5. $\{-\frac{2}{3}, \frac{1}{4}\}$
7. $\{0, \frac{5}{3}\}$
9. $\{1, 2\}$
11. $\{-1, 4\}$
13. $\{-\frac{1}{2}, 3\}$
15. $\{-4\}$
17. $\{-5, 5\}$
19. $\{\frac{1}{2}\}$
21. $\{0, 2\}$
23. $\{0, \frac{5}{2}\}$
25. $\{-\frac{1}{2}, 1\}$
27. $\{-2, 3\}$
29. $\{-2, 3\}$
31. $\{3, 5\}$

Answers to Odd-Numbered Exercises

33. $\{-3, 2\}$

35. $\{2\}$

37. $\{-4, -2\}$

39. $\{0, 3\}$

41. -3 and 2
43. $-\frac{1}{2}$
45. -1 and 2
47. $-\frac{1}{2}$ and 1
49. 6 and 7
51. -3 and 0 or $\frac{1}{2}$ and $\frac{7}{2}$
53. 3 inches by 9 inches
55. 7 inches by 15 inches

Answers 4.6

1. $\langle -\infty, -2 \rangle$

3. $\langle -1, \infty \rangle$

5. $\langle 3, \infty \rangle$

7. $\langle -\infty, -2 \rangle$

9. $\langle -1, 2 \rangle$

11. $[-3, -1]$

13. $\overline{[-2, 1]}$

15. $[-2, 1]$

17. $[-1, 1]$

19. $\langle -3, 2 \rangle$

21. $\langle -1, 4 \rangle$

23. $\langle -3, 3 \rangle$

25. $\langle 0, 2 \rangle$

27. $[-2, 2]$

29. $\langle -4, 2 \rangle$

31. $\{2\}$

Answers 5.1

1. $\overline{\{5\}}$
3. $\overline{\{0\}}$
5. $\overline{\{-2, 2\}}$
7. R
9. $\frac{3}{5}$ $\overline{\{1\}}$
11. $\frac{x+3}{x+2}$ $\overline{\{-2, 2\}}$
13. $\frac{2}{x-2}$ $\overline{\{-2, 2\}}$
15. $x+4$ $\overline{\{4\}}$
17. $\frac{1}{2x}$
19. $\frac{2}{3x^2}$
21. $\frac{2}{x+1}$
23. $\frac{x+5}{x+1}$
25. $\frac{x}{x-2}$
27. $\frac{2x+1}{3(x-1)}$
29. $\frac{1}{x}$
31. $\frac{(x+2)^2}{(x-1)(x+1)}$
33. $2x$
35. $x+2$

37. $6x$

39. $x^2 + x$

41. $\dfrac{x+2}{x+1}$

43. $\dfrac{x-1}{x+3}$

45. $\dfrac{x-1}{x-5}$

47. $\dfrac{x+3}{x+4}$

Answers 5.2

1. $\dfrac{x}{3}$

3. $\dfrac{6x^2}{5}$

5. $\dfrac{-6x-10}{3}$

7. $\dfrac{x}{3x+3}$

9. $\dfrac{3x+1}{3}$

11. $\dfrac{3x+2}{4}$

13. $x - 1$

15. $x^2 - 2x$

17. $\dfrac{x-2}{x+2}$

19. 1

21. $\dfrac{x+5}{x-3}$

23. $\dfrac{x-1}{x+1}$

25. $\dfrac{2x+3}{5}$

27. $\dfrac{2x}{7}$

29. $\dfrac{3x+1}{3}$

31. x

33. $\dfrac{x-5}{5}$

35. $x - 1$

37. $\dfrac{3x+1}{9}$

39. $\dfrac{11x}{6}$

41. $\dfrac{10x+1}{6}$

43. $\dfrac{x+3}{10}$

45. $\dfrac{x+32}{40}$

47. $\dfrac{x-13}{70}$

49. $\dfrac{11}{3x}$

51. $\dfrac{-11}{10x}$

53. $\dfrac{2x+1}{x^2}$

55. $\dfrac{-3x+4}{6x^2}$

57. $\dfrac{3x+1}{2x}$

59. $\dfrac{7x+3}{6x^2}$

61. $\dfrac{2x+3}{x}$

63. $\dfrac{x^2-2x-3}{x}$

Answers 5.3

1. $2x(x + 2)$
3. $2(x + 1)^2$
5. $2x(x + 3)$
7. $(x + 1)(x - 1)(x + 2)$
9. $\dfrac{6}{3(x + 1)}$ and $\dfrac{1}{3(x + 1)}$
11. $\dfrac{3x - 3}{2x(x - 1)}$ and $\dfrac{10}{2x(x - 1)}$
13. $\dfrac{3x - 3}{(x + 1)(x - 1)}$ and $\dfrac{2}{(x + 1)(x - 1)}$
15. $\dfrac{x^2 + 3x + 2}{(x - 2)(x - 1)(x + 2)}$ and $\dfrac{x^2 - x}{(x - 2)(x - 1)(x + 2)}$
17. 1
19. $\dfrac{x + 3}{x}$
21. $\dfrac{3x + 2}{x(x + 2)}$
23. $\dfrac{-x^2 - 3x - 1}{x(x + 1)}$
25. $\dfrac{x + 5}{3x(x - 1)}$
27. $\dfrac{2x - 13}{6(x - 1)}$
29. $\dfrac{3x + 9}{x(x + 5)(x - 1)}$
31. $\dfrac{2x + 3}{(x + 1)(x - 1)(x + 2)}$
33. $\dfrac{x^2 + 6x + 3}{(x + 2)(x + 1)(x + 3)}$
35. $\dfrac{3}{(x - 2)^2(x + 1)}$
37. $\dfrac{x + 1}{x - 5}$
39. 2
41. $\tfrac{1}{3}$
43. $\dfrac{1}{x + 3}$
45. $x + 7$
47. $6x^2 + x + 1$
49. $3x + 4$
51. 3

Answers 5.4

1. $\{-\tfrac{4}{7}\}$
3. $\{\tfrac{1}{4}\}$
5. $\langle -\infty, -\tfrac{1}{7}]$
7. $\{1, 3\}$
9. $\varnothing$
11. R
13. $\{-\tfrac{5}{6}\}$
15. $\{\tfrac{16}{25}\}$
17. $\{-8, 2\}$
19. $\{0, 5\}$
21. $\{2\}$
23. $\{\tfrac{2}{3}, 1\}$
25. $\{-4\}$
27. $\{4\}$
29. $\{-2, 5\}$
31. $\langle \tfrac{18}{5}, \infty \rangle$
33. $\{-\tfrac{3}{4}\}$
35. $\{\tfrac{5}{2}, 5\}$
37. $\{4\}$
39. $\{-2\}$

Answers 5.5

1. 24
3. -9 and -8
5. $-\frac{3}{4}$ and $\frac{9}{4}$
7. 1 and 13
9. 740
11. $564.00
13. 5%
15. 4.5%
17. $\frac{1}{2}$
19. $\frac{3}{9}$
21. 35 hours
23. 6 hours
25. 1.2 miles per hour
27. 40 miles per hour
29. $\frac{12}{7}$ grams
31. 35
33. $2500.00
35. 1.2 quarts

Answers 5.6

1. $\dfrac{3x}{2}$
3. $\dfrac{2}{3x}$
5. $\dfrac{x+2}{5}$
7. $\dfrac{3}{2x+5}$
9. $\dfrac{3x+6}{2x}$
11. $\dfrac{6x}{7x+42}$
13. $\dfrac{x+4}{x-1}$
15. $\dfrac{2x-1}{x-1}$
17. $\dfrac{3x+9}{x-1}$
19. $\dfrac{x(x+5)}{2x-1}$
21. $\dfrac{1}{x-1}$
23. $x+4$
25. $\dfrac{2x+1}{x}$
27. $\dfrac{x-2}{x+3}$
29. $\dfrac{2x-4}{x(x-1)}$
31. $\dfrac{x+1}{x+5}$
33. $x+3$
35. x^2-x+3
37. $(x+5)(x+1)(x-3)$
39. $(x^2+1)(x+1)(x-2)$
41. $x+1;\ 1;\ x+1+\dfrac{1}{x+1}$
43. $x+3;\ -3;\ x+3+\dfrac{-3}{x+2}$
45. $x+3;\ 5;\ x+3+\dfrac{5}{2x-1}$
47. $x^2+1;\ 1;\ x^2+1+\dfrac{1}{x+2}$
49. $x+1;\ 3;\ x+1+\dfrac{3}{x^2+1}$
51. $1;\ 3x+1;\ 1+\dfrac{3x+1}{2x^2+1}$
53. $x+1;\ 2;\ x+1+\dfrac{2}{x+3}$
55. $\frac{1}{2};\ -\frac{1}{2};\ \dfrac{1}{2}+\dfrac{-\frac{1}{2}}{2x^2+1}$

Answers 5.7

1. $\frac{1}{5}$
3. $\frac{3}{32}$
5. 1
7. 5
9. $-\frac{1}{4}$
11. $-\frac{1}{16}$
13. 4
15. 1
17. 16
19. $\frac{1}{9}$
21. 16
23. 4
25. 16
27. -1
29. $\frac{3}{4}$
31. 5
33. $-\frac{1}{8}$
35. $\frac{1}{9}$
37. $\frac{3}{4}$
39. 0
41. $\dfrac{2}{x^4}$
43. $\dfrac{1}{9x^2}$
45. x^2
47. $\dfrac{6}{x^3}$
49. $x + 2$
51. $3x + 9$
53. $\dfrac{4x + 1}{x}$
55. $\dfrac{-2x + 1}{x^2}$
57. $2x^2$
59. $\dfrac{x}{4}$
61. $\dfrac{x + 1}{x - 1}$
63. $\dfrac{x - 1}{x^2}$
65. $x^3 + x^2 + x + 1$
67. $\dfrac{x - 5}{2x + 1}$

Answers 6.1

1. $\{(1, 2), (1, 3)\}$
3. $\{(1, -1), (1, 1)\}$
5. $\{(1, 1)\}$
7. $\{(-1, -1), (-1, 1), (1, -1)(1, 1)\}$
9. 50
11. 5
13. $\{(3, 4)\}$
15. $\{\ \}$

17.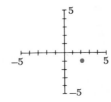

19.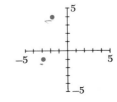

21.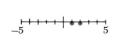
23.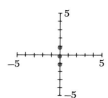

25. $\{(-2, 4), (-1, 4), (0, 4), (1, 4), (2, 4)\}$
27. $\{(2, -4), (2, -3), (2, -2), (2, -1), (2, 0), (2, 1), (2, 2), (2, 3)\}$
29. $\{7, -1\}$ **31.** $\{3\}$
33. $\{16, 1, 2\}$ **35.** $\{-10\}$
37. $\{(1, 2), (3, 1)\}$ **39.** $\{(1, 8)\}$
41. $\{(2, 0), (0, -4), (-2, 0)\}$ **43.** $\{\ \}$

45. 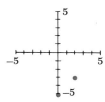 **47.**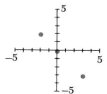

49. 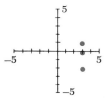 **51.**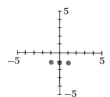

Answers 6.2

1. $3x - y$ **3.** $-x + 2y$
5. $4x$ **7.** $7x$
9. $x^2 + xy - 2y^2$ **11.** $4x^2 + 4xy + y^2$
13. $3x^2 - 11y - 4y^2$ **15.** $3x^2 - 7xy + 2y^2$
17. $(x + y)(x - y)$ **19.** $(x + 3y)(x - y)$
21. $(4x + y)(x - 2y)$ **23.** $xy(2x - y)$
25. $x(x + 2y)(x - 2y)$ **27.** $(x + y)(x^2 + 2)$
29. $\dfrac{-x^2}{y^2}$ **31.** $\dfrac{x - y}{x + 2y}$

354   Answers to Odd-Numbered Exercises

33. 1

35. $\dfrac{x+y}{x-y}$

37. $\dfrac{3x-y}{6}$

39. $\dfrac{x-y}{x^2 y^2}$

41. $\dfrac{3x}{2(x+y)}$

43. $\dfrac{x-y}{xy(x+2)}$

45. $8x^3 y^6$

47. $y+x$

49. $x^2 y^6$

51. $\dfrac{1}{xy}$

53. $x^2 + 2xy + 2y^2$

55. $(x-y)(x^2+xy+y^2)$

Answers 6.3

1. $\{(1, 2)\}$
3. $\{(-1, 1)\}$
5. $\{(0, -3)\}$
7. $\{(0, 3), (1, 1)\}$
9. $\{(0, 0), (\tfrac{2}{3}, 2)\}$
11. $\{(-1, 2), (0, 5), (1, 8)\}$
13. x-intercept is 2; y-intercept is 4
15. x-intercept is $\tfrac{5}{2}$; y-intercept is -5
17. x-intercept is 0; y-intercept is 0
19. x-intercept is 5; y-intercept is $-\tfrac{5}{2}$

21.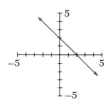

23.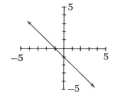

25.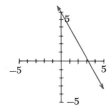

27.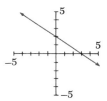

29.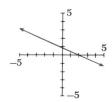

31.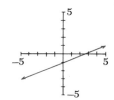

33.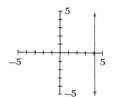
35.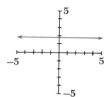
37.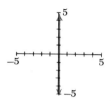
39.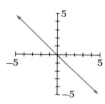
41.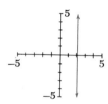
43.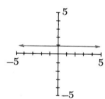
45.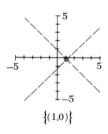
$\{(1,0)\}$
47.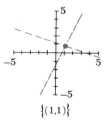
$\{(1,1)\}$
49.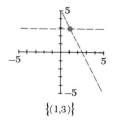
$\{(1,3)\}$
51.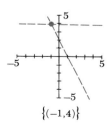
$\{(-1,4)\}$

Answers 6.4

1.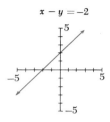
$x - y = -2$

3.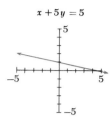
$x + 5y = 5$

5.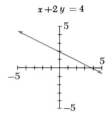
$x + 2y = 4$

7.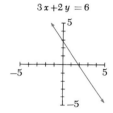
$3x + 2y = 6$

9. $\{(-1, 1)\}$
11. $\{(1, 3)\}$
13. $\{(2, -1)\}$
15. $\{(1, 3)\}$
17. $\{(1, -2)\}$
19. $\{(1, -3)\}$
21. $\{(-2, 1)\}$
23. $\{(-3, -2)\}$
25. $\{(-\frac{1}{3}, \frac{2}{3})\}$
27. $\{(0, 0)\}$
29. $\{(2, -1)\}$
31. $\{(-1, -1)\}$
33. $\{(0, 0)\}$
35. $\{(3, -1)\}$
37. 4 and 8
39. The first number is -1; the second number is 5
41. 29
43. Mary is 12 and Bill is 6

Answers 6.5

1. $\{\ \}$
3. $\{(x, y): x = 6 - 2y\}$
5. $\{(-\frac{1}{2}, -\frac{1}{2})\}$
7. $\{(x, y): y = 3x + 5\}$

9.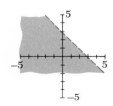

11.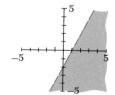

13. 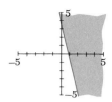 **15.**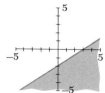

17. 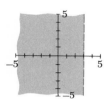 **19.**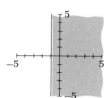

21. 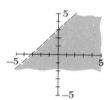 **23.**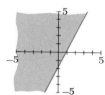

25. 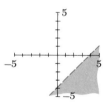 **27.**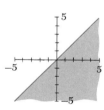

29. 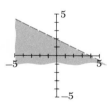 **31.**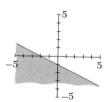

33. 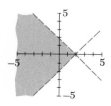 **35.**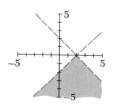

358 Answers to Odd-Numbered Exercises

37. 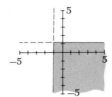   39.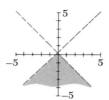

41. 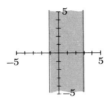   43.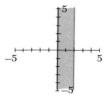

Answers 7.1

1. Domain = $\{2, 3\}$; range = $\{0, 1\}$; function
3. Domain = $\{1\}$; range = $\{5\}$; function
5. Domain = $\{-3\}$; range = $\{1, 2, 3\}$; not a function
7. Domain = $\{3\}$; range = R; not a function
9. $f = \{(x, y): y = 3x - 2\}$; linear
11. $h = \{(x, y): y = x^2 + 3\}$; not linear
13. $f(x) = 3x + 5$; linear
15. $f(x) = 3$; not linear (But the graph is a line!)
17. $f(x) = -x + 5$; linear   19. $f(x) = -x^2 + 3$; not linear
21. $f(x) = x/3$; linear   23. $f(x) = \dfrac{3x - 5}{4}$; linear
25. 11   27. 1
29. -2 31. 4

33. 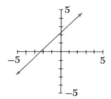   35.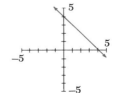

37. 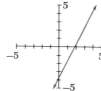 **39.**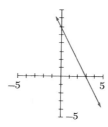

41. Function; domain = $\{0, 1, 2\}$; range = $\{-2, 0, 1\}$
43. Function; domain = $[-3, 2]$; range = $\{2\}$
45. Function; domain = R; range = $\{-4\}$
47. Function; domain = R; range = $[0, \infty)$
49. Function; domain = $[-2, 2]$; range = $[0, 2]$
51. Not a function; domain = $[-3, 3]$; range = $[-3, 3]$
53. Function   **55.** Not a function

Answers 7.2

1. 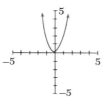 **3.**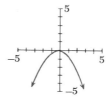

5. 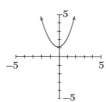 **7.**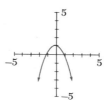

9. 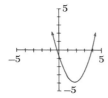 **11.**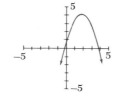

13.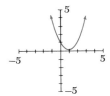

15.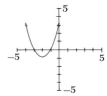

17.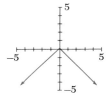

19.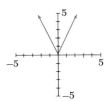

21.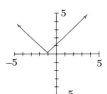

23.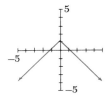

25.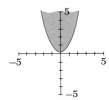

27.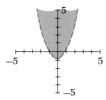

29.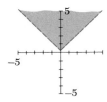

31.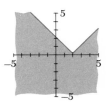

33.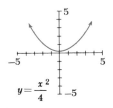

$y = \dfrac{x^2}{4}$

35.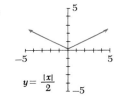

$y = \dfrac{|x|}{2}$

361   Answers to Odd-Numbered Exercises

37.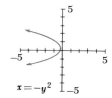

39.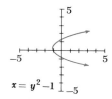

Answers 7.3

1.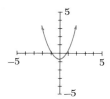

3.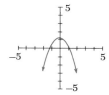

5.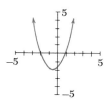

7.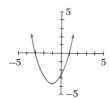

9.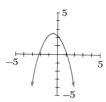

11.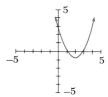

13. $\langle -\infty, 2]$

15. $\langle -\infty, 1]$

17.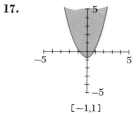
$[-1, 1]$

19.
$\langle -2, 2 \rangle$

21.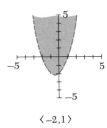
⟨−2,1⟩

23.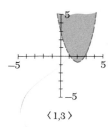
⟨1,3⟩

25.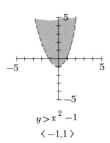
$y > x^2 - 1$
⟨−1,1⟩

27.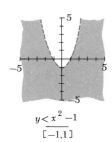
$y < x^2 - 1$
$[-1,1]$

29.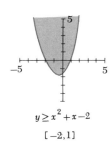
$y \geq x^2 + x - 2$
$[-2,1]$

31.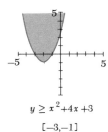
$y \geq x^2 + 4x + 3$
$[-3,-1]$

33. $(0, -4)$
35. $(1, -1)$
37. $(2, -1)$
39. $(-1, -4)$
41. Domain $= R$; range $= [-4, \infty)$
43. Domain $= R$; range $= [-\tfrac{1}{4}, \infty)$

Answers 7.4

1. $\{(3, 1)\}$
3. $\{(1, 1), (2, 2)\}$
5. $\{3, 4\} \times \{2\}$
7. $\{3\} \times R$

363   *Answers to Odd-Numbered Exercises*

9. $\{(x, y): y + 2x = 6\}$ 11. $\{(x, y): y < 2x + 3\}$

13. 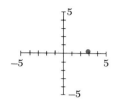   15.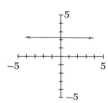

17. 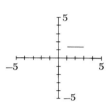   19.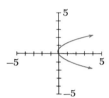

21. Domain $= \{2\}$; range $= \{1\}$ 23. Domain $= [0, 2]$; range $= [0, 3]$
25. Not a function   27. Function
29. $f^{-1}(x) = x/3$ 31. $f^{-1}(x) = x$
33. $f^{-1}(x) = \dfrac{x + 3}{2}$ 35. $f^{-1}(x) = \dfrac{-x + 4}{3}$
37. 2   39. 0

Answers 8.1

1. $-1; 1$ 3. $-\sqrt{5}; \sqrt{5}$
5. 512   7. $\sqrt[3]{-3}$
9. $-2; 2$ 11. $-\sqrt[4]{3}; \sqrt[4]{3}$
13. Rational; 8   15. Rational; -5
17. Rational; 1   19. Irrational
21. Not Real   23. Rational; 0
25. Rational; 0   27. Not Real
29. Rational; $\tfrac{1}{3}$ 31. Rational; $\tfrac{2}{5}$
33. $\{1, \sqrt{3}, \sqrt{5}\}$ 35. $\{-1, -2, -\sqrt{7}\}$
37. $\{-1, 2\}$ 39. $\{2, 3\}$
41. $\{81\}$ 43. $\{-5\}$
45. $\{-\sqrt{3}, \sqrt{3}\}$ 47. $\{\sqrt[3]{2}\}$

Answers 8.2

1. $\sqrt{3}$
3. $-\sqrt{5}$
5. 6
7. -10
9. $\sqrt[3]{4}$
11. 5
13. $\sqrt[4]{10}$
15. -1
17. $(\sqrt{2})^3$
19. 27
21. $(\sqrt[3]{2})^2$
23. 16
25. 1
27. -1
29. 0
31. 0
33. $\dfrac{1}{\sqrt{2}}$
35. $\dfrac{1}{\sqrt[3]{5}}$
37. $\tfrac{1}{2}$
39. $\tfrac{1}{2}$
41. $\dfrac{1}{(\sqrt[3]{2})^2}$
43. $-\tfrac{1}{2}$
45. $\tfrac{1}{4}$
47. 1
49. 6
51. 4
53. 2
55. $\tfrac{1}{3}$
57. 2
59. 12
61. $\{0, -1, \tfrac{1}{3}\}$
63. $\{-1, -\tfrac{1}{2}\}$
65. $\{2, \tfrac{1}{2}\}$
67. $\{2, 3, 4\}$

Answers 8.3

1. $\sqrt{15}$
3. 4
5. 14
7. 6
9. 3
11. $\tfrac{1}{5}$
13. $2\sqrt{2}$
15. $3\sqrt{3}$
17. $4\sqrt{3}$
19. $4\sqrt{5}$
21. $\dfrac{\sqrt{6}}{3}$
23. $\dfrac{\sqrt{10}}{4}$
25. $5\sqrt{2}$
27. $\sqrt{5}$
29. $11\sqrt{2}$
31. $-2\sqrt{3}$
33. $\sqrt{3} + 3$
35. $5 + 4\sqrt{2}$
37. 1
39. $4 + 2\sqrt{3}$
41. $1 + \sqrt{2}$
43. $3 - \sqrt{2}$

365   Answers to Odd-Numbered Exercises

45. $\dfrac{-1+\sqrt{5}}{2}$
47. $-2-\sqrt{3}$
49. $\dfrac{2}{\sqrt{10}}$
51. $\dfrac{3}{\sqrt{3}+3}$
53. $2\sqrt{7}$
55. $\dfrac{2-\sqrt{2}}{4}$
57. $\sqrt{2}-1$
59. $\sqrt{3}+\sqrt{2}$

Answers 8.4

1. $[0, \infty)$
3. $[1, \infty)$
5. $[0, \infty)$
7. R
9. R
11. R
13. $2\sqrt{x}$
15. $5\sqrt{x}$
17. x^2
19. x^6
21. $4\sqrt{x}$
23. x^2
25. $x+\sqrt{x}$
27. $x+2\sqrt{x}-3$
29. $x-4$
31. $x+6\sqrt{x}+9$
33. $|x+1|$
35. x^2
37. $|x^3|$
39. $|x|$
41. $\{10\}$
43. $\{1\}$
45. $\{1\}$
47. $\{2\}$
49. $\{-2\}$
51. $\{4\}$

Answers 9.1

1. 9
3. 36
5. $\frac{9}{4}$
7. $\frac{25}{4}$
9. $\{-\sqrt{5}, \sqrt{5}\}$
11. $\emptyset$
13. $\{-5, 1\}$
15. $\{1-\sqrt{2}, 1+\sqrt{2}\}$
17. $\{-4, 1\}$
19. $\left\{\dfrac{-5-\sqrt{29}}{2}, \dfrac{-5+\sqrt{29}}{2}\right\}$
21. $\{2-\sqrt{3}, 2+\sqrt{3}\}$
23. $\{-6-\sqrt{3}, -6+\sqrt{3}\}$
25. $\{4\}$
27. $\left\{\dfrac{-3-\sqrt{5}}{2}, \dfrac{-3+\sqrt{5}}{2}\right\}$
29. $\{-2-\sqrt{3}, -2+\sqrt{3}\}$
31. $\{1-\sqrt{3}, 1+\sqrt{3}\}$

33. $\left\{-2 - \dfrac{3}{\sqrt{2}},\ -2 + \dfrac{3}{\sqrt{2}}\right\}$ 35. $\left\{-1 - \sqrt{\dfrac{5}{3}},\ -1 + \sqrt{\dfrac{5}{3}}\right\}$

37. $\{-\tfrac{1}{3}, 4\}$ 39. $\left\{\dfrac{1 - \sqrt{13}}{2},\ \dfrac{1 + \sqrt{13}}{2}\right\}$

Answers 9.2

1. $a = 2;\ b = 5;\ c = 4$ 3. $a = 4;\ b = 0;\ c = 5$
5. $a = 1;\ b = -2;\ c = -5$ 7. $a = 1;\ b = 0;\ c = -2$
9. $\left\{\dfrac{-3 - \sqrt{5}}{2},\ \dfrac{-3 + \sqrt{5}}{2}\right\}$ 11. $\left\{\dfrac{-1 - \sqrt{13}}{2},\ \dfrac{-1 + \sqrt{13}}{2}\right\}$
13. $\{-\tfrac{1}{2}, 1\}$ 15. $\left\{\dfrac{5 - \sqrt{13}}{6},\ \dfrac{5 + \sqrt{13}}{6}\right\}$
17. $\{-3, \tfrac{1}{2}\}$ 19. $\{-3/2, 2\}$
21. $\{-1 - \sqrt{5},\ -1 + \sqrt{5}\}$ 23. $\{-2 - \sqrt{2},\ -2 + \sqrt{2}\}$
25. $\{-2 - 2\sqrt{2},\ -2 + 2\sqrt{2}\}$ 27. $\left\{\dfrac{-1 - \sqrt{17}}{2},\ \dfrac{-1 + \sqrt{17}}{2}\right\}$
29. $\{-\sqrt{3},\ \sqrt{3}\}$ 31. $\left\{\dfrac{-\sqrt{5}}{2},\ \dfrac{\sqrt{5}}{2}\right\}$
33. $\{-5, 0\}$ 35. $\{0, \tfrac{5}{2}\}$
37. $\{2\}$ 39. $\{-1 - \sqrt{2},\ -1 + \sqrt{2}\}$

Answers 9.3

1. -23 3. 1
5. Two; rational   7. None
9. Two; irrational   11. One; rational
13. None   15. Two; rational
17. Cannot be factored   19. $(2x - 1)^2$
21. $(2x + 3)(4x - 1)$ 23. $(2x - 3)(3x + 2)$

25. 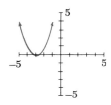   27.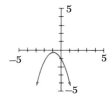

Range: $[0, \infty)$ Range: $(-\infty, 0]$
Vertex: $(-3, 0)$ Vertex: $(-1, 0)$

29.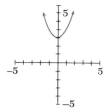

Range: [3, ∞⟩
Vertex: (0, 3)

31.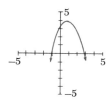

Range: ⟨−∞, 4]
Vertex: (1, 4)

33.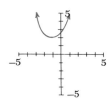

Range: [2, ∞⟩
Vertex: (−1, 2)

35.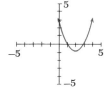

Range: [−1, ∞⟩
Vertex: (2, −1)

37.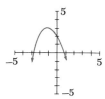

Range: ⟨−∞, 3]
Vertex: (−1, 3)

39.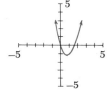

Range: [−1, ∞⟩
Vertex: (1, −1)

Glossary

Absolute value Of a real number, a particular nonnegative real number which corresponds to it. In symbols, $|a| = a$, if a is either positive or zero, and $|a| = -a$, if a is negative. (Note that $-a$ is actually *positive* if a is negative.)

Additive identity The number 0 is called the additive identity. (This is because for every $a \in R$, $a + 0 = 0 + a = a$.)

Additive inverse Of a real number x, the number $-x$, such that $x + (-x) = 0$. If the sum of two expressions is 0, each term is called the additive inverse of the other.

Algebraic expression Any combination of numerals, operational symbols, and variables (such as $3x - 5$) which collectively may represent more than one number.

Axiom A general statement which is assumed to be true and is accepted without proof. (See the real number axioms on pp. 78–79.)

Axis Each of the two number lines used to establish a coordinate plane. The horizontal line is called the x-axis, and the vertical line, the y-axis.

Base *See* Power.

Between If the real number x is both greater than a and less than b, then x is between a and b, and we indicate this in symbols by $a < x < b$.

Binomial Any expression (such as $3x + y$) which consists of exactly two terms.

Cancel To delete a pair of identical factors which appear in a fraction, one in the numerator and the other in the denominator, resulting in an expression which is equivalent to the original.

Cartesian product Of sets A and B, in symbols $A \times B$, the set of all ordered pairs with first component a member of A and second component a member of B.

Closed A set S is closed with respect to an operation (addition, subtraction, etc.) if the result of performing that operation with operands chosen from S is *always* a member of S. *See also* Interval.

Coefficient The numerical factor in an expression (such as -3 in $-3x^2$).

Complement Of set A, the set $\bar{A}$, whose members are exactly those elements from the universe which are not members of A.

Complex fraction A fraction whose numerator or denominator (or both) contains a fraction.

Component In the ordered pair (a, b) the first component is a and the second component is b.

Connective A symbol (such as $=$, $<$, or $>$) used to connect the two expressions in a sentence.

Constant A symbol (such as 3 or a) which may represent only a single number.

Coordinate On a number line the coordinate of a point is the real number which corresponds to that point. In the coordinate plane the coordinates of a point are the components of the ordered pair which corresponds to that point. The first component is called the x-coordinate, and the second, the y-coordinate.

Coordinate plane A plane in which a one-to-one correspondence between its points and the set of ordered pairs of real numbers has been established in a prescribed manner. See p. 217.

Cube root The cube root of a real number a is represented by $\sqrt[3]{a}$, and it is the real number b such that $b^3 = a$.

Degree For the expression ax^n, where n is a whole number, the exponent n is called the degree of the expression. The degree of a polynomial is the degree of its term with highest degree.

Difference The result of subtracting two expressions, as in $x - y$.

Discriminant The value of the expression $b^2 - 4ac$ for the quadratic equation $ax^2 + bx + c = 0$ is the discriminant of that equation.

Dividend In the quotient $x \div y$ or x/y the dividend is x.

Divisor In the quotient $x \div y$ or x/y the divisor is y.

Domain Of an algebraic expression or an open sentence in one variable, the replacement set for that variable. The domain of a relation (or function) is the set of all first components in the ordered pairs of the relation.

Equal Two numerals are equal if they represent the same number. Two sets are equal if they have exactly the same members (without regard to the order in which the members may be listed). Two ordered pairs are equal if they have exactly the same components in the same order.

Equation Any sentence in which the connective is an equal sign.

Equivalent Two algebraic expressions are equivalent if the value of the first expression is the same as the value of the second expression for each allowable replacement of the variables in those expressions. Two open sentences are equivalent if they have the same solution set.

Exponent *See* Power.

Factor Both x and y are factors of the product xy. To factor an expression is to write it as a product. (In a trivial sense any expression may be regarded as a factor of the product of that expression and 1.)

Fraction Any symbol of the form a/b.

Function Any set of ordered pairs of real numbers in which each (distinct) pair has a different first component.

Graph The set of all points which correspond to a set of numbers or to a set of ordered pairs of numbers.

Greater than The real number a is greater than the real number b; in symbols $a > b$, if $a - b$ is positive.

Half plane The set of all points in a plane which lie on one side of a line in that plane. If the line itself is included in the set, the set is called a closed half plane. If the line is not included in the set, the set is called an open half plane.

Inequality Any sentence in which the connective is $<$, $\leq$, $>$, $\geq$, or $\neq$.

Integer Any member of the set $I = \{\ldots, -3, -2, -1, 0, 1, 2, 3, \ldots\}$. The members of $\{1, 2, 3, \ldots\}$ are called positive integers (or natural numbers), and the members of $\{-1, -2, -3, \ldots\}$ are called negative integers.

Intercept In a graph the x-coordinate of any point which is on the x-axis is an x-intercept of the graph, and the y-coordinate of any point which is on the y-axis is a y-intercept.

Intersection The intersection of sets A and B, in symbols $A \cap B$, is the set C such that the members of C are exactly those elements which are in both A and B, simultaneously.

Interval If a is less than b, the set of all real numbers which are between a and b is an open interval and is represented by $\langle a, b \rangle$. If the set also contains the end points a and b, it is a closed interval and is represented by $[a, b]$.

Inverse Of the relation R, the relation R^{-1}, which is obtained by interchanging the components in each of the ordered pairs of R.

Irrational number A real number which is not rational. An irrational number cannot be written as the quotient of two integers, and its decimal form is non-terminating and nonrepeating.

Less than The real number a is less than the real number b; in symbols $a < b$, if $a - b$ is negative.

Linear equation An equation of the form $ax + by = c$, where a and b are not both zero, is said to be a linear equation. Its graph is a straight line.

Linear function A function which is determined by a first degree polynomial expression, $ax + b$ with $a \neq 0$, is called a linear function. Its graph is a straight line.

Minuend In the difference $x - y$ the minuend is x.

Monomial Any expression (such as $3x^2$) which consists of exactly one term.

Multiplicative identity The number 1 is called the multiplicative identity. (This is because for every $a \in R$, $a \cdot 1 = 1 \cdot a = a$.)

Multiplicative inverse If $x \neq 0$, the multiplicative inverse of x is the number $1/x$, such that $(x)(1/x) = 1$. If the product of two expressions is 1, each factor is called the multiplicative inverse of the other.

Natural number Any member of the set $N = \{1, 2, 3, \ldots\}$; also called *counting numbers*.

Number line A line on which a one-to-one correspondence between its points and the set of real numbers has been established in such a way that if point P is to the left of point Q on that line, then the coordinate of P is a number which is less than the coordinate of point Q.

Numeral A symbol (such as 3, $-\frac{2}{5}$, or $\sqrt{2}$) used to represent a specifically designated number.

Numerical expression Any combination of numerals and operational symbols (such as $3 + 4 \cdot 2$) which collectively represents a specific number.

Open sentence A sentence (such as $3x = x + 4$) which contains one or more variables and which becomes either a true or a false statement when each of its variables has been replaced by a member of its replacement set.

Ordered pair A pair of real numbers (a, b) in which one of the numbers is designated as the first and the other as the second.

Parabola Any curve which is the graph of a quadratic function or the inverse of a quadratic function.

Placeholder Any symbol which may represent a number. Both constants and variables are placeholders.

Polynomial Any expression of the form

$$a_0 + a_1x + a_2x^2 + a_3x^3 + \cdots + a_nx^n$$

in which each term is the product of a real number and a whole number power of some variable, or any expression which is equivalent to one of that form.

Power An expression of the form b^a. The base of this power is b and the exponent is a.

Product The result of multiplying two expressions, as in $x \cdot y$.

Quadratic Any second degree polynomial. A function which is determined by a second degree polynomial is called a quadratic function. An equation of the form $ax^2 + bx + c = 0$, where $a \neq 0$, is called a quadratic equation. Similarly, for a quadratic inequality.

Quotient The result of dividing two expressions, as in $x \div y$ or x/y.

Quotient number A positive rational number. The quotient of two natural numbers is a quotient number. The set of all quotient numbers is represented by Q.

Radical sign The symbol $\sqrt{}$ used to designate a principal root, as in $\sqrt{}$ for square root, $\sqrt[3]{}$ for cube root, etc.

Range The range of an algebraic expression is the set of all values the expression can assume after replacements have been made for each variable in the expression. The range of a relation (or function) is the set of all second components in the ordered pairs of the relation.

Rational number Any number which may be represented exactly by a fraction whose numerator and denominator are both integer symbols. The quotient of any two integers (with divisor not zero) is a rational number. Every terminating or repeating decimal represents a rational number. The set of all rational numbers is represented by F.

Real number Any member of the set whose elements satisfy all the real number axioms (pp. 78–79). Every real number may be represented by a decimal, either terminating, repeating, or nonrepeating. The set of all real numbers is the union of the set of rational numbers and the set of irrational numbers and is represented by R.

Relation Any set of ordered pairs of real numbers.

Replacement set The set of all possible replacements for a variable is the replacement set for that variable.

Sentence Any combination of two expressions and a connective (such as $3x = 2$ or $5 \cdot 2 < 12$) which collectively expresses some assertion.

Similar terms In a polynomial expression, any two terms which have the same degree are said to be similar terms. Also, any two terms which are identical or which are identical except for their numerical coefficients (such as $3x^2y^3$ and $-5x^2y^3$).

Solution set For an open sentence in one variable the solution set is the set of all replacements from its domain which result in a true statement. For an open sentence in two variables the solution set is the set of all (allowable) ordered pairs which result in a true statement.

Square root Of a nonnegative number a, any number b, such that $b^2 = a$. The *principal* square root of a is represented by $\sqrt{a}$, and it is the *nonnegative* number b, such that $b^2 = a$.

Statement Any sentence to which one of the labels "true" or "false" can be assigned.

Subset Set A is a subset of set B if every member of A is also a member of B.

Subtrahend In the difference $x - y$ the subtrahend is y.

Sum The result of adding two expressions, as in $x + y$.

Terms Both x and y are called terms of the sum $x + y$. (In a trivial sense any expression may be regarded as a term of the sum of that expression and 0.)

Trinomial Any expression (such as $2x^2 + 5x - 3$) which consists of exactly three terms.

Union The union of sets A and B, in symbols $A \cup B$, is the set C, such that the members of C are exactly those elements which are in either A or B (or both).

Universe If every set under discussion is a subset of some set U, then U is called the universe or the universal set.

Value The specific number which a numerical expression represents. An algebraic expression has a value only if a (numerical) replacement is made for each variable in the expression.

Variable A symbol (such as x or y) which may represent more than one number.

Whole number Any member of the set $\{0, 1, 2, 3, \ldots\}$.

Index

A

Absolute value 47
  with square roots 306–307
Addition axiom for equations 90
  axiom for inequalities 101
  method for finding solution sets 242
  of rational expressions 167–179
Additive identity 46
Additive inverse 24, 46
  of a polynomial 86
Algebraic expression 3
  equivalent 10, 157–160
  in two variables 219
Algebraic statement 37
Associative axiom 38
Axes 217–218
Axiom
  addition 90
  additive identity 46
  additive inverse 46
  associative 38

Axiom (cont.)
  closure 44
  commutative 38
  completeness 79
  distributive 40
  multiplication 90
  multiplicative identity 45
  multiplicative inverse 46
  of positive numbers 79
  of real numbers 78–79
  squaring 308
  substitution 90, 101
  trichotomy 79
Axis of symmetry 274

B

Bar as a symbol of operation order 2
Base 82
Between 73
Binary operation 41

Binomial 126
Brackets to indicate operation order 2

C

Canceling with rational expressions 159–162
  with rational numbers 63
Cartesian product 215
  graph of 217
Check of the solution set of an equation 93
Closed half plane 251
  interval 105
  set 43
Closure axiom 44
Coefficient 83
Commutative axiom 38
Complement 21
Complete factorization 138
Completeness axiom 79
Complex fraction 197
Component 215
Connective 8
Constant 3
Coordinate 72, 235
Coordinate plane 217
Correspondence 14
Counting number 24
Cube root 286

D

Decimal
  nonrepeating 75
  repeating 67
  terminating 25, 67
Degree 83
  of an equation 95
  of an inequality 100
Denominator, restrictions upon 25
Difference 55
  of two squares 132
Discriminant 326
  use with factoring 328
  use with solution sets 327
Distributive axiom 40
Dividend 58
Division of rational expressions 162–167
  as a means of factoring 201–202
  of polynomials 199–204

Divisor 58
Domain of an expression 4
  of a function 14
  of rational algebraic expressions 158
  of a relation 255
  in two variables 220

E

Element of a set 18
Empty set 19
Equal sets 18
Equations 32
  axioms for 90
  checking solution sets 93
  equivalent 89
  identical 97
  left expression of 89
  linear 240
  polynomial 95
  quadratic 143
  right expression of 89
  two variables 230–235
Equivalent expressions 10
  algebraic expressions 157–160
  in two variables 225
  equations 89
Even root 287
Exponent
  integral 208
  laws of 209, 295
  natural number 82
  rational number 293–294
  zero 208
Expressions
  algebraic 3
  equivalent 10
  left and right 89
  numerical 2
  value of 4

F

Factor 39
Factored form 131
Factoring 138–140
  expressions in two variables 226–228
False numerical statement 7
Finite set 23
Function 14
  definition of 259

Function (cont.)
  graphical test of 259
  inverse of a relation 280
  linear 258
  nonlinear 264
  quadratic 265
  zeros of 16, 272

G

General quadratic equation 319
Graphs
  of relations 255, 269
  of sets of numbers 105
  of sets of ordered pairs 217
Greater than 72–73
  or equal to 107

H

Half line 31
Half plane 250
  closed 251
  open 250
Horizontal axis 217

I

Identical equation 97
Identity
  additive 46
  multiplicative 45
Index 208, 288
Inequality 100
  axioms for 101–102
  second degree 154
  in two variables 249–252
Infinite set 23
Integers 24
  addition of 51
  division of 58
  as exponents 207–213
  multiplication of 53
  simplest form of 24
  subtraction of 55
Intercept 235
  technique for determining 236
Interchanging variables 279
Intersection of sets 20
  of domains 159–160
  of graphs 237
  of solution sets 150, 240–244

Interval 100
  closed 105
  open 105
Inverse
  additive 24
  axioms 46
  multiplicative 46
  of a relation 277
Irrational number 75
  proof with $\sqrt{2}$ 288–289

L

LCM 168
  of algebraic expressions 170, 174
  of given denominators 168, 174
  of natural numbers 168
  simplifying complex fractions 197–199
  use with multiplication axiom 186–187
Least common multiple (see LCM)
Left expression 89
Less than 72–73
  or equal to 107
Line 31
  number line 72
Linear
  equation 240
  function 258

M

Member of a set 18
Minuend 55
Monomial 126
Multiplication
  axiom for equations 90
  axiom for inequalities 102
  finding solution sets 181–187
  of polynomials 125–130
  of rational expressions 162–167
Multiplicative identity 45
Multiplicative inverse 46

N

Natural number 1, 2, 24
Negative integer 24
Nonlinear function 264
Number 1
  integer 24
  irrational 75
  line 72
  natural 24

Number (cont.)
  quotient 25
  rational 25
  real 75
Numeral 1
Numerical expression 2
Numerical statement 7

O

Odd root 287
Open
  half plane 250
  interval 105
  sentence 8
Order
  of operations 2
  of rational numbers 72–73
Ordered pair 215
Origin 77

P

Parabola 265
Parentheses to indicate operation order 2
  use with exponents 182
Partial quotient 203
Placeholder 3
Plane 28
  half plane 250
Point 28
Polynomial 83, 85
  additive inverse of 86
  degree of 83, 84
  equation 95
  factored form 131
  factoring 138–140
  long division of 200–204
  prime 138
  as rational expressions 161, 165
  standard form of 85
Positive integer 24
Power 82
Prime polynomial 138
Principal root 287
Product 39, 120
Projection 260–261

Q

Quadratic equation 143
  completing the square 315–318

Quadratic equation (cont.)
  discriminant of 326
  factoring 143
  general 322
Quadratic function 265
Quadratic inequality 150–155
Quadratic theorem 322
Quotient 58
Quotient numbers 25
  addition 44
  multiplication 44

R

Radical equations 305–309
Radical expressions 304
  simplifying 298–302
Radical sign 288
Range
  of an expression 4
  of a function 14
  of a relation 255
Rational algebraic expressions 157–159
  common denominator of 168
  domain of 158
  polynomial as special class 161, 165
  standard forms of 160, 163, 175
Rational numbers 25
  addition of 63
  additive inverse of 64–65
  canceling property 63
  division of 68
  equality of 62
  as exponents 292–295
  multiplication of 64
  multiplicative inverse of 68
  order of 72–73
  simplest form of 66
  subtraction of 7, 68
Rationalize 301
Ray 31
Real numbers 75
  axioms for 78–79
Relation 255
  inverse of 277
Relatively prime polynomials 160
Remainder 203
Repeating decimal 67
Replacement set 4
Right expression 89
Root 286
  principal 287

S

Schematic method 127
Second root 286
Segment 31
Sentence 8
Set 18, 28
   builder notation 106
   empty 19
   replacement 4
   solution 9
Similar terms 84
Simplest form 66
   of integers 24
   of natural numbers 1, 2
   of rational numbers 66
Solution set 9
   sentence in two variables 221, **230**
Square, completing 315–318
Square root 286
Squaring axiom 308
Standard form
   of polynomials 85
   of rational expression 160
Statement
   algebraic 37
   numerical 7
Straight line as a graph 234
Subset 19
Substitution axiom for equations 90
   for inequalities 101
Subtraction of rational expressions 167–179
Subtrahend 55
Sum 39
Summary
   axioms for equivalent equations 90–91
   factoring polynomials 140
   factoring second degree trinomials 135
   graphing relations 269
   nature of solution set members 327
   order of operations of arithmetic 2
   product of two binomials 127
   roots of real numbers 287
   solution set intersections, two variables 244
   solution sets by completing the square 318
   solution sets, equations with square root radicals 309
   solution sets of quadratic equations 146

Summary (*cont.*)
   solution sets of quadratic inequalities 154
   standard form of sum or difference, rational expressions 175
   standard forms of rational expressions 163
   standard set names 26
   theorems for rational powers 295
Superscript 277
Symmetry, axis of 274, 278

T

Term 39
   similar 84
Terminating decimal 25, 67
Third root 286
Trichotomy axiom 79
Trinomial 126
True numerical statement 7

U

Union of sets 20
   of solution sets 143
Universe 20

V

Value of expressions 4
Variable 4
Venn diagram 32
Vertex of a parabola 274, 329
Vertical axis 217

X

x-axis 235
x-coordinate 235
x-form 269
x-intercept 235, 271
   connection with zeros of a function 273
xy-plane 235

Y

y-axis 235
y-coordinate 235
y-form 258
y-intercept 235, 271

Z

Zero
   division by 59
   of a function 16, 272